The Voices of Nature

The Voices of Nature

The Voices of Nature

HOW AND WHY ANIMALS COMMUNICATE

NICOLAS MATHEVON

WITH A FOREWORD BY
BERNIE KRAUSE

ILLUSTRATIONS BY
BERNARD MATHEVON

PRINCETON UNIVERSITY PRESS

PRINCETON & OXFORD

Originally published in French under the title *Les animaux parlent—Sachons les écouter* © 2021 by Humensciences/Humensis

Published in English by Princeton University Press
41 William Street, Princeton, New Jersey 08540
99 Banbury Road, Oxford OX2 6JX

press.princeton.edu

First paperback printing, 2025
Paperback ISBN 9780691239989
All Rights Reserved

The Library of Congress cataloged the cloth edition as follows:
Names: Mathevon, Nicolas, author. | Krause, Bernie, 1938- writer of foreword.
Title: The voices of nature : how and why animals communicate /
    Nicolas Mathevon; with a foreword by Bernie Krause.
Other titles: Animaux parlent. English
Description: Princeton : Princeton University Press, [2023] | "Originally published in French under the title Les animaux parlent—Sachons les écouter © 2021 by Humensciences/Humensis"—title page verso. | Includes bibliographical references and index.
Identifiers: LCCN 2022036324 (print) | LCCN 2022036325 (ebook) |
    ISBN 9780691236759 (hardcover) | ISBN 9780691236766 (ebook)
Subjects: LCSH: Animal communication. | Animal sounds. | BISAC: SCIENCE / Life Sciences / Zoology / Ethology (Animal Behavior) | SCIENCE / Life Sciences / Neuroscience
Classification: LCC QL776 .M37313 2023 (print) | LCC QL776 (ebook) |
    DDC 591.59/4—dc23/eng/20220729
LC record available at https://lccn.loc.gov/2022036324
LC ebook record available at https://lccn.loc.gov/2022036325

British Library Cataloging-in-Publication Data is available

Editorial: Alison Kalett & Hallie Schaeffer
Production Editorial: Ali Parrington
Jacket/Cover Design: Katie Osborne
Production: Danielle Amatucci
Publicity: Matthew Taylor & Kate Farquhar-Thomson
Copyeditor: Jennifer McClain

Jacket/Cover art and book illustrations by Bernard Mathevon

This book has been composed in Arno and Helvetica Neue

To my beloved wife and children, Patricia, Elise,
Pauline, and Baptiste.

To my family and friends.

To my partners in adventure.

To my students and colleagues at the ENES
Bioacoustics Research lab.

To all those who have allowed my research.

To the boys and girls curious about the world around them.

To the naturalists.

To the next generation of bioacousticians.

To Paul Géroudet and Christian Zuber.

To all those who made me dream.

And to Thierry Aubin. Of course.

What is this faculty we observe in them, of complaining, rejoicing, calling to one another for succour, and inviting each other to love, which they do with the voice, other than speech? And why should they not speak to one another?

THE WORKS OF MONTAIGNE (W. HAZLITT, ED.;
JOHN TEMPLEMAN, LONDON, 1369)

CONTENTS

If we ever hope to understand how a living system operates, we'd better learn to listen to its voice. On the other hand, the marvels of natural sound have been made clear on every page of Professor Nicolas Mathevon's book *The Voices of Nature: How and Why Animals Communicate*. With his new publication, Mathevon explores with us one of the more elusive portals, the voice of the natural world in all its varied parts. Professor Mathevon's narrative reminded me of the time in the mid-1970s when I served as a consultant on a doctoral committee headed by the late naturalist Gregory Bateson. The young candidate had called a meeting to discuss his frustration at having failed to answer one question central to the subject of his dissertation. After a short appeal for help by the student, everyone in the room grew quiet. And then Bateson spoke up. "Here's a story that might help," he said, directing his comments to the anxious young man. "At the end of World War I, the great English philosopher, Alfred North Whitehead, was invited to join Harvard's faculty. As a condition of his acceptance, Whitehead insisted that his good friend and colleague, Bertrand Russell, with whom he'd collaborated on the publication of the *Principia Mathematica,* be included as part of the package. During that period, Harvard faculty members were required to give a lecture introducing themselves and their work. Russell chose as his topic a clarification of Max Planck's *Quantum Theory.* Given on a hot August night, the chapel filled with Boston society and the school's faculty, Russell labored for ninety minutes in the sweltering heat. When he finished, exhausted and drenched in sweat, he turned away from the podium and walked slowly to his seat. After a moment of polite applause, Alfred Whitehead stood up, glanced kindly at his colleague, and moved to the podium. In his high-pitched,

English-accented voice, he declared, 'I'd like to thank Professor Russell for his brilliant exposition. And, especially for leaving unobscured the vast darkness of the subject.'"

To some extent the sonic world is not that difficult to understand. Many of us operate intuitively at various levels of success within its borders. But it is difficult to explain because we lack the precise words to express acoustic experience. Most of us have the capacity to hear. And some of us are even gifted with the ability to listen. What makes grasping the biophonic world even more tricky and obscure is that, in the West, we rely primarily on what we see. What we know, what we talk about, the objects we tend to enjoy, are mostly understood through a graphic lens. Yet, our vocabulary is shy of terms that define the elements of acoustic environments which we are all a part of, whether they be urban, rural, or wild.

In *The Voices of Nature*, Nicolas Mathevon meets these issues headon, leading us eloquently along evolutionary paths of communication that have informed us from the time when at least half a dozen of our hominid species appeared on this planet. This is a terrific exposition into the world of natural sounds and the organisms that produce them. Filled in equal measure with accessible narratives of discovery, revelations, and the characters who have observed the patterns that connect, we are transfixed by Mathevon's passionate stories of The Others and the researchers who identified the behaviors as they unfolded. Like a fine novel, this notable work is impossible to put down once you begin.

The sense of wonder that emanates from those dedicated to this topic issues from every page of Mathevon's chronicle. From Peter Marler's studies of white-crowned sparrow dialects to Thierry Aubin's studies of parent-offspring vocal recognition in penguins, the sense of devotion to unraveling the secrets of biophony is inspirational. As the naturalist and professor from Pitzer College, Paul Shepard, once suggested, "We may never fully understand the permutations of the natural world. And perhaps we weren't meant to." I would add to that a quip from the late Professor Kenneth Norris, Emeritus, University of California, Santa Cruz, the marine biologist who discovered how dolphins echolocate. When we were trying to solve the mysteries of singing sand dunes in the American Southwest, Norris responded to my puerile complaint about the incredible amount of time we had spent without a clear result, "Bernie, we're having such a grand time searching for the answers, I hope we *never* find out!" *The Voices of Nature* brings new proximity to that immense journey and our reconnection to the route.

Bernie Krause
Author of *The Great Animal Orchestra:*
Finding the Origins of Music in the World's Wild Places
Wild Sanctuary, Sonoma, California
April 2021

NOTE TO THE READER

This book is accompanied by a website where you can listen to some examples of songs and other sound signals produced by animals: https://mathevon0.wixsite.com/soundjourneybook or https://tinyurl.com/228ns885 or scan this QR code:

Although intended for a wide audience, this book is also written for students and researchers in bioacoustics. If you are one of them, I encourage you to pay attention to the notes (numbered in the text and listed at the end of the book). You will find more details and, above all, a substantial and up-to-date list of papers published in international peer-reviewed journals (around 700 references). An excellent way to start a bibliography on a new subject.

One last message before you dive into the book: Feel free to send me (mathevon@univ-st-etienne.fr) the publications that would have raised your interest and that are not referenced in this book, including yours, of course!

1

Animal chatters

TINBERGEN'S FOUR QUESTIONS

Calanques National Park, near Marseille. It is noontime under the sun of Provence. The heat is intense, the light bright. The garigue smells of thyme, rosemary, and lavender. The background music ("tchik-tchik-tchik") is provided by the cicadas, pressing against the bark of the pine trees, their rostra stuck into the trunk to pump out the beneficial sap. In the deep blue sky, swifts glide like arrows: "Weer!! . . . . Weer!!!" A large locust spreads its colorful wings and flies in front of me, then lands a few meters away. Its long hind legs oscillate rapidly as they rub against its wings, producing a strange chirping sound, like rustling sheets of paper. When the legs freeze, the sound stops. On its branch, a subalpine warbler emits its cheerful ritornello, briskly playing sometimes fluted, sometimes squeaky notes. Suddenly, the warbler falls silent and dives into a bush. It soon comes out to sing again. In the background, far away, some sheep bleat. A dog barks. Later, when the sun has waned and it is getting cooler, the cicadas will stop their relentless concert. Others will take their place, and the entire night will rustle with the song of locusts, grasshoppers, toads, and other nocturnal creatures, in an apparent cacophony. Just before daybreak, the dawn chorus of birds will come alive. The cicadas will remain silent, waiting for the heat to start vibrating their cymbals, those small membranes hidden under their wings, which rattle several hundred times a second. "Tchik-tchik-tchik . . . tchik-tchik-tchik . . . tchik-tchik-tchik. . . ." This is the great concert of life!

The soundscape of the Mediterranean scrubland is unique. There is this multitude of sounds produced by animals, to which are sometimes added the breath of wind in the trees and the sound of waves crashing on the shore. For me, they are associated with all the times I've spent in this region of France. Have you ever wondered why animals make these sounds? Not simply to charm our senses, of course: They are not intended for us. In fact, their purpose is to communicate. To *communicate*— that's a big word! And yet . . . these songs, cries, and other shrill sounds are *signals* that, like our human words, allow them to converse with other animals. What do they say to each other, you may ask? This is what I propose to discover in this book. You are about to enter worlds of sound, some of which are familiar but most of which are completely unknown to you and which you never even knew existed. How could you, since some of them are not even accessible to our ears?

Many animals exploit the sound-transmitting properties of water or air to communicate: to find a partner, to defend a territory, to signal the presence of a predator or food source, to collaborate in hunting, to recognize and interact with members of the group. These communications are essential for many species, including our own. We know this well— we whose articulated language demonstrates an incredible complexity, commensurate with that of our social interactions; we whose simple cries, from the moment we are born, signal our emotions and needs to other humans. The fact that animals are comparable to humans has been demonstrated by a great deal of scientific work over the last forty years. We can no longer set our species apart from other animals: each species has its own biological, ecological, social, and sometimes cultural characteristics that define its own world. Acoustic communication systems are therefore diverse, but all are worthy of interest. They are evidence of the diversity of life.[1]

How are animal vocalizations produced? What information do they contain? Can we understand animal languages? For a long time, the diversity of these sound worlds was difficult to access, but technical advances—such as the tape recorder and then the computer—have changed this. In recent decades, scientists have begun to read the scores of animal concerts and decipher their meaning.

I'm involved in the science that studies animal acoustic communications, called *bioacoustics*. Bioacousticians are working to decipher how animals make and hear sounds, what information is encoded in their sound signals, what this information is used for in their daily lives, and also how their acoustic communication systems have developed over the history of life. We will see that studying the richness and complexity of animal acoustic communication can help us understand how our own communication system works—our words, our laughter, our cries. Bioacousticians are a bit like Champollion, the French historian who deciphered Egyptian hieroglyphics using the Rosetta Stone, a fragment of a stele where the same text is written in several languages. To decode animal languages, many other methods must be used, but the goal remains the same: to decipher their meaning. The sounds produced by animals are signals carrying information whose meaning we are trying to decipher.

Bioacoustics is a discipline rooted in ethology, the science of animal and human behavior. The development of this branch of biology is relatively recent, dating back to the 1960s. In 1973, the Nobel Prize in Physiology and Medicine was awarded to three ethologists. The first was Konrad Lorenz. You may have already heard of this Austrian researcher, who became famous for his experiments on imprinting, in which memory of certain events or individuals is built up very quickly and very early in life. Lorenz discovered imprinting in his observations of geese. If goslings hatch in the presence of a human being, they consider that person to be their mother and follow him wherever he goes.[2] The second Nobel laureate was Karl von Frisch. He discovered the dance of the bees, this unique communication system through which the worker bee, on her return to the hive, can inform her sisters of the whereabouts of new flowers.[3] It is impressively precise: the angle formed between the axis of the bee's walk along one of the honeycombs and the vertical axis corresponds to the angle formed between the direction of the sun and the direction of the flowers when exiting the hive. Simply incredible! But there is more: the frequency of the vibrations of the insect's body and wings contains information about the amount of food provided by the flowers. It is by vibrating that the bee signals it is worthwhile to go on a shopping spree. The third researcher was Nikolaas Tinbergen. Of the three, he is my favorite. Tinbergen spent most of his career studying animal behavior using the experimental method.[4] He invented ways of questioning animals in order to understand the causes and consequences of their behavior. For example, in order to test whether it was the red spot on the herring gull's beak that caused the chicks to beg for food, he offered them various objects (sometimes simple sticks) more or less faithfully reproducing an adult's head and bearing a bigger or smaller spot, and in different colors. He then measured the intensity of the chick's behavioral response—its speed in beating the lure with its beak. Tinbergen thus highlighted the importance of the "red spot" signal in the parent-chick relationship in this seabird species. In addition to being a remarkable experimenter, he sought to formalize scientific research in ethology. He explained that in order to fully understand animal behavior, four questions had to be answered. This method is still valid today,[5] and every bioacoustician keeps Tinbergen's four questions in mind when studying sound communication:

(1) What are the *mechanisms* of the behavior I observe?
(2) What are the *evolutionary causes* that explain the existence of this behavior?
(3) How did this behavior *develop over the course of the individual's life*?
(4) What has the *evolutionary history* of this behavior been over geological time?

Let's look at these four fundamental points in more detail. Let's imagine, for example, that you want to understand why American robins, *Turdus migratorius*, sing in the spring and what the drivers of this communication are.

You first need to understand the *mechanisms* of both the production and reception of signals, i.e., the processes that lead an animal to produce a sound and those that explain a behavioral response to what it perceives—for example, to understand why, when a male robin hears another male robin singing, it responds by singing in turn and sometimes by attacking the intruder. What is it about the song that causes this reaction? First and foremost, there must be particular acoustic characteristics identifying the American robin, which ensure that its song is not confused with that of another animal species, especially another bird species. Second, why is the reaction aggressive? Is it, for example, because the robin is ready to reproduce, and the high level of sex hormones circulating in its blood increases its reactivity? If we want to study these proximal causes of the behavior, we need to describe the properties of the stimuli that provoke the robin's reaction, both external (the intruder's song) and internal (hormonal balances). We also want to understand all the physiological processes, from the reception of the stimulus (How does hearing work?) to the expression of the behavioral response (Why all this agitation? To defend one's territory?). To explore these questions, you can set up experiments in acoustic playback with a loudspeaker placed near where the robin sings, and question it directly: "Is this song a territorial signal for you?"

Once you have addressed the first of Tinbergen's questions, you can turn your interest to the second question: the *evolutionary causes* of the communication. Why has this singing behavior rather than another been favored during the evolution of the species? In other words, how does singing confer advantages that might explain why, once it appeared, it has been retained over time? Does singing increase a male robin's likelihood of being noticed by a female? Will an aggressive individual, singing louder, more often, and for longer than others, be more effective in defending its territory and food resources? These two aspects would increase his reproductive success, i.e., the number of young he fathers and who survives into adulthood. Singing behavior would then be favored by the two facets of sexual selection: intersexual selection—females prefer some singers to others—and intrasexual selection—males drive off insufficiently aggressive colleagues more easily. But beware of the other side of the coin: Doesn't singing like a madman increase the probability of being spotted and captured by a predator such as a hawk or any other bird of prey? You could hypothesize that natural selection may have limited this behavior and favored individuals inclined to sing less loudly. Thus, sexual selection, like natural selection—the two major evolutionary mechanisms identified by Charles Darwin—probably participates in the evolution of communication behavior. You can see that things are complicated and that establishing the evolutionary causes of sound communication is not easy: all behavior is the result of a balance between constraints that sometimes have the opposite effect. You should not forget that evolution is also very much subject to chance (so-called stochastic processes). Your task as an evolutionary biologist will certainly be very difficult.

Let's see if it's easier to answer the third question formulated by Tinbergen: How was this communication behavior *acquired during the life* of our American robin? At birth, the robin chick cannot sing. It simply makes short calls to beg its parents for food. In the weeks following hatching, its brain develops and the chick gradually acquires the ability to produce more complex vocalizations. It is then essential that the young chick be able to hear adult songs, which it will learn by imitating. How are the two types of processes articulated? There are the innate processes (a robin will never sing like a wren; it has a genetic predisposition to sing "American robin") and the acquired processes (the young robin learns to sing by imitating an adult). This is a vast field of investigation. We discuss it in detail in chapter 12.

The fourth question remains, which is by far the most difficult to address: What is the *evolutionary history* of the communication that you are studying? To put it plainly, what are the stages that gradually led from the ancestor of birds—a kind of dinosaur, perhaps emitting dinosaur vocalizations[6]—to a robin singing a song? Quite a story, isn't it? All animal species, including humans, are rooted in the depths of time and share common ancestors. While we are beginning to understand the evolutionary mechanisms of diversification of living species rather well, particularly with regard to their genetic heritage, anatomy, physiology, and morphology, reconstructing the evolution of behavior remains a challenge. How and when did birdsong emerge over the course of evolution? Why in some species is it only the males that sing, whereas in many other species females also vocalize? Is song an ancestral trait in both sexes? Were dinosaurs, the ancestors of today's birds, capable of producing sounds? Did they use them to communicate? Can we imagine a tyrannosaurus "singing" to call his or her partner? Did young tyrannosauruses learn their vocalizations by imitating an adult? When and how did this learning happen? Answering all these questions is difficult, if not impossible, because behavior leaves few fossil traces.[7] My great frustration as a bioacoustician is not being able to listen to and record extinct species. I dream of being able to record baby tyrannosauruses and then have their parents listen to these vocalizations. And to see their reactions! It's obviously unlikely that we'll ever be able to achieve this kind of thing—but who knows? Maybe one day we'll be able to reconstruct "real" dinosaurs from fossil genomes, like in *Jurassic Park*. A Japanese team is trying to revive the woolly mammoth in this way. However, there are scientific methods that make it possible to establish solid hypotheses about the evolution of communications. We'll talk about that too.

Most bioacousticians focus their research on only one of Tinbergen's four questions and therefore do not aim to understand all aspects of the sound communication being studied. But keeping all of these questions in mind provides a fertile framework for thinking. Even when one is interested in relatively simple mechanisms, such as how a sound stimulates the robin's eardrum and is then transformed into nerve potentials that can be interpreted by the bird's brain, it is useful to consider that these mechanisms have a history.

Therefore, to conduct research in bioacoustics, a solid knowledge of biology is required. Most of my PhD students and postdocs have years of study in zoology, anatomy, physiology, neurobiology, ecology, ethology, and evolution. But bioacoustics requires, like many other disciplines in the life sciences, proficiency in scientific fields other than biology. In bioacoustics, there is certainly the prefix *bio* (living), but there is above all the root *acoustic*. Studying animal sound communications requires an understanding of what a sound is. Follow me and hang on tight: we're going on a detour through the physics of acoustic waves. Don't worry; it's not that complicated, and it's important for our journey into the world of animal sound. If, however, physics gives you a serious headache, and after trying to read the following chapter (science does require some effort, after all) you have trouble understanding what I have written, I'll take you straight to chapter 3. There, we start our journey by venturing into the Brazilian rainforest. For now, let's try bravely—a little courage! I first explain what a sound is, how it propagates, and how it can be described.

2

Making circles in water

A SHORT VADE MECUM OF PHYSICAL ACOUSTICS

When I started looking for a place to prepare my PhD thesis—I was fascinated by birds at the time—I was lucky enough to enter what was probably the last laboratory to be housed in a private home. Jean-Claude Brémond was waiting for me there. Mr. Brémond—I never called him by his first name—was director of research at the famous French *Centre national de la recherche scientifique* (CNRS). He told me straightaway: "If you want to work with us, that's OK, but there won't be any research position for you after you've completed your thesis." His words had the merit of being clear, and I was grateful to him for that. As a high school teacher at the time, I was not looking for a paid research position. On the contrary, I was ready to accept anything to satisfy my passion. The walls of the small room where we met were covered with posters of birds, and the atmosphere enthused me. I jumped in with both feet, thinking I would soon be wandering through the woods to listen to the birds singing. It was the beginning of the computer age, and Thierry Aubin—then junior researcher at the CNRS and later my companion in so many expeditions—emerged from a jumble of electronic machines to introduce himself and begin my training. But instead of discussing the latest scientific discoveries about birds and their sound communications, the first lesson Thierry gave me was an acoustics course. I'm not sure I followed all his explanations well: I had a solid background as a naturalist, but was seriously out of my league in physics and mathematics. Even today, after so many years spent observing oscillograms and spectrograms, I am still learning. "A person who wants to work in bioacoustics must have a feeling for sound," Thierry once told me. Never forgetting this maxim, I always ask students who want to work on my research team if they have a feeling for sound. When the students are musicians or music lovers (this is common among bioacousticians), there is no doubt. Otherwise, I am always attentive to their answers. Studying animal acoustic communications requires an understanding of the physics of sound waves and their methods of analysis. To hope to decipher an animal language, one must first understand such notions as frequency, amplitude, modulation, and many others. We discover the essentials in this chapter. If you have some knowledge of physics, the following lines may make you smile. You will find my metaphors approximate, and I hope you will forgive me. But if this is the first time you have dealt with these concepts, don't worry; we'll walk through them together.

To understand the physical nature of a sound, imagine that you are having fun throwing a pebble into a pond and looking at the circles in the water that form and become larger and larger as they move away from the point where the pebble landed. These ripples, which start when the pebble hits the water and then move away from the point of impact in concentric circles, are waves. As a wavelet passes, the water level rises and then falls for a short distance. The water molecules do not move with the wave: they return to their initial position once the wave has passed. It is the waves that propagate through the water, not the water itself. In fact, if you put a small cork on the water, you can see that it moves up and down as the wave passes, but remains at the same distance from the point hit by the pebble. While our circles in the water form and propagate only on the surface of the pond in as many circles, sound waves propagate in three dimensions. To picture this, imagine a loudspeaker—or, better yet, look at one. Its centerpiece is a membrane that vibrates and makes the air in contact with it vibrate. This membrane is the equivalent of a pebble hitting the surface of the pond. The sound waves created by the speaker's membrane are small variations in the pressure of the air (or water in the case of an underwater speaker) that move apart, forming larger and larger spheres, like the ever-larger circles in the water created by our pebble.

The analogy of circles in water, of course, has its limits. While the pebble only disturbs the water's surface for a brief moment, creating a very small number of circles, the vibrating membrane of the loud-speaker produces successions of sound waves. Moreover, the speed of our sound waves is much greater than that of the circles in the water: about 340 meters per second if the sound propagates through the air and much faster if it propagates in water, where sound waves travel at 1500 meters (1.5 km) per second. The speed of sound in air or water, however, is infinitely slower than the speed of light, which is close to 300 million meters per second! This discrepancy between the speed of light and the speed of sound is particularly striking during a thunder-storm. When you see a flash of lightning streaking across the sky, the sound of thunder comes later, with a greater or lesser time lag. In 0.00001 seconds, light travels 3 kilometers, while sound waves need almost 9 seconds to travel the same distance. It is therefore possible to estimate the distance from the impact of lightning by counting the sec-onds between the lightning and the thunder, then multiplying the number of seconds by 340 meters: 9 seconds multiplied by 340 equals 3060 meters, or about 3 kilometers.

A sound is therefore a wave that is transmitted in a propagation me-dium. To exist, sound needs particles: air molecules or water molecules (or solid molecules). There can be no sound in a vacuum. The micro-phones of the Perseverance rover have thus succeeded in recording sound on the planet Mars, since this planet has an atmosphere.[8] How-ever, it is impossible for sound waves to travel from Mars to Earth (or vice versa), since they would have to travel through a vacuum. Intersidereal space, where air is absent, is totally silent. If the propagation medium is air, the sound wave corresponds to a variation in atmospheric pressure. We hear about atmospheric pressure (also referred to as barometric pressure) in weather reports: if the pressure rises, the weather is going to be fine; if it falls, a low-pressure system—i.e., rain—is coming. At-mospheric pressure is the result of gravity acting on the layer of air (the atmosphere) that surrounds our planet. Gravity, the force of attraction exerted by the Earth, gives this layer of air a weight, which exerts pressure on any object on Earth. We are not aware of this because atmospheric pressure surrounds us constantly. Yet it weighs on our shoulders. The only time we perceive it is when we suddenly climb to high altitude, in a cable car, for example, or in a poorly pressurized plane: we then feel this pressure in our eardrums, and our ears can become blocked. This is because the internal ducts of our ears have not had time to adapt to the significant variation in atmospheric pressure due to the change in altitude: the higher the altitude, the thinner the layer of air above us, the lighter its weight, and the lower the atmospheric pressure. While at sea level (altitude = 0 meters) it is slightly above 1000 hectopascals (hPa), it barely reaches 900 hPa at 1000 meters and 700 hPa at 3000 meters. At the peak of Mont Blanc in the Alps (4810 meters), the atmospheric pressure is around 550 hPa. In short, let's just remember that the air around us is under pressure and that this pressure varies by a few hundred hectopascals depending on the altitude. It also varies with the temperature and humidity of the air: the difference in atmospheric pressure between a rainy day and a sunny day is in the tens of hectopascals.

In comparison, the order of magnitude of air pressure variations due to the passage of sound waves is ridiculously low. Thus, the human ear only begins to perceive them when they reach 20 millionths of a pascal (Pa). If we take the pascal as a unit of measurement (100 times less than 1 hectopascal), the variation in atmospheric pressure due to a change in altitude of 1000 meters is on the order of 10,000 Pa; the variation caused by the passage from bad to good weather is around 40 Pa; and the variation caused by the passage of sound waves begins to be audible when it is equal to 0.000002 Pa! On the other hand, when the pressure differential caused by the passage of sound reaches 20 Pa, hearing becomes unbearable and even painful. If we take up the analogy of the pebble in the pond, the variations in atmospheric pressure due to variations in weather conditions would correspond to variations in the average water level in the pond on the order of a meter, while the height of our waves spreading out from the impact of the pebble and representing sound waves would be on the order of a millimeter. Our ears—and the ears of all animals, including insects—are therefore ultrasensitive sensors of variations in atmospheric pressure. Moreover, when caused by atmospheric conditions, these changes occur slowly—in contrast to pressure variations due to the passage of sound waves, which are very fast. The ears are also therefore ultrafast sensors: we are able to detect oscillations in atmospheric pressure due to the passage of sound waves between 20 and 20,000 times per second! Our senses are particularly acute when these variations are repeated between 200 and 10,000 times per second.

If you're feeling like this is a little over your head, don't panic. The important thing to remember is that a sound is a wave of pressure variation that is transmitted by vibrating the air, and that these weak vibrations are perceived by our ears. Hence the two essential notions that define a sound produced by an animal: its intensity and its frequency. The *intensity* of a sound—also known as its amplitude—is the power at which we hear it: it can be of low intensity (we have difficulty hearing it; when someone whispers, for example) or high intensity (when someone shouts). This intensity corresponds to the variation in atmospheric pressure we have just been talking about. Again, let's use the analogy of the pebble in the pond: a small pebble will create small ripples while a larger one will create larger ones. The height of the wavelets is the amplitude of the pressure variation. We have seen that the human ear perceives pressure variations between 0.000002 and 20 Pa. Below that, we hear nothing; above that, the sensation is painful and we are unable to tell if the sound becomes louder or not. In order to take measurements, a logarithmic scale was invented, which spreads these variations between 0 and 120 decibels (dB). A differential of 0.000002 Pa compared with the local atmospheric pressure corresponds to 0 dB and therefore defines the threshold of minimum intensity audible to the human ear. At the other end of the scale, 120 dB corresponds to a pressure variation of 20 Pa. Beyond that, the intensity of sounds is such that it destroys the hair cells of our inner ear, at the risk of leaving us irremediably deaf. A remarkable property of the decibel scale is that an increase of 6 dB corresponds to a doubling of the value of the pressure differential. Thus, when we go from 0.000002 to 0.000004 Pa, the intensity of sound increases by 6 dB. Unfortunately, this dB scale was constructed for the human species. Not all animals have a sound detection threshold at 0.000002 Pa (i.e., 0 dB): some are certainly more sensitive and capable of detecting sounds at lower intensities (negative decibel values), others are less sensitive and only hear sounds above 0 dB. This does not prevent decibels from being very useful for reporting the intensity of a sound, however, and they are widely used in bioacoustics.[9]

Now let's take a little detour to the aquatic environment. We already know that sound waves propagate faster in water than in air. This is due to a difference in resistance to the passage of the sound wave between the two propagation media. Because water molecules are closer to each other than air molecules, they collide with each other more quickly, facilitating the transmission of pressure variations and thus increasing the speed of propagation of the sound pressure wave. In fact, the denser the propagation medium, the faster mechanical vibrations, such as sound, propagate. It is not for nothing that, in Western movie scenes, the gangsters listen for the arrival of the train they plan to attack by pressing their ears against the railroad track. They can hear it coming long before an airborne sound reaches them, because the vibrations propagate much faster in the metal than in the air, at more than 5000 meters per second (compared to the 340 meters per second of sound waves moving through the air). Water as a propagation medium lies in between. It is less dense than metal but denser than air: sounds move through it at about 1500 meters per second. In short, the decibel scale is different from one medium to another. When we work on the aquatic environment, we cannot use the minimum threshold of sound intensity for the human species that has been measured in the air (the 0.000002 Pa discussed earlier). This is another reference point that we use, independent of the human species: 0.000001 Pa, which therefore corresponds to 0 decibels underwater. Yes, all this is very complicated and not very rational. Let's just remember that a measurement in decibels is very beneficial for getting an idea of the intensity of sounds, and that measurements in decibels taken in air and in water are not easily comparable.

We now know that a sound wave is a variation in pressure, which propagates at a speed of 340 meters per second in air and 1500 meters per second in water, and whose wave amplitude defines the intensity of the sound. We also know from experience that a sound can be more or less high-pitched or low-pitched. This characteristic defines the *frequency* of the sound, which in turn corresponds to the rhythm of pressure variations. If this frequency is rapid, in other words, if the pressure alternately decreases and increases at a rapid rate, the frequency of the sound will be high and the sound will be high-pitched. If this rhythm is slower, it will be low and the sound will be lower. Let's take a last look at our analogy of the pebble in the pond. If we throw not just one but several pebbles one after the other, we will create a succession of circles in the water. If our pebbles are thrown very close together, the distance between the circles in the water—the wavelength—will be short; the little cork floating on the water will bob up and down very often. A short wavelength is synonymous with a high frequency. Conversely, if the pebbles are thrown in a more nonchalant way, the distance between the rings will be greater and the small cork will wobble less often. The wavelength is long, which is synonymous with a low frequency. I have already said above that our ear is particularly capable of detecting variations in air pressure when they are repeated between 200 and 10,000 times per second and that the human ear can perceive sound waves with a frequency between 20 and 20,000 cycles per second. Each animal species that communicates through sound has a different range of frequency sensitivity, measured in hertz (Hz). For example, elephants hear very low sounds: they perceive frequencies below 20 Hz, which are inaccessible to us (they are called infrasounds for this reason). Bats, on the other hand, produce extremely high-pitched sounds (up to 100,000 Hz for some species), which are also inaudible to humans (they are called ultrasounds). There is no difference in physical nature between infrasounds, sounds, and ultrasounds; these designations just indicate a difference in human hearing ability.

If you've followed me this far, bravo! Now you know what a sound is: a pressure wave that propagates, in water or in air (or in metal or other solid media), with a certain intensity and frequency.[10] You are now perfectly armed to enter the world of animal acoustic communications. Follow me. We're about to begin our journey in a South American rainforest, the *Mata Atlântica*, where a small bird, sweetly named the white-browed warbler, awaits us.

3

The warbler's eyebrows

WHY DO BIRDS SING?

Morro Grande Reserve, Atlantic Forest, Brazil. A first song reaches our ears: notes that are at first high-pitched, then gradually fall in a regular modulation. A few seconds later: the same song, muffled and indistinct this time. It is a reply. The second bird must be a good 50 meters away and the dense vegetation of the tropical forest greatly reduces the range of its vocalizations. This November morning, two white-browed male warblers engage in a territorial dispute. Thierry and I are the attentive witnesses. The second individual approaches, his more intense song further riling up his opponent. The latter is singing at the top of his lungs, perched a few meters high, then he flies resolutely toward the intruder. The battle is won, the newcomer flees. In birds, as in many other species, the owner of a territory usually wins, because his motivation to keep his property is unparalleled: I'm here and I'm staying put! In the following days, white-browed warblers will have to deal with a new intruder, and by no means the least formidable. We'll arrive armed with a tape recorder and loudspeaker and begin aural jousts with them . . . we'll learn a lot.

Why study this greenish warbler from the Atlantic Forest, the *Mata Atlântica*, on the east coast of Brazil? The idea had been given to us by Jacques Vielliard, then professor at the University of Campinas in Brazil, whose knowledge of the birds of South America and their songs was unequaled. We were interested in the problems that the forest environment poses to the sound communication of birds—dense vegetation being an obstacle to the propagation of acoustic waves. We were looking for an abundant species that was easy to find, producing a song that was simple and short enough that its acoustic structure could be studied without too much difficulty. The song also had to produce a frank (i.e., aggressive) territorial reaction when the bird heard an opponent. We wanted to observe the response of birds to songs that were recorded and then broadcast from a loudspeaker.

"*Basileuterus leucoblepharus*," said Jacques in his slow, modulated voice; the white-browed warbler, as it's commonly known in English. This was the model we needed. "A bird absolutely typical of the Atlantic Forest." The die was cast; our bird would be the warbler. And the work began. We spent long hours in the rainforest, first observing the warblers to learn about their territories and their behavior, then recording their songs and testing them with signals emitted from our loudspeaker. This is how Jacques, Thierry, and I were going to unlock the secrets of their sound world.

The Morro Grande Reserve is only an hour and a half from the megalopolis of São Paulo. But what a contrast! Here, the Atlantic Forest, of which there is almost nothing left, is still splendid. When we first arrived, it was late afternoon. Night falls early in the tropics, and the forest was already rustling with the sounds of the night. Once we got out of the car, Jacques solemnly said to us: "Gentlemen, welcome to the *Mata Atlântica*." I was very impressed: it was my first experience in the rainforest, and my first steps were very hesitant. I imagined a hostile environment, where you could step on a dangerous animal at any moment, where any number of unknown threats were waiting for you. The following morning, in the midst of the dawn chorus, Jacques taught me that the characteristic feature of the tropical forest is its biodiversity and not its density of animals. Reassured, I listened, and within a matter of minutes, we spotted an incredible number of different bird species, each represented by a small number of individuals. The same is true of insects, snakes, and any other species living in the tropical forest: an extreme diversity of species but a small population of each.

We got to work. First, we observed and recorded male warblers. The bird is discreet in appearance, but Jacques was right: the white-browed

White-browed warbler

warbler is one of the most common species at Morro Grande. Its song is simple and short: a few very high, tenuous notes, almost difficult to make out, then a succession of small notes becoming lower and lower, with a very regular descent—"twiiii—twiii—twi—twuu—twu—tuu—tu"—all for about 10 seconds.

Why do birds sing? For their pleasure, one might think; and it is quite possible, insofar as the act of singing seems to fulfill an irrepressible need. However, scientific observations and experiments over the last 50 years have shown that singing has two essential functions: attracting sexual partners and repelling competitors.[11] Sometimes the males are the only ones who sing. Sometimes it is the females that sing. Or females and males vocalize together, in duets so synchronized that we don't know which one produces which note. Songs are indeed communication signals; that is, they contain information for other birds of the same species. Like our words, the songs represent a means of calling out to fellow birds. They allow the American robin or Eurasian wren to identify itself as an American robin or a Eurasian wren, and even to say, "My name is So-and-So." This information is encoded in the acoustic properties of the song. In principle, a song consists of a succession of notes, alternating with silences, according to a particular rhythm and sequence. The notes are characterized by their frequency or pitch (high or low), their amplitude or intensity (high or low), and their variation or modulation during emission (i.e., the frequency or intensity of a note can move). To decipher this coding, the analysis of the acoustic structure of the vocalization is an essential step. However, only experiments consisting in emitting the vocalization from a loudspeaker allow us to really understand the role of the acoustic signal, the information it carries, and how it is coded. It's a bit like asking animals questions.

In the white-browed warbler, males establish territories about 100 meters in diameter, where they live with a female, and which they defend against other males. In the forest, where the birds cannot be seen beyond a few meters, it is through song that the male signals his presence, and therefore his territory, and repels possible competitors. However, in the tropical forest, where the plants intertwine in a profusion of leaves, branches, and creepers of all sizes, it is not easy to make his song heard. These obstacles considerably hinder the propagation of sounds by absorbing their energy and reflecting sound waves in multiple directions, creating echoes and distortions. In such an environment, a song suffers significant changes as it is diffused far away, and the degradation of its structure worsens with the distance of propagation. This is nothing extraordinary: we know that it is difficult for us to understand what someone in a distant room or on another floor is saying. In order to evaluate these signal changes imposed by propagation through the forest, we placed a loudspeaker and a microphone at different distances from each other, and at different heights in the vegetation, at the respective positions of a transmitter bird and a receiver bird. Then we played warbler songs from the loudspeaker, while recording them with our microphone after they had propagated through the forest environment. These experiments gave a clear result: at a distance greater than 100 meters in the forest, the first half of the warbler song (the very high notes) became totally inaudible because the high-pitched sounds were not transmitted well. Only the second half of the song could be recorded. And that's not all: the silences were filled by echoes due to the reverberations of the sound waves on the vegetation. How can a male warbler still identify another male of his species at this distance?

Animal sound signals carry two broad categories of information. The first category is related to the stable characteristics of the singer: his belonging to a species or to a group, his morphology and anatomy, his sex, his weight, his age, his past experiences, etc. This information is described as *static* because, even if it sometimes varies with the individual's age, it forms a kind of identity card for the issuing individual. I sing who I am. The second category is related to the state of the moment; for instance, the degree of motivation to defend one's territory or to seduce a partner, the need to warn of the presence of danger, etc. I sing my state of mind in the here and now. This is *dynamic* information, the value of which is likely to vary rapidly in the signal. Our studies of the warbler focused on static information: we wanted to understand how male warblers could signal both their species ("I am a white-browed warbler") and their individual identity ("I am Marco") despite the extreme constraints on sound transmission in a forest with particularly dense vegetation.

The first step in our experiments was to understand how a warbler recognizes over a long distance that a song is being produced by another male of its species.[12] In other words, how can the acoustic structure that has largely been degraded by propagation through the rainforest still convey enough information to be recognized as a warbler song? Displayed on a computer screen, the spectrogram of a birdsong resembles a musical staff, which is very useful for visualizing its structure. The notes of a warbler song seem to hang on a taut wire from a very high note with a frequency of about 8000 Hz to a much lower one around 2000 Hz (if you haven't read chapter 2 on the structure of a sound, just remember that the frequency of a sound corresponds to its pitch—high, medium, or low). This imaginary thread follows what a mathematician or a mountain dweller would call a slope; here it's quite regular—a *glissando*, as the music lovers say. Since the notes descend from high to low, the song is said to follow a descending frequency modulation slope. By comparing the songs of several males that we had recorded, we easily noticed that they share the same construction rule: the songs of all the warblers follow the same slope. We then hypothesized that this slope carried the information "I am a white-browed warbler." To test this idea, we made computer-generated artificial songs that mimicked the warbler's song, but whose frequency slope was either less steep than in the original song or, on the contrary, more pronounced. The acoustic playback experiments could then begin.

We positioned the speaker in the middle of a male warbler's territory. Thierry operated the tape recorder where the prepared sounds were stored, and I held my binoculars fixed on the area around the loudspeaker. Then we played various songs (the original and the modified ones) and observed the bird's reactions. They exceeded all our expectations. Indeed, in response to a natural song, recorded from a warbler living a few kilometers away, the bird that owned the territory would fly directly to our loudspeaker, singing and screeching loudly. Conversely, as soon as the loudspeaker played a modified song, with even a slightly different slope—no reaction. The bird continued to go about its ordinary business, scornful of our calls. On the other hand, its attention increased sharply again when it heard an artificial whistle in a frequency that followed the slope of the natural song, even without alternating notes and silences. In other words, a male warbler does not need to hear the detail of the notes of the song to understand that a rival is present and is trying to take away his territory. Everything in our results indicated that the slope of the frequency modulation of the song is the criterion, the only criterion, used by white-browed warblers to acoustically recognize their fellow creatures. We had understood the rule followed by this bird to identify members of its species. Pleased as punch, we had found the species signature code of a rainforest bird!

Species identity, however, is only one piece of information among many encoded in the bird's song. We then decided to investigate how warblers differentiate between their neighbors and strangers. Let me explain: when we play back to one warbler the song of an individual recorded at a distance, a bird it has never had a chance to hear before, the tested individual responds very aggressively, whereas if it is a song from a closer territory (for example, on the edge of its own territory), its reaction is greatly attenuated. It has recognized its neighbor, the one who occupies the territory next door and has no designs on its own domain. This recognition between neighbors is frequently found in

Black redstart

birds and other animals. Tudor Draganoiu, a researcher at the University of Paris Nanterre, has studied it in a bird species common in France: the black redstart *Phoenicurus ochruros*. These little birds love old stones and establish their territories in small villages in the Massif Central and the Alps. They are easy to observe because the males do not shy away from singing from the rooftops. Their song consists of several successive musical phrases: the first are rapid successions of notes, while the last one sounds like rustling newspaper. Tudor has shown that redstarts from the same hamlet have similar melodies—they share common phrases—and they tolerate each other. They can spot a stranger from his song and then forcibly chase him away. This is known as the dear-enemy effect: it allows the birds not to worry unnecessarily, and to reserve the energy allocated to the defense of the territory for really perilous situations; in other words, for when an unknown intruder arrives and seeks to establish himself. A neighbor whose territory is stable does not represent a great danger.[13]

In the case of our warblers, we first had to establish whether their song has a signature that could differentiate one individual from another. The analysis of the acoustic details of the recorded songs was very instructive: not only did they all follow the species-specific rule of descending pitch, but the descending slope of the notes was also punctuated by small breaks here and there. In other words, the gradual descent of the frequency of the notes—from the highest to the lowest—is in fact not perfectly regular; sometimes one note that follows another is a little lower than might be expected if the regularity were perfect. Remarkably, the songs of the same individual all have the same irregularities, although they differ from one male to another. They could therefore seriously be suspected of bearing the individual signature of a male and allowing recognition between neighbors—an interesting hypothesis that still needed to be tested.

The second stage of our playback experiments would confirm this. When our loudspeaker, placed at the edge of the tested bird's territory, emitted the song of its neighbor, our warbler continued to indulge in his warbler routine. However, if we changed the irregularities of slope in this song, the reaction was immediate: no longer recognizing its usual neighbor, the bird rushed to threaten the intruder. I'm always amazed when a playback experiment provokes an animal's reaction. Isn't it incredible to be able to enter into the intimacy of a nonhuman world in this way?

In this second stage of our exploration, we had discovered something exciting: the irregularities of slope that bear the individual signature are only found in the first part of the male's song, the very part that is transmitted only at short distances—a few dozen meters at the most. A bird will therefore be able to locate each of its close neighbors individually but will be unable to precisely identify males further away, since the information that encodes their individual identities has disappeared, absorbed along with the high notes by the tropical vegetation! The individual identity of the white-browed warbler, therefore, is private information: only the closest neighbors, and probably females in or near the territory of the singing male, have access to it.

Another bioacoustician, Frédéric Sèbe, who has since become a close colleague, was to show some time later that the irregularities of the songs of neighboring warblers sound more similar than those of songs from distant individuals. In other words, the closer the males' territories, the more similar the individual acoustic signatures. This is reminiscent of the shared phrases of the black redstart. In the warbler, we do not know if this resemblance between neighbors is the result of an imitation of a signature, which would allow the structure of the songs of individuals living next to each other every day to converge, or if it is the result of genetic proximity. Perhaps the males of this species tend to establish their territory close to their hatching site and then find themselves close to a brother or a cousin, whose vocal characteristics they share. As for the possible interest of this sharing of vocal signatures, it remains a mystery. One can imagine that it facilitates recognition between neighbors. It seems difficult to test these hypotheses on the warbler of the faraway Atlantic Forest, but we can hope that the French redstarts, less difficult to access, will let us in on their secret.

In our warbler, the coding of the two pieces of information—species identity and individual identity—is thus adapted to the constraints that the tropical forest imposes on the transmission of sound waves: if any listening warbler, even a distant one, is interested in spotting my presence, then he (or she) can, but only my neighbors need to know that I am Marco.[14] We have also shown that a warbler can locate the position of a transmitting individual based on the extent of the changes in its song during transmission. Indeed, the longer the propagation distance, the fewer notes remain in the song. After 20 meters, let's say that, from the original song "twiiii—twiii—twi—twuu—twu—tuu—tu," we only hear "twi—twuu—twu—tuu—tu." At 100 meters from the loudspeaker, only "twuu—twu—tuu—tu" is audible. And Marco knows whether or not to start a fight with the emitter!

Therefore, despite the strong propagation constraints to which it is subjected, the warbler's seemingly simple song contains information enabling these birds to manage their social relationships.[15] This example illustrates how important it is to break through the mechanisms for encoding and decoding the information present in the songs. This is how the mysteries of a sound-based communication system can be deciphered, and how we can understand which environmental constraints have shaped this system over the course of evolution.

Numerous studies have sought to test whether the sound signals emitted by animals, particularly the vocalizations of birds, are precisely adapted to the environment in which they live. The results of these investigations were somewhat contradictory.[16] It was thought that species living in forest environments produce lower notes, degrading less quickly than high notes. Their songs should also be poorly modulated, i.e., made up of notes whose frequency varies slowly, making them less sensitive to the echoes caused by the reverberations of sound waves on trunks and branches. Yet a forest bird like the wren produces a very high-pitched song, made up of a succession of very fast trills. It therefore degrades as it spreads through the forest. To put it another way, the hypothesis of "acoustic adaptation" (which we discuss in chapter 17) was only partially supported. To make sure, Thierry and I asked Jacques to propose a list of birds in the *Mata Atlântica* that represented the different layers of the forest, for which he would have quality recordings in his archives. As Jacques had spent his life recording birds, he had a great sound library and plenty to satisfy us!

After a few days spent preparing our tape recorders, we were ready for new adventures, new propagation experiments. The forest was going to ring with the songs of the brown tinamou, the squamate antbird, the white-browed warbler (again and again), the white-shouldered fire-eye, the swallow-tailed manakin, the giant antshrike, the surucua trogon, and the hooded berryeater.[17] This list is not a random miscellany: it ranges from species that spend most of their time on the ground or at very low altitudes to those that live in, and even way up high on, the canopy. Our hypothesis was that the song of each one propagates better, and at a longer distance, at the height from which the bird usually vocalizes (otherwise,

Hooded berryeater

Surucua trogon

Giant antshrike

White-browed warbler

White-shouldered fire-eye

Squamate antbird

Swallow-tailed manakin

Brown tinamou

why choose to stand there?). As with what we had found in the literature, the results were not obvious to interpret. Take the brown tinamou: it walks on the ground in search of its food and does not perch to sing. Yet the range of its song would be greater if it sang at a height of several meters. Quite the reverse is true for the hooded berryeater, which lives at the very top of the canopy, and it is indeed here that the greatest range of its song is obtained (we hired a tree climber to place the loudspeaker there). In short, some birds sing at positions where their melody spreads efficiently over long distances; others do not. This is probably because a bird's position in the forest strata does not depend solely on acoustics. Many other factors come into play: Where is its favorite food? Where are the predators? Where is it sunny more often? Where is it windy? And who knows what else? Sometimes, however, it seems easy to explain why a species emits a signal that is difficult to hear from a distance. For example, swallow-tailed manakin males gather in assemblies called leks, in small forest clearings. There, females join them—they know the place! Their courtship ritual combines singing and dancing, with the males hopping on branches. No need to be heard from afar. On the contrary, by emitting songs of relatively low intensity, the manakins probably avoid announcing their whereabouts to predators. Clever, those manakins.

Two birds that are very common in Europe were going to teach me that there are other ways to improve communication in the forest other than emitting songs with an optimal acoustic structure to disperse them. Let's take a look at the cute Eurasian wren *Troglodytes troglodytes* and the Eurasian blackcap *Sylvia atricapilla*. The former belongs to a rich French research tradition. Jean-Claude Brémond, the man who introduced me to the world of bioacoustics, as I stated in chapter 2, had cleared up the coding of the identity of the species, showing that male wrens are very sensitive to the rhythm of the notes, i.e., the speed at which they are produced, and that this is what enables them to recognize other male wrens.[18] Michel Kreutzer, from the University of Paris Nanterre, had also shown that each wren population has a particular dialect, with acoustic properties different from the songs produced by other populations, similar to regional patois and accents in humans.[19] As for me, I decided to study the behavior of this bird in the presence

Eurasian wren

of songs propagated over a great distance; in other words, songs whose acoustic structure is very degraded.

The wren is a tiny forest bird. Imagine a small ball of feathers, with a fine beak and a small, erect tail, sneaking through the bushes with elegance and precision. The male sings surprisingly loudly for its size and actively defends a territory about 50 meters in diameter. I made a very simple experiment: I placed a loudspeaker on the ground, in the middle of a wren's territory, and then broadcast the vocalizations of another wren that was unknown to him, having been recorded far from his "home." Two versions of this unfamiliar song were played in succession: one was what in science is called a *control stimulus*, i.e., an unmodified version supposed to provoke a strong response; the other was an *experimental stimulus*, i.e., the same song but recorded after propagation over a long distance in the forest. Each time (because the experiments must, of course, be repeated), the tested bird responded to both versions by vocalizing in response. However, the exploration behavior that followed was very different. In response to the undisturbed control song, the birds came very close to the loudspeaker, some landing squarely on it, frantically searching for their invisible opponent. On the other hand, when they heard the experimental stimulus—the melody coming from far away—they preferred to perch several meters above the ground to sing. Why such a difference? I made two hypotheses. First, like the white-browed warblers, the wrens must have perceived information about the distance separating them from the individual transmitter: the control song (not degraded by the propagation) was to be interpreted as that of an intruder who had entered their territory, while the song propagated over a long distance was attributed to a more distant bird. Second, perching to respond to a distant individual might improve the quality of the signal and the information to be transmitted into the distance.[20] Both hypotheses had yet to be confirmed.

The wren's song is powerful and high-pitched. The hypothesis that it is seriously degraded during long-distance propagation could be tested by experiments of signal propagation in the forest. The results proved quite interesting: the degradation of the acoustic quality of the song depended strongly on the height of the loudspeaker placed in the tree, and especially on the height of the microphone. When the loudspeaker was located high up in the tree, the song was certainly received with greater intensity and underwent fewer transformations (fewer echoes and better preservation of high notes), but these improvements were much better when the *microphone* was perched several meters high. Here one must keep in mind what a forest environment typically inhabited by wrens looks like. Starting from the ground and moving to the treetops, there are several strata: first, an area of fairly dense bushes, with a few shrubs and trees in early stages of growth; then a relatively clear space between the top of the bushes and the underside of the canopy; and, finally, a more or less dense area where the tree branches intertwine. The wren spends most of its time in the bushes, not very far from the ground. Perching when it sings improves the range of its vocalizations: the song will be heard from further away. But the improvement is mainly in the quality of its listening: perched high up, it can more easily perceive another wren chirping in the distance.[21]

Eurasian blackcap

A few years later, together with my colleague Torben Dabelsteen, a professor at the University of Copenhagen, we were to confirm these results by working on another species of European forest bird, the black-cap. Torben is a specialist in sound propagation issues. By making very precise measurements, we showed that when a blackcap moves from a perch at a height of 4 meters to a perch at 9 meters, the horizontal range of its song increases by about 25 meters. In other words, perching a few meters higher is equivalent to getting the blackcap's song 25 meters closer to a possible receiving bird. Note that this 25-meter gain from perching at a greater height applies to all directions. Since the diameter of a black-cap's territory is between 50 and 100 meters, this increase is enormous and has a considerable advantage.[22] Let's do a little math to get the ideas straight. Let's say that the song of a blackcap perched 4 meters high can be heard up to 50 meters from the bird. The area over which the song will be heard around the warbler will be equal to 3.14 (the number pi) multiplied by the square of 50 (remember your college lessons: the area of a disc is equal to the product of the number pi multiplied by the square of the radius of the disc), i.e., $3.14 \times 50^2 = 7850$ square meters. When the blackcap perches at 9 meters, the range of the song (thus the radius of the disc) increases by 25 meters, as we've said. The area over which the song is heard then becomes $3.14 \times (50 + 25)^2 = 17,662$ square meters. That is more than double the initial surface of 7850 m^2. This means that a blackcap that climbs 5 meters into a tree more than doubles the area in which its song is heard. Playback experiments further showed that the blackcap is able to identify a song of its species even if it has been particularly degraded during its propagation in the forest.[23]

Acoustic communication thus makes it possible to exchange information even if we cannot see each other, which is particularly interesting in an environment with dense vegetation. Clearly, while the echoes created by the reverberation and absorption of the energy of sound waves by vegetation can greatly modify the acoustic structure of signals, birds have developed strategies during their evolution that overcome these drawbacks: perch high up, emit a song with high intensity or propagation-resistant characteristics, use information coding that counteracts or even uses signal degradation.

However, forest environments present yet another difficulty for acoustic communications: competition for sound space, when multiple species—not only birds but also insects and amphibians—use sound to communicate. A real mess! Don't they get in each other's way? How do they navigate this huge jumble of sound? This question has been the subject of many studies and, it must be said, of a certain amount of speculation.[24] The acoustic avoidance hypothesis has thus had a good run and is still being actively discussed (along with the acoustic niche hypothesis[25]). It's the "Let's not get in each other's way" hypothesis. Animals might talk to each other on different frequencies, with different rhythms, even at different times of the day or night, to avoid interference from equally voluble neighbors. This hypothesis has been verified in some animals, particularly crickets, whose noisy stridulations are more or less acute depending on the species.[26] But if you've ever listened to the dawn chorus of multiple species of birds singing together just before sunrise, you'll certainly agree that acoustic avoidance is not a priority for them. In fact, there are several strategies for transmitting and receiving sound signals that allow animals to identify the signal they are seeking in the midst of the ambient cacophony. I'll take you far away from the rainforest, to much colder climates, to see how seabirds use acoustics to communicate in the terribly noisy environment of their breeding colonies.

4

Cocktails between birds

NOISE AND COMMUNICATION THEORY

Hornøya Island, Sea of Barents, Norway. From the lighthouse where I'm staying, you can see the furious sea. To reach the gulls' cliff, we have to cross a windswept plain, where angry gulls dive on me as if they wanted to open my skull; probably convinced that I'm getting too close to their young. Once I reach the cliff, the white birds' antics become dizzying. Nests overflow from every crevice. The cliff is bristling with them. Everywhere, seagulls are nesting. The season will be short, and woe to the adults who are late in breeding as their young have no chance of survival. The incessant screeching drowns out the sound of the waves. Here one individual—female or male?—vociferates as it flies toward the rock. It comes to relieve its partner, who responds just as loudly. How do these gulls find their way through the multitude of nests in this cliff building? No doubt the birds have a precise knowledge of the topography. But chances are that their partners' calls help them find their way, despite the surrounding noise. This is what I came here to study, at the farthest reaches, with Thierry, my usual partner in crime.

In this chapter, we discuss a concept that is essential for all acoustic communication—the *noise*—and see how animals manage to exchange information through sound signals in spite of it. All communication involves three elements: an emitter, a signal (sound waves), and a receiver. The emitter, or *sender*, produces the signal in which information is encoded. The signal propagates in the environment while being transformed to a

Gull

greater or lesser degree. As we have already seen, the sound waves propagating through a forest reverberate on the vegetation, creating echoes; the energy of high-pitched sounds is more easily absorbed than that of lower-pitched sounds. The signal is then received by the receiving individual. The receiver hears the signal and decodes its information. Once this information has been integrated into its brain, it will potentially be able to react, for example, by attacking the intruder or responding to its partner. This succession of steps from the sender to the receiver conditions all communication and constitutes the chain of information transmission. It was an American, Claude Shannon, who in 1948 formalized these concepts by establishing the mathematical theory of communication (typically known as information theory).[27] In scientific language, contrary to what is sometimes thought, a theory is not a hypothesis or something that has not yet been proved and could have competing explanations. Instead, a scientific theory is a system that explains a part of the world around us. It is, in a way, the result of a large body of scientific research whose observations and results are consistent. A scientific theory can certainly evolve as a result of new discoveries, but if it is accepted, it is because no knowledge or experience has ever come to oppose it at its foundation. In the life sciences, two theories are particularly well founded: the cell theory, which states that all living beings are made up of cells and that every cell comes from a mother cell (the corollary of which is that spontaneous generation does not exist and that all animals, including humans, share a common ancestor), and the theory of evolution, which explains that living beings change over time, under the combined effect of chance (e.g., genetic mutations) and selective constraints favoring one characteristic or another (natural and sexual selection).

The mathematical theory of communication was not originally intended for biologists. It had been established to provide a framework for the telecommunications research that was taking off at that time. The aim was to characterize both qualitatively and quantitatively the deformations undergone by any signal during its transmission in a propagation channel. More specifically, the aim was to understand how a signal can carry information from a transmitter to a receiver in a noisy channel, and thus to implement strategies for encoding information in telecommunication signals in order to transfer messages as efficiently and quickly as possible. The information can be seen as a cube of a certain volume that must pass from the transmitter to the receiver. At the time it is produced by the sender, the signal corresponds to a certain volume of information. During its propagation in the environment, the signal is deformed. When it reaches the transmitter, the volume of information it encodes has decreased: the volume of the cube has become smaller. In a way, the information cube is stripped down during propagation. This reduction in the amount of information is the result of noise. To understand this, let's take a concrete example. Imagine that you have a breakdown on the side of the road and you call your mechanic and say, "My car has broken down at the side of road 2 at mile marker 103." If the telephone transmission is unsatisfactory, your mechanic may hear "grrr ... M ... car ... broke down ... grrr ... road ... 103." Your mechanic will understand that your car has broken down and you are calling for help, but the original information has not been transmitted in full. The amount of information you sent has been reduced by two things: on the one hand your words have been distorted, and on the other hand some parasitic sounds ("grrr") have been added. These two types of modifications, which in nature correspond to signal modifications due, for example, to reverberations on vegetation and the vocalizations of other animals, contribute to the noise of the signal. In fact, they *are* the noise.

Shannon's goal was to understand noise in order to imagine ways to encode information in telecommunication signals to limit its impact. Let's take the example of the roadside breakdown. How can you make your mechanic hear you despite the noise from the telephone transmission? Shannon's theory shows that three strategies are possible. The first one is obvious: you can repeat the message until you are sure that the mechanic understands it. Emitting the same signal several times in order to improve the transmission of information in a noisy environment is to practice what is called information redundancy. Since the noise of the telephone line is irregular, each time you repeat it, a different part of the signal will reach your mechanic—e.g., first that your car has broken down and then what your location is. A second strategy is to talk louder, or even to shout into your phone. Increasing the strength of the signal makes it stand out against the noise. Finally, you can change the tone of your voice, by making it higher-pitched, for example, especially if the crackling on the line is in the low register. Avoiding sound frequencies occupied by noise is a strategy we often use when the propagation channel is congested, for example, when talking to someone at a noisy party. By using these three tools, you try to make your signal stand out from the noise, which in acoustics jargon means increasing the *signal-to-noise ratio*. These three strategies—redundancy, increasing the intensity, and modulating the tone of the voice—play on the three components of the sound signal: time (increasing the duration of the signal), amplitude (increasing the strength of the signal), and frequency (changing it to avoid the frequency band occupied by the noise). Shannon put all these processes into mathematical equations, and it became possible to make predictions about the coding strategies to be employed based on the characteristics of the noise and the information to be conveyed. I won't take you that far, don't worry. But we will see that nonhuman animals grapple just as much with the problem of noise from acoustic signals during their communications, and that they, like us, develop strategies in accordance with the mathematical theory of communication. This theory actually applies to the entire living world. It is particularly useful for the study of animal and human acoustic communications.[28]

Most seabirds—seagulls, gulls, albatrosses, shearwaters, auks, penguins, and others—breed in colonies of hundreds, thousands, even millions of individuals. Finding a place to nest is not easy when you spend

Black-headed gull

most of your time in the middle of the ocean—or even in the ocean for penguins, which lead a decidedly aquatic life. Seabirds therefore find themselves nesting on the same coasts, sharing the available space. Nesting together limits the risks. While a predator would not hesitate to pounce on an isolated bird, it will hesitate to venture into a colony where individuals are densely packed. Another characteristic of most seabird species is that individuals form breeding pairs. They are said to be monogamous and in principle particularly faithful, at least during the same breeding season, if not for life—so much so that raising young in the marine environment requires unfailing parental cooperation. Being able to recognize your partner among the other birds in the colony is therefore essential. Sometimes, identifying one's young can also prove very useful, as in some penguins where the young gather in a kind of nursery and the risk of confusion is high. Scientific research carried out in the field, as close as possible to the animals, has shown that acoustic communication plays a primordial role during this recognition between partners and between parents and their young.

It was with Isabelle Charrier, my first PhD student and now director of research at the *Centre national de la recherche scientifique* (the famous CNRS), that I began my investigations into parent-offspring recognition in a gull species, the black-headed gull *Larus ridibundus*. Like other gull species, the black-headed gull forms breeding colonies in which each pair establishes its nest sometimes only a few meters from the nest of the neighboring pair. The black-headed gull is quite eclectic in its choice of habitat and is easily found inland, where it establishes breeding colonies on ponds. It was in France, on a pond in the Forez plain, that Isabelle and I studied it. The doors of the pond had been opened to us by Jean-Dominique Lebreton. Director of research at the CNRS, member of the Académie des Sciences, erudite, and passionate about nature and animal behavior, Jean-Dominique had been studying the demography of the black-headed gull for many years and was delighted to see ethologists interested in his favorite bird. We wanted to record the calls of the parents of young gulls and then test with acoustic playback experiments whether the young were able to recognize their parents from their calls. This meant getting as close as possible to the animals. How could this be done? Jean-Dominique had a perfectly successful technique: "You're going to have to build a floating blind. And squeeze into a wetsuit. You will go unnoticed and be able to approach the nests." Imagine our construction: a wooden and polystyrene frame, covered with branches and leaves, with small openings allowing us to observe outside and to pass a pole on which our microphone was fixed. I ordered two diving suits from the administration of my university, who questioned me about the purpose of the operation, suspecting me of organizing a holiday by the sea. We built a blind large enough to house two people. Walking carefully on the bottom of the pond, we could finally approach the birds. "Careful, not too close, though—you mustn't disturb the broods," warned Jean-Dominique, always concerned about preserving nature.

The pond of the Ronze is a magical place. It was a rare privilege for me to enter it incognito, hidden beneath my blind. As for the flora, the water lilies, pondweed, and reeds were reflected in the water and enchanted the eye. A multitude of different species of birds gathered there: coots, teals, mallards, and redheads paddled across the surface. Some grebes too, including the little black-necked variety, whose red tufts and vermilion eye make it look like an operetta marquis. And gulls by the hundreds. Making an unholy racket. To wait near a nest for one of the parents to come back to feed the young, record its calls, then go near another little gull family, such was our first step. The next day, we came back with a camouflaged loudspeaker fixed on the blind, to play parental and nonparental calls a few meters away from the young in "our" nest from the day before, hoping to provoke different reactions that would prove they could distinguish their parents' voices from those of other adults. This merry-go-round had to be repeated day after day—true as it is that, in science, each experiment must be repeated and repeated again for the result to be validated. Busy with my university classes, I could only go to the field once or twice a week, and it was Isabelle who had to provide the bulk of the effort. Within a few weeks, her patience was rewarded with success: we obtained experimental proof that young gulls recognize their parents by voice. Above all, we showed that, in the hustle and bustle of the colony, redundant information is necessary for optimal voice recognition. Sometimes a particularly sensitive chick would be able to recognize its parent if we made it hear a single call. Half of the chicks tested needed three calls. However, when the parental call was repeated at least four times, 100% of the chicks tested responded to the voice of their parent while ignoring the voice of another adult in the colony. The behavior of our gulls was consistent with the mathematical theory of communication.[29]

If the noise of a gull colony is already a major constraint to communication, this constraint can be much more extreme. This is the case with king penguins. To my regret, I have never worked on king penguins. That would have required missions of several months in the Southern Hemisphere, which my position at the university made difficult. I also did not want to stay so long away from my family. So I experienced the penguin adventure vicariously. But Thierry told me so much about it, and I've read so many publications and seen so many films that I almost feel like I've been there! Come with me, let's go to Crozet Island, in the middle of the Indian Ocean, to discover how the king penguins use acoustics to recognize each other. This is certainly the most beautiful example of communication in a noisy environment that nature gives us to observe.

Penguins form a unique and very special family of birds, the Spheniscidae. Their peculiarities? They live only in the Southern Hemisphere, are unable to fly, and spend most of their time swimming underwater—sometimes at very great depths—to fish. They only go ashore to reproduce, which entails nuptial parades, mating, brooding, feeding, and protection of the young against the cold for the species closest to the South Pole. There are about 20 species of penguins. Even if you have never been particularly interested in these birds, you certainly know the large penguins: the king and the emperor (*Aptenodytes patagonicus* and *A. forsteri*, respectively). In these two species, the adults and their young form huge colonies, which can number thousands of birds—up to a million for the largest colony of king penguins. But the penguin parents are faithful and exclusive: they look after only one young at a time. The question of recognizing family members is therefore crucial. First, during the incubation period and the first days posthatching, when one of the two parents is brooding or staying with the newly hatched young, the other is at sea, restoring its health. When it returns, it must find its partner in the colony. This is a difficult task because large penguins do not make nests: the parent can walk around with the egg or chick on its legs, warm in a fold of skin. Later, when the young have gained a little independence and can join other penguins in a kind of mobile crèche, each parent has to be able to find the chick and feed it. If recognition is not forthcoming, the outcome is certain: the chick will die of hunger. The penguin pair can't rely on the other parents. Long ago, barbaric experiments in which penguins had their beaks taped shut to prevent them from vocalizing showed that adult penguins who had been rendered dumb were unable to recognize their partners even if they passed each other. In adult penguins, recognition between partners in a couple or between parents and chicks is only vocal. The problem with this acoustic communication is obvious: at any given moment, there are dozens of birds in the colony looking for their partner or young, shouting at the top of their lungs. The background noise is frightening. It can exceed 95 decibels, which is considerable (between a lawn mower and a jackhammer). Moreover, since this noise is produced by the penguins themselves, it is obviously not possible to avoid it by pitching the frequencies of their voices higher or lower. Nor is it possible to squawk more loudly: everyone is already giving their all! For the emperor penguin, finding the right chick can take more than two hours—a real challenge. And they succeed![30]

Let us return for a moment to the chain of information transmission that characterizes all communication: an emitter, a signal, and a receiver. To communicate in noisy environments, each element of the chain is decisive. With our previous example of the breakdown on the highway, we know that the sender can adopt signal production strategies that will facilitate the transmission of information. Are such strategies found in penguins?[31] One of the scientific studies led by Thierry on this subject was published in a prestigious British journal under the title "How Do King Penguins Apply the Mathematical Theory of Information to Communicate in Windy Conditions?" It has met with great success in the world of bioacoustics and beyond. When the wind is strong—and it is true that it's windy under the latitudes where penguins live—the difficulty of communicating in the colony increases considerably. This wind contributes to the noise of the signal. To counteract it, the king penguins practice redundancy of information. Their song is a succession of syllables, all of which carry the individual identity of the emitter. Researchers have measured wind speed and counted the number of syllables in the penguins' songs. And, hold on tight, the number of syllables in the song of the king penguin is proportional to the wind speed! It's hard to believe. More precisely, as long as the wind speed remains below 8 meters per second, the penguins produce songs of between 4 and 6 successive syllables. Above 8 meters per second, the songs get longer, increasing to 10 syllables when the wind blows at 9 meters per second and easily reaching 14 syllables when the wind speed exceeds 11 meters per second. The relationship between the number of syllables and the wind speed is linear, meaning that the number of syllables can be predicted just by knowing the wind speed.[32]

When vocalizing, penguins use many other strategies to counter the noise of the colony and the resulting confusion. For example, when one individual starts to squawk, all its neighbors within 7 meters fall silent. This courtesy rule limits interference and reduces signal noise. In addition, penguins vocalize by adopting body postures that facilitate the transmission of their signals. For example, the king penguin points his beak toward the sky and extends his neck as far as it will go: the sound waves then propagate above the bodies of the other penguins and avoid being absorbed by the masses of feathers along the way. The emperor penguin uses his chest as a reflective mirror: he sings with his beak lowered against his torso, which reflects the beam of sound waves in the direction of his partner.

The adaptations of penguins to communication in noisy environments cannot be limited to the behavior of the emitter. Penguin song is a signal with interesting characteristics in the face of this constraint. The analysis of the temporal and frequency structure of the song, together with playback experiments with vocalizations that have been previously modified, show a very precise individual vocal signature. This signature is coded differently depending on the species of penguin. In the king penguin, it is the frequency modulations that count. "Wrooinnn—wrooin wrooin wrooin wrooin—wrooin wrooin"—the song of the king penguin sounds like a trumpet blast. On the spectrogram, the frequencies rise and fall several times, rapidly, with a rhythm and acceleration specific to each individual. If you artificially make the song lower- or higher-pitched, it does not alter the recognition. On the other hand, the slightest change in the frequency modulation, and the signal of the partner or parent becomes an unknown voice.[33] If the receiving bird is close, half a single syllable is sufficient for recognition: "wroo. . . ." As with the black-headed gull, redundancy of information is important: the penguin trumpets and trumpets again and again, repeating its simple name: "wroo . . . wroo. . . ." The emperor penguin is a bit of an exception: in his case, three successive syllables are necessary. In these two penguins, the individual signature is coded by the temporal dynamics of the song: frequency modulation in the king's case,[34] amplitude modulation in the emperor's case.[35] Another remarkable fact: both species produce two-voice signals.[36] I'll try to explain the phenomenon in a simplified fashion.

To produce a vocalization, an instrument is required. As in the world of music, there are two main categories: wind instruments, such as the trumpet, where a breath of air makes a flexible structure vibrate; and friction instruments, such as the violin, where solid structures rub or collide, making them vibrate as well. In both cases, the vibrations are transmitted to the air, producing sound waves. Birds and mammals have wind instruments: the syrinx and the larynx, respectively. Other modalities of sound production may exist in these two groups of animals; for example, some birds have feathers that vibrate during nuptial flights and produce sounds.[37] But the larynx and syrinx are the two main vocal instruments of mammals and birds. Larynges and syringes are very complex organs: cartilage, bone, muscles, and the nervous system work together to vocalize. In simple terms, they are first of all membranes placed in the path of the air flow coming out of the lungs—membranes that will be stretched by muscles to a larger or smaller extent, allowing the emitted frequencies to vary. We see this in more detail in the next chapter. For the moment, you should know that the larynx of mammals and the syrinx of birds differ in one essential point: they are not in the same place. While the larynx is located high up on the respiratory tract (at the Adam's apple), the syrinx is placed very deep, just at the exit of both lungs, at the junction of the two primary bronchi—the tubes that carry air into the lungs.[38] In some birds (e.g., the songbirds), there is a functional originality of the bird syrinx: it is a double organ. With two systems of vibrating membranes, controlled by two nerve commands, the syrinx is a musical instrument that can emit two different sounds at the same time. This is in theory.[39] In practice, in most songbirds, either the two voices of the syrinx are perfectly tuned because its two parts are coupled, or only one of the two semisyringes emits a sound while the other remains silent; sometimes the two semisyringes alternate. One could say that there is only one voice.

In large penguins, the two semisyringes produce two sounds of different frequencies at the same time. This is called the two-voice phenomenon. These two frequencies are not very far apart, of course, but they are far enough apart to be noticed on a spectrogram and to play a role in encoding the individual identity of the transmitter. It is in the emperor penguin that the two-voice phenomenon is most remarkable. The two frequencies are close enough to generate a bizarre phenomenon called beats, which results in a regular modulation of the signal amplitude: "WO-wo-WO-wo. . . ." I will not dwell here on the physical explanation of the beats. Just know that they are well known to guitarists: they get them by simultaneously plucking two strings emitting very close notes. In the emperor's case, the rhythm of the beats is characteristic of each individual. It is the double syrinx that controls the vocal signature. This coding of individual identity information is extraordinarily reliable: while intensity modulations managed by variations in the flow of air leaving the lungs would probably be fluctuating, the beats depend directly on the anatomy of the syrinx of the transmitter bird. Since each individual has a unique anatomy, the beats are unique. In addition, an individual's beat remains more or less intact during propagation, because the two frequencies that create it reach the receiving individual with little distortion and thus restore the entire initial beat to its original level. Beats carry the signature of the emitter far and wide with great reliability.[40]

Large penguins therefore have emission strategies and a sound signal that are well adapted to the constraints placed on individual vocal recognition by the noise of the colony. On the signal reception side, they are also very efficient. Chicks have shown an extraordinary ability to extract information from a noisy signal. In playback experiments, the call of one of the parents was mixed (as a DJ might) with the calls of other adults, mimicking ambient noise. This controlled-signal noise allowed researchers to set a very precise signal-to-noise ratio; that is, the emergence (or disappearance) of the parent's voice against the cries of other adults could be precisely controlled. The results of the experiments are breathtaking. King penguin chicks were able to recognize their parent's voice even if its intensity was 6 decibels lower than that of the ambient noise.[41] In other words, even if the noise level is twice as loud as the parent's call, the parent is still heard and identified by its chick. By way of comparison, we humans would have great difficulty identifying a call as soon as its intensity falls to the level of the noise. This astonishing ability to extract the signal from the noise—the cocktail-party effect—is based on the ability to identify the frequency modulations of the parental call lost in the middle of the heterogeneous noise. It ensures the survival of the king penguin chick.

The two large penguins are the ones facing the most important constraints for recognition between partners and between parents and young: terrible ambient noise, birds constantly moving around the colony, and no visual cues. Other species of penguins, such as the Adélie penguin *Pygoscelis adeliae*, form less dense breeding colonies, where each pair incubates its egg in its nest, which is often a simple, small depression in the ground, but a nest nonetheless—a clearly identifiable home—where the young stays and is fed by its parents until it becomes independent. In these species, adults returning from a feeding trip to the ocean will not have much difficulty finding their partners or young. They first return to their nest, probably based on their visual knowledge of the area. It is only on arriving at the nest that they start to vocalize. The colony is therefore much quieter, and the ambient noise is much less pronounced than that of the colonies of large penguins. The same experiment that I have just told you about on the king penguin—the one with the parent's song mixed with vocalizations from other adults—was carried out on the Adélie penguin. Under these constraints, the Adélie penguins are much less successful than their royal counterparts. As soon as the parent's call no longer emerges from the noise (i.e., a signal-to-noise ratio of 0 decibels), the chick becomes unable to recognize its parent's voice. The natural constraint is lighter, as is the sensory performance.[42]

There is much more to be said about penguins, such as the fact, for example, that a king is particularly gifted at finding the exact location of his or her singing partner or parent.[43] Are these differences in the reliability of vocal recognition between species with different levels of risk of confusion found in other animals? Early studies have compared species of American swallows nesting in large, dense colonies with solitary swallows where each pair breeds in isolation. Individual vocal signatures are much more pronounced in the calls of colonial swallows.[44]

The results of Isabelle Charrier's work on acoustic communications in seals are in the same vein: the more densely a species forms its breeding colonies, with a high risk of confusion between individuals, the more reliable the vocal signatures of mother and young are and the more effective the individual recognition.[45] The transition is now in place: in the next chapter, we will observe the communication between mother and young in pinnipeds. We are about to go deeper into the complex world of acoustic communication between parents and their offspring.

5

Family dinner

PARENT-OFFSPRING COMMUNICATION

Igloolik, Foxe Basin, Arctic Province of Nunavut. After pushing the snow clouds southward, the icy wind from the pole has died down and the low Arctic sun is timidly reborn. We will be able to set sail. Aboard a small motorboat, a derisory walnut hull in the immensity of the ocean, I am with Isabelle in search of walruses, these enormous animals with long tusks resting on the blocks of the fractured pack ice. Isabelle studies marine mammals, and she is an outstanding field researcher. "The objective of the mission is very simple," she told me. "To find females with their young and test if the young calves are able to recognize their mothers by voice." But nothing is easy in this inhospitable region, starting with the journey to the camp. It required several flights from France, with the size of the plane becoming progressively smaller at each stopover. In Iqaluit, the gateway to Nunavut, we definitely left the heat of the Canadian summer behind us. As we flew over Foxe Basin, the first ice floes appeared. Finally, we reach Igloolik and its ice pack— an immensity of white. Only the airstrip is clear. Another hour of snowmobiling to the cabins, then ... 13 long days of Arctic summer follow, confined in our cabins, completely isolated from the outside world by the storm. The Arctic is untamable. It's the most hostile environment I've ever been in.

Walruses, seals, elephant seals, fur seals, and sea lions form a group of marine mammals called pinnipeds.[46] Marine, yes, but not through and through. Unlike whales and dolphins, pinnipeds must go ashore to reproduce. On land . . . or sometimes on the ice, like the walrus when in the Arctic spring the pack ice is just beginning to break up and the mainland shores are still inaccessible. Some pinnipeds, especially eared seals, form large colonies of tens, hundreds, and even thousands of individuals, similar to the penguins we discussed in chapter 4. The young face the same problem: finding a parent who will feed them. In pinnipeds, being mammals, it is the mother who suckles her young, and therefore who feeds, whereas in penguins, both female and male feed the chick. As in the case of penguins, the difficulty depends on the size and density of the colony. It is therefore highly likely that the reliability of recognition between mother and young will be higher in pinniped species, where the risk of being separated from their young is greater. This is the hypothesis that Isabelle is testing as she travels the world to study and compare recognition systems between mothers and their young in these animals.[47] Let's take the case of the walrus first. With these pieces of floating ice drifting on the ocean as the only places to give birth, the risk of the mother being separated from her calf in just moments is significant. All the more so since she is forced to make an occasional deep-sea dive to feed. Although her young are able to swim from birth and try to follow her everywhere, they regularly lose sight of one another and have to find each other again because the young walruses are in great need of their mothers. They continue to suckle for two to three years after birth. We were in Igloolik to test the hypothesis that a walrus mother must recognize her cub's voice accurately.

Igloolik. *The place where there are igloos.* Originally a place where Inuit families met temporarily during their nomadic life. Imagine a single gravel airstrip, a few dozen houses, a kind of grocery store–warehouse, a small church, a curious building resembling a flying saucer pompously named the Igloolik Research Center, and, a little higher up, a few large fuel tanks, probably replenished just once a year. Igloolik, with its 1500 inhabitants and its dirt streets, is not a megalopolis. The surroundings don't have a very pronounced topography. The city is plain, without any particular charm, in a rather desolate landscape. The nearest village, Hall Beach (Sanirajak), is 70 kilometers away as the crow flies. This is definitely the Arctic: sled dogs tied up at the entrances of houses or wandering in the streets and some impressive polar bear skins on drying racks, waiting to be tanned, testify to this. And the omnipresence of white. Not the familiar, jagged white of a mountain with its bare rocks and forest patches, but the boundless, flat, infinite white of the frozen sea blending with the horizon.

Brad Parker is waiting for us on the tarmac. His stocky build, large hands, and trapper's demeanor exude a roughness mixed with a surprising cynicism. He does, however, inspire confidence. In any case . . . we have no choice. The Arctic is another planet. The rainforest may be impressive when you discover it, but life is easy there. In the far north, life is a daily challenge. After ordering us to throw our bags in the back of his pickup, Brad invites us to have a solid breakfast at his place. "There're still a lot of ice floes this year. Walruses won't be easy to find. Plus, the weather here is unpredictable. Bad most of the time. And sometimes a day when you're rewarded beyond your wildest dreams. The advantage is that the day never goes to bed." He adds, "We'll stay in my huts. We won't come back to town. So eat up!" A little later we glide over the ice pack in a sled pulled by a snowmobile, our legs protected by a reindeer hide. We head for the huts. "The greatest danger is the polar bear. He doesn't fear man. When you go out, look everywhere." The year before, a hungry bear had knocked down the wall of the cabin where Brad cooks. "Pakak . . . ," Brad said, pointing to the man who would be our guide, "Pakak was once saved by another hunter. A polar bear had pinned him to the ground and opened his back with its claws." And then to underscore his point he added, "just as easily as that—crash! Like peeling an orange." A polar bear's claws are like daggers, each up to ten centimeters long. Pakak had some impressive scars.

We learn from the dictionary that the Latin name of the walrus, *Odobenus rosmarus*, comes from the Greek *odo*, which means "tooth," and *benus* ("I walk"). This name was built on a legend: it was long believed that this animal moved across the land thanks to its long tusks. The tusks are mainly used as weapons during fights between males. The walruses also use them as a tool in the search for food and to climb onto a block of ice when coming out of the water, which is very handy for getting a

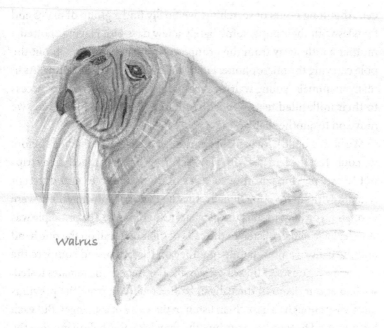

Walrus

good grip. A walrus eats mainly mollusks from the seabed around 80 meters deep, and more rarely at a depth of between 200 and 500 meters. An adult female walrus weighs up to 1300 kg, males 1800 kg, and newborns 85 kg, which in itself makes for a beautiful baby.[48]

Taking advantage of a day of good weather where the sun undulates endlessly around the horizon, we go in search of walruses. Pakak is at the controls of the motorboat. Minimalist equipment: sandwiches, a rifle with a few cartridges, no navigation system. We suggested Pakak use our GPS. After taking a quick look at it, he got rid of it, disdainfully. We will never understand how Pakak finds his bearings in this universe without landmarks. The point is, he knows where he's going. So we trust him absolutely and we focus on finding the walruses, eyes glued to our binoculars. What a sight! The Arctic spring is a magical season: here, a flight of snow geese, a magnificent bird adapted to the coldest regions; over there, the double breath of a bowhead whale, then the head of a bearded seal emerging from the waves. And on this strip of pack ice, a female polar bear with her two cubs. The walruses are playing hard to get. After long hours of searching, we finally find a group of males and females with their pups, some barely a few days old. Having spotted a mother a little away from the group, we approach and stretch out the pole carrying the microphone. First, we have to record the young. As in many mammals, young walruses regularly ask their mothers for access to their milk-filled teats. Once the first one's calls are in the box, we move on to another baby.

We had to build up a bank of recordings of several individuals before we could test the females with playback experiments. It took many trips out to sea, playing with the vagaries of the weather, to find different females with their young. Finally, the big day arrived when there were enough recordings to start the playback experiments. The principle was simple: have females listen to the calls of their own calf on the one hand and the calls of a non-family member on the other, and compare the mothers' responses to these two signals. For each of the females tested, we had at our disposal the calls of babies recorded from the previous days: easy enough to have them listen to the voice of a stranger. But each time, we had to start by recording the female's baby, before moving the canoe back a good distance so that Isabelle on her laptop could prepare the signals we were going to emit from our loudspeaker. Since the walruses were resting on drifting blocks of ice, it was indeed impossible to find the same females twice in a row. Another constraint: It was unthinkable to kidnap the little one and replace it with our loudspeaker. Walruses are dangerous animals, and they would have plunged into the water if we had gotten too close. We were going to play back the sounds from the canoe. By dint of tenacity, and almost surprisingly, it was a success. Of the 13 females that we managed to test in two successive missions, all of them responded to the calls of their young from our loudspeaker. With variations, of course: Some females just turned their heads in our direction; others barked loudly. On the other hand, they remained motionless as they listened to the calls of an unfamiliar pup. Only one female reacted to it by taking a quick glance. The evidence was there: walrus mothers recognize their young by voice without fail.[49] Two field trips were necessary to obtain the experimental demonstration. Isabelle stayed a good month each time. Thierry and I took turns. I can hear you from here: all that for this? It's true that we would have liked to explore the recognition mechanisms in walruses in greater detail. But the severe conditions in the Arctic decided otherwise. During the second stay, Thierry and Isabelle had even come close to dying there. Their guide didn't have Pakak's know-how and had been caught in the trap of monstrous ice packs! This memorable episode is told in the film *Bonjour les morses* that I invite you to watch.[50]

Our walrus experiments supported Isabelle's hypothesis: when there is a high risk of separation between the mother and her young, speech recognition is reliable. Is this reliability lower when the risk is less? With their varied social structures, pinnipeds are a good model for answering this question. Some species are solitary: the female gives birth to her young alone on a beach or on a floating ice floe. This is the case in many earless seals, such as the bearded seal *Erignathus barbatus*, for example, where the female stays permanently with her pup for several weeks. On the other hand, most eared seals form colonies, the density of which varies from species to species. There, females suckle for several months. During these long lactation periods, they alternate between stays on land, where they look after their young, and trips to sea, where they feed and replenish their reserves. What are the results obtained by Isabelle by comparing the mother-juvenile recognition systems in these different situations? First of all, how reliable is the individual vocal signature? In solitary seals, such as the Weddell seal *Leptonychotes weddellii*, the harbor seal *Phoca vitulina*, or the Hawaiian monk seal *Monachus schauinslandi*, the voice of a pup is rather variable and is therefore not easy to recognize vocally. On the other hand, the voice of a small walrus, a small elephant seal,[51] a small Australian sea lion *Neophoca cinerea*, or a small subantarctic fur seal *Arctocephalus tropicalis* (both species breeding in colonies) is well differentiated and facilitates its identification. It should also be noted that, in species where it has been tested, recognition is often mutual: mother and young are able to identify each other vocally. Another interesting criterion is the speed at which this recognition is established. Investigations into the implementation of mother-juvenile recognition are rare. Only four species of eared seals have been studied, with very different results. It takes between 10 and 30 days for a Galápagos baby sea

Fur seal

lion *Zalophus wollebaeki* to learn how to recognize its mother's voice, compared to 10 days for a Galápagos baby fur seal *Arctocephalus galapagoensis* and between 2 and 5 days for the subantarctic fur seal. The greatest temporal amplitude was observed in the Australian sea lion: if mothers can recognize their cubs 10 days after birth, the cubs take several weeks![52] This is curious, and still unexplained.[53]

The subantarctic fur seal or Amsterdam fur seal *Arctocephalus tropicalis* lives in the Indian Ocean and breeds on small islands in the area (including the island of Amsterdam, which is one of the French Southern and Antarctic Lands). Isabelle spent nine months studying this species as part of her doctoral thesis. After the birth, a mother fur seal from Amsterdam Island stays for a few days with her newborn but has only one thought in mind—to go back to the ocean alone to feed. She has been fasting for several weeks, and there is not much to eat in the immediate vicinity of the island. She will leave her calf for two to three weeks. Unbelievable but true—two to three weeks when the baby is on its own in the colony with no food to eat. After that time, the mother returns. The reunion is an amazing sight to behold, with mother and baby calling each other repeatedly. You have to see this little black ball clawing its way across the rocks that separate it from its mother to understand the sheer strength of its motivation. Once the little one is finally there, the mother sniffs it, probably as a final olfactory check. All this happens very quickly; within 10 minutes after she touches the shore, the mother is in contact with her cub. These reunions are crucial: a lack of recognition means no feeding, and the death of the young. Recognizing the mother's voice is a vital imperative.

Playback tests carried out by Isabelle show that a newborn baby reacts to the voice of any female within hours of birth and that it must learn to recognize its mother. The most surprising result was when, on Isabelle's return to France, we compared the number of days before the mother's recognition with the time of her departure to the ocean. The two were correlated: the faster the baby learned, the sooner the mother abandoned it and vice versa.[54] In other words, a fur seal mother seems to pay extreme attention to her baby's ability to recognize her: "Do you recognize me? Well, I can leave you alone now."[55]

What happens as the length of the mother's absence increases, you might ask? To find out, Isabelle played the barks of female fur seals to pups when their mothers had abandoned them for a short or a long time. The results of her experiments show that the motivation of the pup to shout in response to the barking of females increases with the time of separation.[56] After being left alone for a day, one in five pups react, while if two weeks have passed, three out of four call out in hunger! What if they hear the voice of an unknown female? No reaction at the beginning, of course, but up to 30% of the pups cry out in response after two weeks. It's certainly not because the cubs have become unable to recognize their mother, as borne out by the dramatic reunion when she finally returns. But it is because they are willing to do anything to get milk, including begging for milk from any nipple. In short, they are too hungry to wait, and they let it be known.[57]

The call of the baby fur seal is a *signal*. It carries information not only of the emitter's identity but also about its state of satiety from the sender to a receiver, the mother fur seal. I'm not hungry—I'm not crying. I'm a little hungry—I yell moderately. I'm as hungry as a wolf—I'm screaming all the time. We know that with our own babies, don't we? Listening to this signal, the mother will change her behavior: calling in turn, getting into position to suckle her young, for example. These observations lead us to a slightly more theoretical reflection. In order for a communication system between a transmitter and a receiver to be preserved by natural selection, the information transmitted must generally be of interest to the receiver. Otherwise it would stop responding. It is not necessary for this information to be systematically interesting or truthful. But it does have to be interesting enough for evolution to have selected a receiver that responds to it. In fact, as a rule of thumb, if the transmitter-receiver pair do not find a mutual interest in making the communication work, it will disappear. And let's be clear: the signal has to be honest *enough* for communication to continue.[58]

This hypothesis of *honest communication* is not easy to test. Indeed, it requires being interested in several elements that are not necessarily easy to grasp. Let's stay with our juvenile begging calls, which are signals that are widely used not only in our walruses and seals but in many mammals and birds. First of all, do these food-begging signals accurately reflect the state of the transmitter? In our baby fur seals, this seems to be the case. Experimental tests have been carried out in birds, where it is easier to artificially manipulate the amount of food ingested by the offspring. Rebecca Kilner, a professor at Cambridge University, conducted famous experiments more than 20 years ago showing that the begging calls of the chicks of the Eurasian reed warbler reflect their need for food. Since then, other studies have confirmed this honesty of information in other bird species.[59]

Kilner was interested in the roles of the begging calls in parent-offspring interactions in birds, and how the cuckoo *Cuculus canorus* might exploit them—you know, that bird that lays its eggs in other birds' nests. We'll talk more about that later. So Kilner's model bird was the Eurasian reed warbler *Acrocephalus scirpaceus*. The warbler chicks solicit their parents by opening their beaks wide, displaying the red background of their throats, and cheeping insistently. In response, the parents shove lots of small flying insects and delicious caterpillars down their throats. In a first experiment, Kilner and her collaborators temporarily removed chicks from their nest and placed them in a small, experimental box.

Kilner then fed them until they stopped begging, then kept them warm but without feeding them further. For the next 2 hours, every 10 minutes she mimicked the arrival of a parent by tapping the box, which inevitably provoked the chicks' begging behavior. The results were clear: the longer the chicks had gone without food, the more loudly they would chirp. You want numbers? Here are some. After 20 minutes of waiting, a 3-day-old chick cheeped at the rate of 2 calls every 6 seconds. After 100 minutes, 8 calls. The phenomenon was even more sensitive with older chicks (6 days old). They gave less than 15 calls after 20 minutes of waiting compared to more than 35 after 100 minutes, more than twice as many. In both the youngest and oldest chicks, the number of calls was proportional to the time spent waiting. But Kilner didn't stop there. She had to check that it was hunger that caused the chicks to call more, not the stress of absent parents or any other reason. She measured the amount of food needed to stop the calls of isolated chicks and found a strong correlation between the amount of food and the number of calls. It is probably safe to assume that the begging calls of reed warbler chicks honestly reflect their nutritional needs. However, other studies have shown that this correspondence between nutritional requirements and the intensity of begging behavior is not as rigid as it seems. It is thus possible to train chicks to chirp more or less according to the promptness of the parental response.[60] If they are fed on their first call, they will soon learn that there is no point in forcing their voice to get rewarded. On the other hand, if we wait for them to chirp insistently before feeding them, they will quickly get into the habit of expressing their hunger vehemently. Honest, but not stupid.

Why not beg as loudly as possible right away? After all, having a tantrum as soon as you feel the slightest pang of hunger would allow you to get more from your parents, and faster. You might as well lie a little or a lot about your condition: you'll eat more and get fatter faster than your brothers and sisters. In other words, why stay honest? Scientists who have investigated this question by mathematical modeling have shown that begging behavior can only remain honest if it has a certain cost. Although I am not a proponent of the use of these anthropomorphic terms, I must confess that they are powerful metaphors. It is well understood that the temptation to dishonesty will be greater if it costs nothing and can be profitable! Where could the cost of begging come from? It is in fact of two kinds: a direct energy cost (screeching requires energy) and an indirect cost due to the predators that may be attracted by the cries.

How can we estimate the direct energy cost? A first approach is to put a chick in a respirometer, a device that measures its oxygen consumption. The results of various studies show that this consumption is multiplied by only 1.3 when the chick emits calls compared to resting, which does not seem considerable.[61] Since time is money, time spent begging can also be considered a cost. Video recordings of house wren nests have shown that chicks beg only 4 to 10 times per hour and that each round of cries lasts 4 to 7 seconds. This makes 16 to 70 seconds spent begging per hour, or 0.4% to 2% of the time.[62] The results of all studies show that the energy cost of the begging calls is rather low. Becoming dishonest by chirping more, therefore, would not cost much.

But that measurement does not take into account the other side of the cost: increased predation. A nest, even well hidden, if full of young, bawling chicks, is like a lighthouse in the middle of the night, and some predators apparently understand this. Be aware that, in birds, predation by other birds, mammals, or snakes is responsible for 80% of brood mortality.[63] Do begging calls have anything to do with that? To answer this question, experiments have been carried out that involve placing small loudspeakers in artificial nests to mimic the presence of young chicks begging. Most of these studies have concluded that the cries of the young attract predators.[64] It is even likely that their impact is underestimated. Predators don't just listen; they watch. The more the chicks solicit their parents, the more the parents tend to feed them and thus increase their comings and goings, probably making it easier to locate the nest.[65]

As noted above, chicks can change their begging behavior depending on external conditions. It has been shown that they can lower the sound level of their calls and make them more high-pitched when the number of predators increases, making them somewhat less noticeable.[66] Sometimes they even decide to remain silent if they hear a predator approaching.[67] And if they don't, their parents can order them to shut up or at least turn it down.[68] But we are not at the end of our surprises. It has been suggested that chicks may instead exaggerate their calls despite the risk of predation, forcing their parents to respond to their requests as quickly as possible.[69] A bit like an angry child throwing a tantrum to demand a piece of candy at the supermarket, parents will give in faster so as not to attract attention.

As you can see, the begging calls are honest . . . to a certain extent. The young are able to modulate them to get more or to limit the danger of being eaten by a predator. Another important element to take into account is that very often a nest houses several siblings who may compete for food provided by the parents. Who will receive the most pudding? Studies indicate that competition within nests can be accompanied by an increase in the intensity and number of calls, with each one exaggerating its real needs in an attempt to tilt parental choices in its favor. The correlation between the chick's state of satiety and its motivation to call becomes less reliable.[70] An experiment with the yellow-headed blackbird *Xanthocephalus xanthocephalus*, a North American bird, has shown that a chick begs longer if it is placed in a nest with another hungry chick rather than with a full chick.[71]

As parents of large families know, children can also join forces to get what they want. In our study of black-headed gulls, Isabelle and I highlighted such a strategy.[72] From our floating blind, we observed the begging behavior of gull chicks in nests of one, two, or three chicks. The first interesting result was that the number of calls per hour decreased with increasing clutch size, from more than 7 calls per hour when the chick was alone to less than 3 calls per hour in three-chick nests. When only one chick was present, the parent regurgitated all the more easily as the chick screamed loudly. But only 17% of the chick's calls were followed by parental regurgitation. In nests with several chicks, the number of regurgitations was proportional to the number of chicks crying at the same time: 23% of the cries were followed by regurgitation when only one chick was crying compared to 87% when both were crying. The group effect was even more pronounced in nests with three chicks. If only one of the three siblings was begging alone, it could motivate the parent in only 13.5% of cases. In pairs, 55% of the begging sequences induced regurgitation. And when the whole tribe cheeped together . . . 100% of their collective begging was satisfied. United we stand. However, it is not certain that this behavior is generalizable to many species. In black-headed gulls, the parents do not feed the young directly; they regurgitate on the nest floor, and the young must then retrieve what has fallen out of the parental beak. Competition between chicks is therefore more likely to occur after parental regurgitation than before. In birds where the parent feeds directly in the bill, there may be less incentive to cooperate.

Interactions between chicks are not always limited to the crucial moment of the parent's arrival. Surprisingly, chicks of many species call out when the parents are absent, even though they are easy prey for predators. Sometimes this can be a misinterpretation, as the chicks hear a noise or feel a vibration that makes them think a parent is arriving. The temptation to make themselves known as early as possible is sometimes great. For example, in the tree swallow *Tachycineta bicolor*, the first chick that calls out is the one that will be best fed.[73] It is also possible that the chicks are talking to each other. When competition between chicks is intense, it may be more interesting to agree on sharing resources rather than spending a lifetime fighting. Alexandre Roulin, professor at the University of Lausanne in Switzerland, his colleague Amélie Dreiss, and their collaborators have thus supported the *sibling negotiation hypothesis*.[74] This hypothesis suggests that, in the absence of the parents, chicks inquire about their mutual motivation to obtain food by calling out. By listening to the calls of the other chicks, each one then decides its chances of actually having access to food when the next parent returns to the nest. Alexandre's favorite study model is the barn owl *Tyto alba*. This large owl, with its pale plumage and silent flight, frequently nests in buildings.

A barn owl nest has an average of four chicks, with a maximum of nine. What a party! With their hooked beaks and sharp claws, barn owl chicks are very well equipped to engage in some serious battles. Yet they're quite friendly to each other, smoothing their plumage, warming each other, and sometimes even exchanging food. There's no physical struggle over who gets the next meal. Instead, they negotiate. How's that? In the absence of their parents, a particularly hungry chick produces many long-lasting cries. If its siblings are less food starved, they remain silent when the parent arrives. It is a so-called iterative process: at first, everyone begs; then as the negotiation progresses, some chicks withdraw while others exaggerate their cries. Sort of a poker table for owls.

The chicks pay attention to the details of the calls. If a hitherto dominant individual gradually gives shorter and fewer cries, those that were previously silent try to get the upper hand.[75] Moreover, the one who ultimately dominates the negotiation is the one who forces the others to progressively issue shorter and less frequent calls. To achieve this, it starts by imitating the duration of the calls of the others, who respond by reducing the duration of their own calls. If this is not enough, the hungry chick then increases the rhythm of its calls; and, in principle, if the others are less determined to eat, they give up. When the owl parent enters the nest, it brings only one prey—a small rodent, for example. This prey cannot be shared among the chicks and will therefore be swallowed by only one of them, the one that won the negotiation. Imagine if the little owls fought beak and claw for the vole. It is easy to understand the value of sitting down at the negotiation table to decide beforehand who will eat. The chicks that were not fed on this occasion will be fed next time.[76] The little barn owls know how to wait.

With this example, we can see that the life of the nest is not limited merely to communication between a chick and its parent. In reality, parents and offspring form a real communication network—a concept that we discuss in detail later as it applies to most animal communication systems. The chicks have an intimate knowledge of their siblings. They know that Peter is hungry, while Paula is full, etc. A few studies have shown that the chicks' calls have vocal signatures, and playback experiments have revealed that the chicks recognize each other.[77] In some species, parents also use these signatures, for example, to share the responsibility for feeding their children. Do you remember Tudor Draganoiu, the man who showed that black redstart males have similar songs when they are neighbors? Tudor also studied the relationship between parents and their young. He found this sharing of the brood in his favorite bird.[78] Two or three days after leaving the nest, the chicks are still unable to feed on their own. A young redstart is then fed preferentially by either its mother or father. Tudor has shown through playback experiments that parents recognize which young they are dealing with on the basis of its calls.

Communication between parents and young starts very early on. It has long been known that bird embryos can inform their parents of their temperature by calling through the eggshell:[79] "I'm too hot!"—the parent stops brooding; "I'm too cold!"—it gets back on the eggs. When I discuss crocodiles in chapter 8, we will see that the embryos signal acoustically to their mother and brood mates when they are ready to hatch. What about the parents? Recently, Mylene Mariette, a researcher at Deakin University in Australia, showed that adult zebra finches *Taeniopygia guttata* talk to their chicks when they have not yet hatched.[80] And what do they tell them? They tell them the outside temperature. When it is very hot and dry in the Australian desert, an adult zebra finch is stressed. It then emits a series of little calls, the *heat call* as Mylene described it, especially when it is in the nest incubating its eggs. Mylene and her collaborators had the excellent idea of playing these calls from a loudspeaker to brooding zebra finch embryos. She noticed that their growth was slowed down, resulting in smaller adults than usual; in other words, birds better adapted to withstand harsh weather conditions and poor food resources. When they were able to reproduce, these birds produced more offspring than control individuals who had not heard the heat calls when they were hatched. Unbelievable, isn't it? We don't know, of course, if the zebra finch parents are deliberately calling "It's too hot!" to their eggs—it's even possible that the heat calls are an unintended consequence of altered respiratory activity due to heat stress. Nevertheless, these sounds alter the growth of the embryos, allowing them to prepare for more difficult living conditions than expected.

In the nest's communication network, children talk to parents, they talk to each other, parents talk to them. And do the parents talk to each other while they're both busy feeding their brood? When Mylene was a postdoctoral researcher in our laboratory, she took part in an unprecedented experiment. In the zebra finch's nest, female and male alternate their presence, each one brooding in turn, so the test involved delaying the male's return to the nest (we had a system to capture the male when he came to the area of the aviary where the food was). This delay altered the vocal exchanges that characterize reunions between partners in this species.[81] More precisely, the time that the female would then stay out of the nest was predicted by the rhythm of the duo. The more the female sped up the pace of the exchanges, the less she would stay in the nest waiting for the male the next time. It is likely that these exchanges of information between parents, these kinds of negotiations, are also commonplace. Observations in natural conditions with the great tit have shown that the female in the nest signals her hunger to the male outside the nest with her calls.[82] Presumably, this is taken into account by the male when it has to decide that it is time to take over from his partner.

While communication signals generally need to carry reliable (honest) information in order for the communication to continue (if her baby cries wolf too often, the parent will eventually stop responding), they can also be a means of manipulating the receiving individual rather than simply informing him or her. This concept of a manipulative signal was put forward as early as the 1970s by Richard Dawkins and John Krebs, both professors at Oxford University.[83, 84] The main idea is this. When communicating, the "purpose" of the sender is often to get something from the receiver (food, copulation, etc.). To do this, it distorts its signal by exaggerating the information it carries. Dawkins and Krebs suggest that, in response to this escalation, the receiver will become increasingly resistant to the received signal over time. This arms race, as it is called, between the emitter and the receiver is reflected on the evolutionary scale of a species by the progressive development of increasingly extravagant signals and a growing resistance to react to them.[85] You can see that this conception of animal communication as an endless spiral, an unstable process, is opposed to the stable balance proposed by the signal seen as honest that we were talking about earlier. We return to these important notions later when we look at the mechanisms of the evolution of acoustic communication. For the moment, let us say that the two points of view can be reconciled by assuming that a communication is likely to be fairly honest and stable when the transmitter and receiver have a common interest (which is the case for our chicks and their parents), whereas it is likely to be a manipulation accompanied by an arms race when the degree of conflict between transmitter and receiver is very high. This second situation is typically encountered when one species of bird has its young raised by another species. This is a very instructive case study for understanding the concept outlined by Dawkins and Krebs.

Parasitic birds represent about 1% of all bird species. Some are specialized: the female always lays her eggs in the nest of the same species. Others are eclectic: the brown-headed cowbird, for example, can entrust its eggs to more than 200 different species of birds.[86] This method is accompanied by multiple strategies to facilitate the integration of the eggs, and then the young parasites, into the host nest.[87] First of all, one thing is obvious: when a parent bird feeds a young bird that is not of its species, the interests are divergent. In short, the parent wastes time and energy—even more if its young have been killed by the parasite or die of starvation—while the young freeloader gains its food, survival, and growth. The young parasite is therefore the archetype of the manipulator, and the adoptive parent will have every interest in arming itself against this manipulation. How to manipulate? One method is to imitate the begging signals of legitimate children;[88] another is to reproduce the signals by exaggerating them.[89] Although few studies have compared the acoustic structure of the calls of parasites and parasitized, it is estimated that the calls of parasitic chicks mimic the calls of chicks of the host species in one out of two parasite species.[90] Let's take the example of the Horsfield's cuckoo *Chrysococcyx basalis*, an Australian bird that is in the habit of having its young raised by the superb fairy wren *Malurus cyaneus*. The cuckoo chicks' calls sound similar to those of the fairy wren chicks. However, some chicks are better at imitation than others. Unskilled imposters will soon be spotted by their hosts, and the adoptive parents of a Horsfield cuckoo abandon their nest in 40% of cases, leaving the cuckoo to perish. Better still, if another parasite, the shining bronze cuckoo *Chrysococcyx lucidus*, tries its luck in a fairy wren nest, it never succeeds in forcing its adoptive parents to feed it and is abandoned every time. The reason is simple; this unfortunate cuckoo emits a begging call that is very different from that of the fairy wren chicks.[91] Not content to imitate, parasitic chicks often exaggerate the intensity of their cries of begging. In a way, they lie (a little) about their condition; the adoptive parents get trapped, believe they are dealing with a particularly hungry chick, and redouble their energy to feed it.

In this arms race between parasites and their hosts, host species develop strategies to limit the impact of the parasite.[92] These strategies are not limited to paying attention to the acoustic quality of the calls of the chicks present in the nest. They also include, for example, harassment of parasitic adults to prevent them from laying eggs in the nest and the recognition and elimination of parasitic eggs once laid. Some strategies can be really subtle. A few years ago, Mark Hauber, my colleague when I was a visiting professor at Hunter College in New York, now professor at the University of Illinois, together with Diane Colombelli-Négrel, Sonia Kleindorfer, and their collaborators, made a very unexpected discovery: the parents of some birds teach their children a password while they are still in the nest.[93] Here's the story.

Once again, the host-parasite pair formed by the superb fairy wren and the Horsfield cuckoo served as a study model. After recording the sound activity of parasitized and nonparasitized fairy wren nests throughout the nesting period, they first discovered that the fairy wren female produced an *incubation call* during the days before hatching (so it's not just zebra finches that talk to their eggs!). The scientists then compared the acoustic structure of these calls with the acoustic structure of the begging calls that the chicks emit once they have hatched. They found strange similarities. The calls of the chicks in one nest were more like those of their own mother than those of another fairy wren female—in other words, there are family accents in the fairy wrens. To test whether this resemblance had a genetic component, scientists exchanged fairy wren eggs between different nests at the very beginning of the brooding period. When they hatched, the fairy wren chicks called out like their adoptive mother. So they *learn* this call when they hear it while in the egg. The next step was to see if the parents paid attention to this family signature when they had to feed their young. Playback experiments showed that this was indeed the case. Fairy wren parents respond much better to the calls of their own offspring than they do to those of foreign chicks.

Let's describe the course of events for clarity. In an unparasitized nest, the chicks hatch after a two-week incubation period. Five days before hatching, the female begins to make her incubation calls. In the days following hatching, the nest becomes silent again. The parents feed the young, who beg by opening their beaks without making any audible sounds. Three days after hatching, the chicks begin to vocalize. Their calls resemble their mother's incubation call. They have learned the password. Both the mother and the father, who has also learned to recognize the acoustic key, continue to feed their young until they fly away. In a nest where a cuckoo has laid its eggs, things are different. If the female begins to make her incubation call about ten days after laying, the cuckoo chick hatches on the twelfth day of incubation, three days earlier than fairy wren chicks. It takes advantage of this situation by removing their eggs from the nest. Faced with a nest with only one chick left, the female fairy wren stops making hatching calls. A few days later, when the cuckoo chick starts to call, it will have some difficulty imitating the call of its adoptive mother. It must be said that the lesson will have been short, heard for only two days behind its shell. Perhaps it is less genetically prepared for imitation than legitimate children. If the fairy wren parents are not convinced by the young cuckoo's vocalizations, they will abandon the nest, leaving the cuckoo chick to its sad fate. I told you that fairy wrens abandon their parasitized nest in 40% of cases. Errors in passwords are probably to blame for this. We can make the hypothesis that the next step in the arms race between the superb fairy wren and the Horsfield cuckoo will be for the fairy wren to further refine the acoustic correspondence between the maternal incubation call and the legitimate chick's call, and for the cuckoo chick to hone its study skills.

Teaching your own children a password—this is amazing, to say the least! And it's not only birds that talk to their young before they're born.[94] A study suggests that the female bottlenose dolphin *Tursiops truncatus* vocalizes while she is pregnant and that her young imitate her calls once born.[95] Might she teach them the vocal signature that will enable them to recognize each other among other dolphins?[96] This provides us with a nice transition. Let's give up the birds for a while and go back to the ocean. The whales and dolphins are waiting for us there . . . vocalizing, of course.

6

Submarine ears

UNDERWATER BIOACOUSTICS

Somewhere in the Foxe Basin, between Igloolik and Rowley Island. We're still with Pakak. He shuts off the engine near a block of ice. He lets us drift a few yards and throws the grappling hook on the ice. The hook slips and finally snags. Here we are, moored. The sea is calm. Leaning on the edge of the canoe, Isabelle lets the hydrophone cable slip through her fingers. The little black capsule dives into the icy water. We quickly lose sight of it. Five meters, ten meters . . . "Take the headphones," Isabelle says to me, "listen . . . they're there." I obey. Immediately, lancinating songs reach my ears, like the wind blowing in the trees: "Twwuhuhuhoohou hoohoooooooooooooooooooooooooo. . . ." Songs, more songs, endless songs—almost disturbing. Long melodies, at first high-pitched, then descending to the low register, both regular and modulated in trills, like someone gradually losing his breath while blowing into a musical pipe. When one melodic line ends, another has already begun. Or several. Some close and strong, others like an echo in the distance. "These are bearded seals," Isabelle whispers. "What you can hear is the song of courting males." A memory surfaces of an old picture book from one of my grandmothers, with a drawing of a seal singing, head down, its body bowed. Sometimes, they say, the seal songs are so loud that you can hear them even when you are out of the water.

The bearded seal—*Erignathus barbatus* by its Latin name—is a sedentary, solitary inhabitant of the great north. When we see it, it is usually

Bearded seal

lying on a block of ice, waiting for time to pass. It has a large body disproportionate to its small head—and incredible whiskers. In the air, they dry out and bend back. But in the water, this imposing comb spreads out harmoniously to probe the bottom in search of crabs and mollusks. During the breeding season, each male defends an underwater territory. Isabelle had hypothesized that the bearded seal's song plays the same role as that of the birds: to ward off intruders. How would we test this idea? With underwater playback experiments, of course. The method was obviously appealing, but not seeing what the seal does underwater made the operation uncertain, to say the least. We could only record a possible vocal response of the seals to our signals and observe their behavior on the surface. It wasn't much, but we decided to try it anyway! From the canoe, we sent our underwater loudspeaker down sixteen times, a good 3 kilometers apart each time to make sure we would encounter different individuals' territories. Almost every time, the males in the vicinity began to sing less often as they heard the song from the speaker. Also, in half of the experiments, we saw a seal head appear on the surface of the water near the boat, as if the local occupant was trying to spot the intruder on the surface that he hadn't been able to see underwater.[97] It is certainly difficult to say from these experiments alone that singing allows the bearded seal to defend its territory. The fact remains that our playbacks changed their behavior, apparently encouraging them to patrol their territorial waters.

These experiments underscore the difficulty encountered in bio-acoustic research in underwater environments. There, underwater, it is difficult to directly observe the behavior of the animals.[98] Studying the acoustic communications of seals, whales, dolphins, fish, and various shrimp[99] in the sea requires special methods, such as using underwater sensors (hydrophones, recorders, GPS), sometimes placed on the bottom, or even directly attached to the animals if they are big enough. Until recently, due to a lack of suitable technology, the sea was thought to be mostly silent. In the 1950s, didn't Jacques Cousteau make a documentary on the seas and oceans called *The Silent World*? Believe me, the opposite is true. The oceans are teeming with sound-producing beasts. In fact, if you had the courage to read chapter 2 of this book, the one in which we made circles in water, you already know that the aquatic environment is particularly conducive to the propagation of sound waves. You surely still have in mind that they move very fast—much faster than in the air (something like a kilometer and a half per second)—and that they travel very far. The songs of some whales can be heard from dozens, even hundreds, of kilometers away. This property of the marine environment has interesting biological consequences. For example, the coral reefs, inhabited by myriads of shrimp and fish, all chattering, rasping, squeaking, and growling, produce a background sound that can be heard several kilometers away. Attracted by this very particular hubbub, fish or larvae of various animals looking for a place to set up home find the reef of their dreams much more easily than if they had looked for it randomly in the blue immensity. It has even been shown that each type of reef has its own unique sound signature, which varies with the time of day and seasonal cycles. But, you may ask, how has it been proved that fish use the sound of the reefs to get there? Here's the story.

First of all, be aware that most of the myriads of fish that live as adults on coral reefs spent their childhood in the open sea. During this larval phase, the fish are more or less left to drift until they reach an age when nomadic life begins to weigh them down. Then they decide to actively swim toward the reef. In the first of several studies, New Zealand scientists, with the knowledge that light attracts fish, deployed two light traps at sea 500 meters apart. Next to one of the traps they placed an underwater loudspeaker that emitted the sound of a coral reef. By repeating the experiment several nights in a row, the scientists found that the loudspeaker trap attracted on average more small sardines than the silent trap. The sound reinforces the light-attracting effect.[100] Another study went even further. In this instance, the scientists built 24 small patches of artificial reefs by placing pieces of dead reefs on the sand at a depth of 3 to 6 meters. On 12 of these fake reefs, they placed loudspeakers that broadcast recordings of reef noise, previously recorded on living reefs. These recordings consisted mainly of sounds produced by fish as well as the snapping of shrimps as they close their claws at high speed, which usually represents the dominant noise of the coral reefs.[101] What superpractical instruments for stunning or killing prey![102]

Most of the fish arrivals were at night. They all arrived in greater numbers on the sonorized reefs. Scientists renewed the experiment by sonorizing some of their reefs either with only shrimp snapping (high-pitched sounds) or only fish sounds (lower-pitched sounds). They found that some species of fish preferred to settle on "shrimp" reefs while other species chose "fish" reefs.[103] The arriving fish seemed to choose their destination according to the sound environment. These discoveries have interesting applications for reef conservation. It has been shown that fish can be encouraged to come to damaged reefs by using loudspeakers that emit the background sound of a healthy reef.[104] We come back to fish later. For the moment, I propose to focus on the giants of the seas: the whales and their cousins.[105]

Whales, dolphins, and porpoises make up the large group of mammals known as cetaceans. They comprise two very distinct categories: toothed whales (odontocetes: dolphins, porpoises, and sperm whales) and baleen whales (mysticetes: the "real" whales).[106] You don't need to be gifted to understand that toothed whales and baleen whales can be easily differentiated by looking into their mouths. If, instead of finding nice, tidy teeth, you see long, stiff bristles, you are looking at a baleen whale, whose considerable baleens (those huge brushes, if you will) allow the whale to get its meager sustenance (fish, shrimp and other crustaceans, etc.) by filtering it out of huge quantities of seawater.

As I mentioned before, cetaceans are mammals. Like you and me. Their ancestor was terrestrial and must have lived a bit like the hippopotamus, about 50 million years ago. The transition from land to water was accompanied by several important anatomical and physiological changes, such as the acquisition of a hydrodynamic body shape, hair loss to limit water resistance during swimming, and a thicker skin with a strong layer of fat to increase insulation.[107] Yet, cetaceans kept their air breathing, which means they must regularly come to the surface to breathe in and out. The famous whale exhalations are blasts of air accompanied by water vapor expelled from their lungs through slightly special nostrils—the blowholes—which are located on the top of their heads. What is the point of knowing this to understand acoustic communication, you ask me? Well, cetaceans, having kept their mammalian respiratory system, have also kept the vocal apparatus: the larynx and its vocal cords. In terrestrial mammals in the human species, for example, it is the air coming out of the lungs that makes the membranes of the vocal cords vibrate. These vibrations produce sound. So far, nothing extraordinary. But here's the rub: the production of sound is usually accompanied by the exhalation of air from the mouth. Try to talk in front of a mirror and you'll see mist forming on the mirror, a sure sign that you are exhaling air. But when a whale sings, you won't see a single tiny bubble of air coming out of its mouth or blowholes.

Olivier Adam, professor at the Sorbonne University in Paris, is a specialist in whale acoustic communications. Let him explain the theory he and his colleagues are defending: "A whale, like the humpback whale, which is known for its long vocalizations, has a special anatomical feature, a kind of bag—the laryngeal sac—whose opening connects just above the vocal cords."[108] Let's visualize it: first, the lungs, which are extended by a pipe, the trachea; at the top of the trachea, the larynx with its vocal cords;[109] after the larynx, the main pipe continues to the blowholes; last, a secondary pipe escapes just after the larynx and opens into the laryngeal sac. "This is the mechanism we propose. The air comes out of the lungs, vibrates the vocal cords, and then goes into the laryngeal sac instead of coming out through the blowholes. The air can then return from the laryngeal sac to the lungs, and the cycle starts all over again. Such a mechanism makes it possible to produce very long songs without blowing air out of the body or breathing in air." Like ingenious closed-circuit bagpipes.

In dolphins, the mechanisms of vocal production are better known, the more reasonable size of these animals making them easier to study in captivity.[110] These mechanisms are really strange. Just look! Like all toothed whales, dolphins have two pairs of special anatomical structures placed in the animal's nose: the monkey lips. Why are they called monkey lips? Because these structures actually look like primate lips, with their oblong shape and folds. The two pairs of lips are connected to the nasal passages and vibrate as air passes through them, producing sounds. In some species, only one of the pairs (right or left) produces sounds, while the other is silent. In other species, one of the pairs produces very high-pitched clicks used for echolocation (the identification of prey and obstacles through their echoes; we'll talk more about this later), while the other emits whistles for communication between individuals.[111] These sound vibrations are then transmitted to a kind of ball of fat, the melon, which focuses the acoustic waves in a directional beam. It is the melon that gives the domed shape to the front of the dolphins' heads. Note that dolphins also have a larynx and vocal cords, like other mammals, but, amazingly, it is still not established whether they use them for vocalization. In short, baleen whales sing with their larynx, a bit like we do, while dolphins and other toothed whales whistle with their noses thanks to specific anatomical devices. What do whales say to each other while singing and dolphins while whistling? This question interests hundreds of scientists and the general public as well. Yet our knowledge on the subject is very limited.

The vocal repertoire of baleen whales varies greatly from one species to another.[112] Some, like the blue whale *Balaenoptera musculus* or the fin whale *B. physalus*, are quite discreet and their vocalizations are rather stereotyped. Others, like the humpback *Megaptera novaeangliae* and the bowhead *Balaena mysticetus*, are very talkative.[113] The songs of the humpback, made famous by Roger Payne and Scott McVay, can last for hours and have infinite variations.[114] Let's take a closer look at this mythical animal.

The humpback whale is a baleen whale, present in almost all seas and oceans, which can reach a respectable size of 15 meters long and a no less respectable weight of 25 to 30 tons. During the summer, it feeds on small fish or krill, the shrimp-like crustaceans that abound in the icy waters of the Arctic and Antarctic. As winter approaches, humpback whales migrate thousands of kilometers to tropical waters where they breed. It's here, in these warm waters, that the males' songs reach their full expression. Endless melodies that can stretch for hours. Each song is a succession of short notes and long moans, "Oohooohoooohoooohoooohooo," lasting up to 8 seconds, whose pitch varies between 30 and 4000 Hz, making them perfectly audible to our ears. These notes and moans are not emitted randomly, in a disorderly manner. On the contrary, the humpback whale's song is perfectly structured, like a kind of musical ritual. It is organized in cycles of about half an hour. Each cycle contains roughly eight themes that follow one another in a precise order. When the eighth or final cycle is over, the whale stops singing, pauses for a moment, and then starts a new cycle from the beginning. Each theme is itself structured in sentences, which are successions of notes. The number of phrases per theme varies from two to about twenty.

Males from the same corner of the planet ocean sing the same songs at the same time. However, the songs vary from place to place, much like the regional dialects of birds.[115] Sometimes the regional dialect changes abruptly. In 1996, Michael Noad of the University of Brisbane and his colleagues noticed that, in the humpback whale population they were studying on the east coast of Australia, two males were singing a different song from the others, with other phrases and themes.[116] And get this: a year later, most of the 112 males they recorded at the same place were singing this new tune, which must have been catchier than the old one. Noad realized that this new song was similar to the one that had been produced for years by the whales on the west coast of Australia. One can assume that some individual from the West, lost in his humpback whale thoughts, had taken the wrong route back from the

Antarctic Seas, and had brought a new and fashionable hit to the East. A cultural revolution of sorts. Do you realize that this event means that humpback whales imitate each other? So they are capable of vocal learning, just like us humans. This is a topic we'll talk about later.

Although it is unpleasant for a researcher to admit it, we do not know precisely why whales sing.[117] Of course, since males sing mainly in the breeding grounds, it can be assumed that their vocalizations serve to charm females and repel competitors, as in birds. But that remains to be seen. To test this hypothesis, as we did for the bearded seal, scientists have tried to provoke behavioral responses from the whales by having them listen to songs emitted by underwater loudspeakers. Success has been mixed. Faced with a playback of a song of their species, some humpback whales flee, while others approach the loudspeaker.[118] It's hard to interpret this. A currently popular hypothesis suggests that song plays a complex role in the interactions between males. Maybe they are in vocal competition? Another hypothesis: Several males singing together could stimulate the receptivity of females. By joining his voice to that of others, a male would increase his chances of mating. The males' chorus melting the females' hearts! Nothing is certain, though. Some even suggest that songs are not communication signals but rather sonar signals, used by whales to probe their environment.[119] The songs are indeed intense enough to reverberate off the bottom, even at great depths, or off other obstacles, such as rocky shores. It is estimated that a whale could hear echoes of its song from more than 5 kilometers away. Perhaps they are building an acoustic image of their surroundings by singing. In terms of acoustic reverberations, the ocean is quite a cathedral.

It is worth noting that whale vocalizations could be very useful for probing the ocean floor! A recent study reports that part of the energy of the sound waves produced by these animals is transmitted to the ground and makes it vibrate. These waves then become seismic waves and propagate in the ocean floor (remember your geology classes in high school: this floor is the oceanic crust); they can be recorded by seismographs, the ultrasensitive devices humans have placed all over the world to detect the smallest of earthquakes. These seismic waves provide information about the organization, structure, and movements of the earth's crust. Using whale songs to understand the geology of our planet. Who would have thought it?[120]

The incredible songs of male humpback whales should not make us forget other vocalizations, however discreet they may be. Isabelle Charrier, Olivier Adam, and their student Anjara Saloma have been working for several years on the calls exchanged between mother and calf.[121] Recording tags were placed on the backs of the animals using a long pole. Not so easy. But very interesting, because the tag records not only the sounds but also a lot of other information, such as the depth or the speed of the animal's dive. Analyses of these recordings show that mother and calf vocalize especially after a dive, when they come to the surface to breathe and rest. Their calls are of very short duration and low amplitude. Nothing is known about the information that is transmitted—the secrets that the mother and her young exchange remain to be deciphered.

With dolphins, things are (a little) clearer. The vocalizations of these sociable animals play a major role. They allow individuals of the same band to recognize each other and facilitate collective decision making, such as whether to go this way or that.[122] Dolphins are known for their whistles, aren't they? Well, you should know that each dolphin has its own unique whistle, which is recognized by the other dolphins in its close circle. When a dolphin whistles, it's a bit like announcing its first name.[123] Sometimes dolphins imitate the whistles of their fellow dolphins, perhaps as a way of addressing one another.[124] Dolphins have a complex social structure where whistling as a means of address is very useful.[125] For example, males, who form long-term alliances by cooperating to seduce females, recognize and call each other when they whistle.[126] Similarly, dolphin mothers whistle when they wish to call back their calf.[127]

There is one toothed whale whose vocalizations send shivers down the spine if you live under the sea. It's the killer whale. The *killer* whale! *Orcinus orca*—the top of the top predators. One of the most socially, culturally, and cognitively complex creatures on the planet, the killer whale is pure intelligence. Killer whales often live and hunt in packs, sometimes with very elaborate techniques, such as cooperating to rock the block of ice on which an appetizing seal is found until lunch falls into the water. Orcas are divided into several ecotypes, i.e., populations with different lifestyles, that feed on different prey. In the Pacific Northwest, resident killer whales eat salmon, transient killer whales eat marine mammals (such as sea lions), and offshore killer whales eat sharks![128] In the Atlantic, one ecotype has a predilection for other cetaceans, another is more eclectic. In the Antarctic, five ecotypes have been identified: one has a preference for the minke whale *Balaenoptera acutorostrata*, another for the seal, yet another revels in the penguin, and the last two are especially fond of fish. Killer whales are real foodies and very picky about their diet.[129]

Orca ecotypes form well-separated populations that do not reproduce with each other. And, as you might expect, each has its own vocal dialect.[130] We also know that killer whales can imitate others: the transmission of dialects is most certainly cultural.[131] The vocal repertoire of killer whales has three main types of vocalizations: clicks, whistles, and pulsating calls. The clicks are extremely short and are used for echolocation. The whistles are highly modulated, varying between high and low pitches. Pulsed cries are very fast trills, which the human ear hears as a short call of about a second. Let's admit it outright, once again: we don't understand what the killer whales are saying to each other.[132] The only certainty is that acoustics play a major role in the cohesion of social groups and when killer whales cooperate to hunt. They also know how to keep quiet when necessary. Thus, killer whales that hunt marine mammals are less talkative than others. It must be said that their prey has good ears.

Charlotte Curé, one of my PhD students, now a researcher at Cerema,[133] and her colleagues recently conducted playback experiments demonstrating that humpback whales can distinguish between the vocalizations of different killer whale ecotypes. The humpback whales approach an underwater speaker that emits herring-eating orca calls as if it were a bell announcing a meal. On the other hand, the humpback whales avoid a speaker emitting whale-eating orca vocalizations, which is a very good idea.[134] Similarly, Charlotte and her team have shown that

pilot whales *Globicephala melas* in Norway are able to distinguish between killer whales from different populations just by listening to their
voices.[135] Being able to distinguish one from the other helps pilot
whales determine whether they can relax or should prepare their de-
fenses before the arrival of the great predator. Even the sperm whale, as
gigantic as it is, can fall victim to killer whales. In fact, the sperm whale
is very careful: if it hears the calls of killer whales from a loudspeaker, it
stops feeding.[136] Instead of diving as deep as might be expected, the
sperm whale flees at full speed or regroups with other kin members,
perhaps hoping to scare off the terrible predator.[137] United we stand!

Let's talk about the sperm whale: *Physeter macrocephalus*, the big-
headed blower. It's a truly unique animal—the largest toothed whale on
the planet; the *biggest brain* on the planet; and an incredible nose,
which, if it isn't shaped like Cyrano's, is much bigger than Pinocchio's;
up to 30% of its body volume! You read correctly: a third of the sperm
whale's body is just a nose. Inside this nose, the space is essentially occupied by a large pouch filled with a rather special oil, the spermaceti,
which changes consistency with temperature. It was for this oil, widely
used in cosmetics, for example, that sperm whales were once hunted.
For a long time, it was believed that spermaceti allowed the sperm whale
to adapt its buoyancy to the depth of its dive.[138] But in the 2000s, another hypothesis was validated. This pocket with its spermaceti is . . . a
musical instrument!

This enormous nasal device is capable of generating sounds—simple
and short clicks—which are extremely powerful.[139] Two hundred and
thirty-six decibels.[140] Enough to stun any squid. As with dolphins, it is
the famous monkey lips that produce the sound. Caricaturing the situation, here's a brief description of how it works. The valves of the monkey lips, connected to the nasal duct, vibrate as air passes through. The
vibrations are transmitted back to the spermaceti pouch and spread
through this oily substance to the skull bone. You should see the shape
of this bone: it's a huge parabola that reflects the sound waves . . .
forward. The spermaceti then acts a bit like glass lenses do with light; it
concentrates the waves into a beam directed in front of the animal. The
sound waves are thus sent forward into the water—"click-click-click."

Sperm whales vocalize to locate and perhaps stun their prey, to navigate the depths of the ocean, and to maintain social ties with their fellow creatures.[141] For echolocation, they use simple directional clicks emitted in regular series. When they find their prey, the series ends with an acceleration of the clicks, causing quite a stir. Acoustic communications between individuals rely on more elaborate sound productions, such as codas, which are short series of clicks: "click-click-click . . . click-click-click."[142]

If we zoom in on one of the clicks constituting a coda, we can see that clicks are not simple sounds. A click has a very short duration, around 100 to 200 milliseconds. Our ear hears it as a single sound ("click!"), yet that click is actually a burst of pulses.[143] (I hope you are following me: a coda is a series of clicks; a click is a series of pulses.) These pulses are echoes of the click produced by the monkey lips. Imagine the enormous spermaceti as a kind of cathedral, in which every click reverberates, creating pulses. Since the sound waves move very fast in the spermaceti (around 1800 meters per second), there is no chance for our ear to discern these echoes; we only hear a single "click"!

The time interval between the pulses of a click depends on the volume of the spermaceti, just as the return time of the echoes in a cathedral depends on the size of the building. By measuring this interval, we can therefore estimate the body size of the animal.[144] As female and male are of very different sizes, it is possible to know the sex of the animal by looking at the pulses that make up the clicks.

A coda has between four and a dozen clicks, all occurring over a period of a few seconds at most. Scientists usually annotate the clicks in a coda by looking at the time between two successive clicks. For example, the annotation "2 + 1 + 1 + 1 + 2" means that the sperm whale made two clicks close together, took its time to produce the next three, and ended with a short burst of two clicks: "click-click . . . click . . . click . . . click . . . click-click." In the early 2000s, Luke Rendell of the University of St. Andrews in Scotland noticed that sperm whales belonging to the same group all produce the same kind of codas.[145] They seem to share the same Morse code, if you will. Rendell suggested that there are acoustic clans among sperm whales. A few years later, it was hypothesized that these codas might instead be used by females and juveniles, perhaps to signal their identity.[146] Again, we have to admit: we still don't understand what the sperm whales are saying to each other.[147]

Perhaps you feel a little frustrated that you haven't learned more about the languages of dolphins and whales. I have, of course, considerably summarized what we know. Nevertheless, our knowledge is still limited: it is extraordinarily difficult to study these animals in their natural environment. However, new technologies, combined with modern methods of analyzing long sound sequences, give us reason to be optimistic. Perhaps we are on the verge of unlocking the secrets of the humpback, the killer whale, and the sperm whale. For now, let's get our heads out of the water. The elephant seals are waiting for us on the beach. There, in our aerial element, it's easier to explore what animals are saying to each other.

7

The tango
of the elephant seals

VOCAL SIGNALS AND CONFLICT RITUALIZATION

Año Nuevo Reserve, near Santa Cruz, California. Jaws open, the two giants clash. Each one wants to push the other back, to bite it, to finish it off—in a word, to kill it. Their flaccid trunks are torn on all sides. Blood flows. The fight is incredibly long: they have been fighting for half an hour already! Their chests stuck together, exhausted, they stop for a few seconds. Time for a breather. Then one of them swings back and picks up momentum, determined to strike again. The enormous daggers of his canines cut into the opponent's thick skin, who steps back for the first time. He will yield. A few more assaults and off he goes, turning tail and running away. A two-ton mass crawls across the beach. The winner is not finished; it's still chasing him. Another bite, a violent bite, on the rump. Then, triumphant, he stops, stands up, and—head high, his trunk thrown backward—launches a hoarse cry, regular as a metronome: "Clork ... clork ... clork. ..." The loser will never fight back again: he will run away.

The first time I visited the elephant seal colony in Año Nuevo, it was raining hard. It was February. Imagine a low sky, a cloud-covered horizon, the Pacific Ocean marbled with large rollers, and hundreds of huge seals on the beach. In *Mirounga angustirostris*, the males are as big as pickup trucks, up to 4 meters long. The females are smaller, 2½ meters at most. Sexual dimorphism (the fact that females and males are not the same size and do not look alike) is obvious at first glance. And here and there, big black tootsie rolls of more than a hundred kilograms each: the babies of the year. Northern elephant seals come to spend the winter in Año Nuevo to breed, give birth, and mate. The adults migrate from their fishing grounds in southern Alaska, arriving here in November–December and leaving again in mid-February. They fast throughout this time. A highly effective weight loss diet: A male or female elephant seal can lose up to 40% of its weight in a single breeding season.[148]

The competition for what is trivially called "access to females" is quite tough between males. Less than 1% of them are able to reproduce in their lifetime. On the beach, relations of dominance are first established through physical confrontations, often of fairly short duration. They are maintained during the breeding season mainly through ritualized behaviors, including characteristic postures (the seal raises its body), vocalizations, and ground tremors (when an individual hits its chest against the sand). Battles such as the one I just described are rare: they only take place between two dominant (alpha) males, both equally motivated to win the competition. The outcome can be fatal.

Why am I interested in elephant seals? As usual, due to a mixture of chance, luck, and curiosity. That year, I was a visiting professor at the University of California, Berkeley, hosted by my colleague Frédéric Theunissen from the Department of Psychology. Together we were conducting a research project on hyena communication, which I will tell you more about later. My visiting professorship was funded by the Miller Institute, a wonderful foundation that invites scientists from all over the world to spend a year at Berkeley. The competition is strong here too, but the reward is great: a year devoted to research, with no constraints other than to attend the Tuesday lunch with the other visiting professors. I decided to take advantage of this opportunity to visit other research laboratories in the region. That's how one day I took the bus to Santa Cruz, a pretty town south of San Francisco well known by surfers. On arrival, Ron Schusterman was waiting for me. Ron had founded the Pinniped Cognition & Sensory Systems Laboratory to explore the cognitive abilities—as scientists call *intelligence*—in pinnipeds. Ron was a pretty amazing guy. "Nic, I have to get my breakfast first. I hope you don't mind." Instead of visiting a lab, I found myself in front of an amazing 80-year-old who quietly started telling me science stories over coffee and pancakes. I liked this unconventional situation, of course.

"I'd like to introduce you to Colleen Reichmuth—my favorite student, who took over my lab. She's great. She's my spiritual daughter." The old man's eyes twinkled. He was happy to have entrusted his life's work to someone he had confidence in. Ron is gone today, and Colleen and I often remember him.[149] Science is also about transmitting wisdom.

The Pinniped Laboratory is a unique structure where seals, sea lions, and sea otters are trained to respond to sound signals. This method is called operant conditioning. Here's the principle of it. You may be familiar with nonoperant conditioning, or Pavlovian conditioning. It was first discovered by the Russian researcher Ivan Pavlov in the late 1800s in his famous experiment with dogs. When meat is presented to a hungry dog, it starts to salivate. In Pavlov's study, conditioning consisted of associating the presentation of a piece of meat with the ringing of a bell. After a few trials where the bell was rung at the same time the meat was presented, the dog salivated as soon as the bell rang, even if the meat did not follow. The dog was conditioned to the bell stimulus. The response to this stimulus is a reflex over which the dog has no voluntary control. It cannot do anything about it; it salivates. In operant conditioning, the aim is to provoke in the animal a voluntary response to the stimulus: the animal must complete an action, such as pressing a button or moving in a certain direction. For example, Colleen's animals learn to hit a target with their muzzle when they hear certain sounds. To bring them to this behavioral response, the seal or sea lion is taught that, if it reacts "correctly" to the sound stimulus, it will be rewarded with a nice fish. Once the seal is trained, it becomes possible to ask questions, for example, to test its hearing. We can see that, if the sound level of the stimulus is lowered too much, the seal stops responding and therefore stops hitting the target. By systematically exploring different loudness levels at different frequencies, Colleen creates audiograms for the animals being tested. If you've ever been to an otolaryngologist, you probably had to listen to sounds of varying intensity to measure your hearing. At the end of the test, your results are plotted onto a curve that shows your hearing ability, ranging from low- to high-pitched sounds. This curve is your audiogram. Using this method, Colleen's goal was to explore the sound world of pinnipeds. When I told her about the elephant seals I had seen at Año Nuevo, she was enthusiastic: "Let's do some field experiments! Fantastic, especially since we already have a connection with the Año Nuevo Reserve; and one of my students, Caroline Casey, will certainly be willing to help us." That same evening I contacted Isabelle Charrier, my former PhD student, now senior researcher at the CNRS and an expert in marine mammals. We had always said that one day we would like to work together again, and her great knowledge of field work with pinnipeds was essential to the success of the project. So a new, small team of passionate scientists—Colleen, Caroline, Isabelle, and me—was formed. We were going to work for several years to explore the sound world of elephant seals (and we still do!). Once again, chance encounters were an essential ingredient in my life as a researcher.

Elephant seals are so large that they have no fear of humans. As long as we didn't get too close, our presence on the beach was mostly tolerated. This made it easier to make recordings and playback experiments—to a certain extent, of course. We had to be very careful because females can become aggressive and try to bite if you get too close to them. Some of them chased us over long distances. And the bite of an elephant seal is more than annoying: their jaws are powerful and their saliva contains bacteria that our immune systems are not equipped to handle. As for the males, they're so big that they can kill a human with a single butt of their head.

The vocalizations produced by males are made up of a succession of clicks, several per second. It has long been known that these signals play a role in the interactions between males: it is common to observe that, when an alpha male begins to vocalize, the males around him beat a hasty retreat. Our first hypothesis was that these calls carry information about the strength and power of the individual transmitter: it is easy to imagine that the strongest males must have the lowest voices or produce

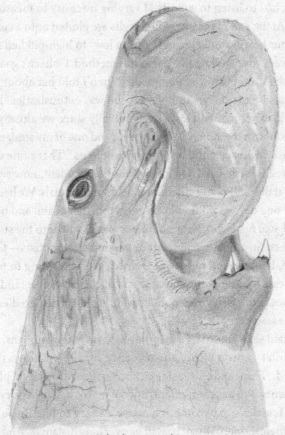

Elephant seal

their calls at a faster rate. An alternative hypothesis was that males produce vocalizations with marked individual characteristics and that they learn to recognize each other by their particular voice. Between the two hypotheses a decision had to be made—or we had to reconcile them.[150]

It was a considerable amount of fieldwork, which required several consecutive missions. Each year, about 40 males were given a mark—a number painted on their lower back—which allowed them to be identified throughout the breeding season. But since the painted marks disappear at sea, Caroline would attach a numbered tag to some of them on one of the rear flippers, hoping to find them the following year. I'll let you imagine Caroline, crouching, approaching one of these large animals from behind. Painting the back of an elephant seal or attaching a tag to its flipper is no small business—not a sport I'd recommend! Each bull was carefully photographed and measured and its voice recorded several times during the season. Tagged males, when found, were recorded over several years.

Isabelle and I had the idea that the most impressive males would be those with the lowest voice pitch, as size and pitch are two characteristics often linked in the animal kingdom. This is indeed what our measurements showed. In addition, the bigger the individual, the faster the rhythm of its clicks. The hypothesis that males can show their strength and thus their ability to win a fight by their voice alone became plausible. We still had to check that the males at the top of the hierarchy were also the biggest.

Estimating the hierarchical rank of the males is not obvious. It takes many hours of observation to describe the interactions between individuals and to record the outcome. Are you familiar with the Elo rating system, named after the chess player who developed it? It is used to rank a group of individuals from the strongest to the weakest, after competitions involving only two players at a time. We used this method to evaluate the dominance rank of male elephant seals. At the beginning of the breeding season, each bull received an Elo score of 1000. Then the Elo scores were modified, up or down, at the end of each male-male interaction. After a few weeks, the scores stabilized: some individuals—the alpha males— were at the top of the rankings because they had won almost every fight they had ever had. Further down in the rankings were the beta males, then the peripheral males who had systematically fled from any opponent.

To test the hypothesis that the males' vocalizations carry information about strength and ability to win a fight, we looked for possible correlations between the acoustic structure of the clicks (their rhythm, pitch) and the males' Elo score. I was very surprised to find nothing conclusive. A very dominant alpha male could have a slow or, on the contrary, a very fast rhythm, a deep or rather high-pitched voice. It was impossible to predict the hierarchical level of an elephant seal just by listening to its voice.

Experiments with acoustic signal playback made it possible to find out for sure. We had males listen to voices of unknown males—voices that we had recorded a hundred kilometers south of Año Nuevo, in the region of Piedras Blancas. We had modified the characteristics of these voices, making them lower or accelerating their rhythm in order to imitate the voice of ultradominant male alphas. The results of the experiments were disappointing: the males mostly ignored our signals or seemed to respond randomly. We were back to where we had started. Now we had to test our second hypothesis: elephant seals probably had to learn to recognize the voices of their fellow elephant seals.

For Caroline, the one of us who spent the most time in the field, there was no doubt that each male had a different voice from the others. Just as you can recognize a familiar person when you hear them on the phone, Caroline could identify each of the males on the beach by hearing them vocalize. Our acoustic and statistical analyses confirmed it: the rhythm of the clicks and the pitch of the voice, higher or lower, were enough to distinguish one individual from another. Even better, the vocal characteristics of each individual were preserved from one year to the next. A male keeps his voice . . . for life. All that remained was to ask the animals the question, Can you recognize each other by voice alone, the way Caroline can?

Asking such a question to an elephant seal is not simple. Caroline was utterly sure of herself: she recognized the males, she differentiated between them very well just by their voices. So we started playback experiments. We played the vocalizations of familiar individuals, who were neighbors on the same beach and with whom our test bull had interacted over the previous weeks, to beta males. For each male tested, two trials were conducted. In one, we made the male listen to the voice of a dominant, i.e., an individual in front of whom the male would back down during an encounter. In the other, we played the voice of a submissive. Imagine the scene: a male elephant seal lying on the beach, the loudspeaker barely 10 meters in front of it. The voice of a dominant reaches its ears, and suddenly the huge mass rises abruptly, turns around, and runs like mad in the opposite direction from the loudspeaker. When we played the voice of a submissive, the animal took its time, reacted only after a while, and then approached the loudspeaker seen, or rather heard, as a weak male. Sometimes, there was even a little breakage and a few welds needing repair in the evening when we returned from the colony. In any case, the results were clear: an elephant seal can recognize Peter, Paul, or James by its voice alone.[151]

Like my teammates, I was now curious to understand the acoustic basis of this recognition. In particular, do elephant seals rely on the rhythm of the successive clicks that make up their calls to identify their fellow bulls? Using the same protocol, we tested 10 males with vocalizations of dominant individuals whose clicking rhythms had previously been modified by making them faster or slower. As long as the change in rhythm remained very small, the males continued to run away from the stimuli: they always recognized their dominant. But as soon as the rhythm was accelerated or slowed down by 10% of its initial value, their behavioral response became more hesitant, or even disappeared altogether. They would lie motionless, seemingly ignoring vocalizations whose sender they no longer recognized. This experimental result is conclusive; it shows that elephant seals memorize and use the rhythm of their fellow elephant seals' calls to identify them. Each individual must remember the voices of several dozen others. What a memory— an elephant's memory, of course![152]

One year, on the beach of Año Nuevo, it was claimed anecdotally that an alpha male was particularly powerful and aggressive. The following November, this individual did not return, and the other males were able to spread their harems over a larger area. A few days later, a new male arrived, poked his head out of the water, and started to vocalize "rlurk . . . rlurk . . . rlurk . . . ," and all the others bolted. It was the alpha male from the previous year who was cleaning up the beach with his calls . . . remembered by his fellow males. If the memory of elephants in Africa and Asia is legendary, the memory of Northern elephant seals is just as brilliant.

Remember: one of Nikolaas Tinbergen's four questions concerns behavioral *development*. How does behavior develop over the course of an individual's life? It is understood that baby bull elephant seals are unable to make the characteristic clicking sounds of the adults. The babies do call out a lot, but they sound like sheep bleating. Also, the male babies are not looking for a fight. In fact, once weaned, they even have a tendency to huddle together. How does the hierarchy of dominance play out? When do they start using their voices to recognize each other? With all these questions in mind, I wasn't finished with elephant seals.

Hierarchies of dominance—the fact that individuals living in groups do not all have the same access to resources—are a characteristic shared by most animals, including humans. What is a resource here? Anything that can be valued by individuals, like food resources. For example, we observe a hierarchy of dominance being set up in groups of hens for access to seeds distributed by a farmer. Sexual partners can also be resources. I grant you that using the term *resource* to designate individuals with whom it would be possible to mate is not the most elegant, but this metaphor from the world of economics has the merit of clarity: if the same resource is desired by many, it can be a source of competition and lead to the establishment of hierarchies of dominance. In elephant seals, the resource prized by the male is the female. This does not mean that female elephant seals do not also desire resources and just sit around idly. For example, they might be competing with each other for a place on the beach that is sheltered from storms, where their young would not be threatened by the ocean. Some are very aggressive toward their peers when they feel they are too close to their young. Moreover, let's not generalize; in animals, including humans, it is not always a case of males competing for females. The animal world is much more complex than that. We will see further on that the situation can be totally reversed, as in spotted hyenas, for example. The female elephant seals themselves may choose to let this or that male approach them. We lack sufficient data here to say for sure.

To understand the establishment of the hierarchy of dominance of male elephant seals and the role of acoustic communication in this development, our small French-Californian research team had its work cut out. We decided to study in parallel the age-dependent changes in space use, social interactions, and vocalizations. The males were classified into five categories: individuals aged four, five, six, seven, and eight years and older. Why start with four-year-old males? Because it is only at this age that young males return to the breeding colony. Before that, they are somewhere in the ocean. Long hours of observation began. The study would last four years.

This fieldwork allowed us to get a clear vision of how social networks work in elephant seals. Our data, obtained through the observation of 1352 interactions involving 207 males, shows that social relationships evolve with age. Younger males come into contact with a smaller number of fellow bulls than older males. In addition, they favor contact with individuals of their own age class. In other words, they avoid the old bulls, probably because they fear them. Moreover, interactions between young males often involve two individuals who have never met before, while older males regularly interact with the same rivals. In short, whereas younger males form fluctuating networks of social interactions, with many unknown individuals, mature males navigate a stable social network made of familiar individuals.

What role does voice play in these interactions? From our earlier work, we knew that adult males have voices with a strong individual signature. And I've told you about our experiences showing that individuals recognize each other vocally. What about younger individuals? Were their voices also different? Were they also able to recognize each other? Our analyses showed that the vocal signature appears gradually, stabilizing when the animal reaches seven years of age. This is, in a way, the age of reason for the elephant seal. Younger males have a voice that is not very personalized, so it is not possible for them to recognize each other with certainty. It is only with age that the individual vocal signature becomes clearer.[153]

In our elephant seals, acoustic communication therefore plays a key role, allowing each seal to navigate within its social network, by adjusting its behavior according to the individuals it encounters and dealing with conflicts mostly by voice rather than in bloody fights. Here's Pierre talking! He's stronger than I am and scares me stiff. I have to get away fast! There, I hear Paul. He's a loser. He doesn't scare me. As soon as he hears me, he'll run away! The life of an elephant seal is no bed of roses: only 5% of male elephant seals survive to adulthood. But without this vocal ritualization, the loss of lives in adulthood would probably threaten the survival of the species.

8

The caiman's tears

ACOUSTIC COMMUNICATION IN CROCODILES

Nhumirim Ranch, Pantanal, Brazil. Standing at the back of the Toyota, Zilca lit the marsh with a powerful torch. Undisturbed by the bumps in the muddy and rutted road, she never took her eyes away from the beam of light. *"Aqui! Uma mae jacaré com seus pequenos!"* Shining like stars, the eyes of the caimans betray their presence. I came to South America, in the Pantanal, near the Bolivian border, to study the acoustic communication between mothers and their young in the jacaré caiman—a species of crocodilian that is abundant in this region, where a wooded savanna, swamps, and magnificent sunsets are accompanied by an incredible diversity of fauna. Nhumirim is a *fazenda*—a farm, with cattle and cowboys—as well as a research station. I have known Zilca for some time, but I had never met her in person before this fieldwork. A few years ago, our first online conversation was an unforgettable moment. To communicate with her, I used a language translator found on the internet, as I don't speak Portuguese and Zilca is not comfortable with English. Since our exchanges were only in writing, I didn't realize until much later that she was a woman. The stereotype of Crocodile Dundee made me think of a big, tough guy when talking about crocodiles. My first memory after landing at the Corumbá airport is of the wrinkled face of a small woman, with her gentle smile and sparkling eyes: Zilca Campos! Sharing the life of this warm and fascinating person during a field mission was a rare human and scientific opportunity.

Reaching the Nhumirim Ranch means a seven-hour trip, mostly on sandy roads. Zilca had organized everything down to the last detail. Once we left the hills of Corumbá, the landscape became flat. The Pantanal savanna is the floodplain of the Paraguay River. During the dry season, from October to April, water is only present in the main rivers and isolated lakes of the Pantanal. But the rainy season floods most of its 100,000 hectares. In Nhumirim, there can be more than a hundred lakes in wet years. Conversely, "some years, it almost doesn't rain and everything is dried up," explains Zilca. This year, luckily, the lakes are full, ensuring good conditions for the caimans.

"This will be your house," she says. "It hasn't been occupied for a while. I lived there for so many years!" With my colleague Nicolas Grimault from the CNRS and our doctoral student Leo Papet, I entered the modest house, delighted by its breathtaking view of a large marsh. Its name, Fauna, was not undeserved. A few days after our arrival, we saw two long snake molts stuck in the grooves of the kitchen door. The animals had probably come out of the cupboard during the night, disturbed by our presence. We finally found one of them coiled up against a wall. The other had evaporated. Every morning, until we decided to tape up all the doors and windows, a new snake came into the house.

The jacaré caiman (*Caiman yacare*) is emblematic of the Pantanal. You have likely already seen them in photographs, either with a colorful butterfly on their snout or throwing a catfish in the air to gobble it up head first, thus preventing it from getting stuck when it passes through the esophagus. It is a sister species to the spectacled caiman (*Caiman crocodilus*), the most common caiman found from Mexico to Argentina. "The number of nests found in Nhumirim varies considerably from year to year. Sometimes we've found more than 60. And sometimes none. The last few years have been the worst," Zilca tells me. "We don't know why these huge variations exist. The decrease in rainfall over the last 10 years may be part of the reason." Zilca is a researcher at Embrapa, a Brazilian research institute, and has been working on the ecology of the jacaré for several decades. She is passionate about this animal. She continues: "The female jacaré lays between 20 and 30 eggs in the middle of the rainy season, from December to February. As with all caimans and alligators, the nest is a mound of vegetation. The eggs take about three months to mature—three months during which the female remains in place, providing protection from predators. The mother defends the nest and takes care of the young." Parental care is a common behavioral characteristic of archosaurs, a group that includes crocodiles, birds, and extinct dinosaurs and pterosaurs. By doing research on crocodiles, I feel like I'm opening a window to the past, a window to these extraordinary animals.[154]

Acoustic communication is a remarkable feature of the mother-young relationship in crocodilians.[155] Females and their young produce vocalizations. In a few species, fathers also participate in watching over the young. In the gharial (*Gavialis gangeticus*), a crocodilian with a very long snout (of which there are still a few rare populations in India), the males group together to watch over newly hatched young while the females go about their business.[156]

When they are ready to emerge, mature crocodilian embryos emit vocalizations that act as signals to their siblings and mother.[157] It was with Amélie Vergne, one of my most passionate students, that we demonstrated this experimentally. The idea came to me during a family visit to a crocodile zoo. On the edge of one of the ponds where dozens of huge Nile crocodiles (*Crocodylus niloticus*) were basking, a sign indicated that crocodiles use sounds to communicate . . . which I didn't know. A little upset by my lack of knowledge, but above all terribly intrigued, I began to gather all the scientific publications on the subject—only to find that no one had really studied the question. Fortunately, Luc Fougeirol, the founder of the zoo *La Ferme aux Crocodiles* in Pierrelatte, France, opened his doors to me to begin the research.

The principle of our first experiment was simple: go into the crocodile pond, move a mother away from her nest (Nile crocodiles lay their eggs in the sand), bury a loudspeaker in place of the eggs, leave the pond, wait for the mother to come back, and play hatching calls from the loudspeaker (or noises of the same duration but without the acoustic structure characteristic of baby crocodile vocalizations). Imagine the scene: two zookeepers with shovels hitting the water to keep away a mother trying to defend her nest, while the zoo veterinarian is digging the hole to bury the loudspeaker. It was a bit chaotic.

Most of the females we tested quickly returned to their nests. In response to the playbacks of the hatching calls, the females began to dig in the sand above the loudspeaker, whereas they remained motionless when we played other noises. It was the calls of the hatching young crocodilians that led to the first maternal care behaviors. In another experiment, where we played the same sound stimuli to eggs about to hatch, we saw that the calls of babies led to the hatching of their siblings.[158] In the wild, everyone hatching together is probably an important advantage: the mother is there to help and to protect the last ones that come out.[159]

In crocodilians, the mother's help doesn't stop at the nest. Have you ever seen animal films that show a crocodile with youngsters in its mouth? The mother (or father in some species) is carrying her young to the river or body of water. After hatching, she stays with them for several weeks. The young crocodiles, which measure about 20 centimeters, are extremely vulnerable to predators such as large birds or carnivorous fish. Cannibalism is also widespread, so one crocodile parent will not tolerate any other adults near its young.

In order to study the acoustic communications between the crocodile parent and its young in the days following hatching, it was necessary to go into the wild with free-living animals. In a zoo, this is not possible because the density of animals is too high, and the young, if left in the basins, would not survive for long. I asked Luc for advice on the best places to approach crocodiles. "Go to Peter Taylor's house," he said. "He is an American who has a field station in Guyana, on the banks of the Rupununi River. He is an expert on the black caiman. You'll see, that caiman is a very big one—a real man-eater." Then he added, "Going to Taylor's isn't easy. So, for a start, you could go to the Kaw marshes. There are black caimans there too." The Kaw marshes: the mythical, virtually inaccessible, and miraculously preserved stretches of swamp in French Guiana. I contacted Eric Vidal, then at the University of Marseille, who had a research program on the Agami herons of French Guiana, precisely in Kaw. With his usual kindness, Eric invited me to join one of his missions, asking me in exchange only to make some recordings of herons. I called Thierry Aubin, who, not losing any opportunity to get close to a wild animal in its natural environment, was immediately willing to accompany me.

It was my first helicopter experience. We took off from the parking lot of the caiman camp hostel, on the road from Roura in the direction of the Kaw marshes. A few minutes later, after diving over the marsh below, the helicopter approached a very small aluminum platform floating on the water. A few cable lengths away, another platform of about 10 square meters protected by a tarpaulin would be our camp. Lost in the middle of the Kaw swamps, we were ready to begin our experiments on the black caiman, the amazing *Melanosuchus niger*.

It was, of course, by boat that we approached our first female. I didn't have a recording of baby caimans, so I decided to play baby Nile crocodile calls; we would see the result. During the first broadcast, the female's response was impressive: she leaped forward, roared, and charged at our boat. Needless to say, we immediately stopped our calls. We were full of hope when we returned to the platform to prepare our signals for the next day's experiments. It was quite something: even coming from a different species of crocodilian, the screams of babies provoked an intense maternal reaction. It opened up new perspectives. But the next day, we were disillusioned. Same individual, same test: no more answers. Yesterday's female ignored us.

This phenomenon is well known to ethologists. It's called habituation: when an animal classifies a sensory stimulus as one to ignore due to lack of reinforcement. Baby crocodile calls in the absence of real babies were indeed of no interest to the female. What was surprising was how quickly the mother caiman became accustomed to our signals. But what was even more surprising was that the other adults in the lagoon that we subsequently tested immediately showed a very weak reaction to the calls of the babies. Let me be very clear: all the black caimans who had witnessed our first experiment and had probably heard our signals had integrated in a very short time that calls coming from a loudspeaker placed on a boat were not worthy of interest. Clearly, we had not taken enough precautions. We had underestimated the animal, its sensory integration capabilities, and its memory. Duly noted. This is a great risk of ethology: not realizing that the animal you are studying has a keen knowledge of its environment   much better than yours—and that it is capable of understanding very quickly that your presence results in negligible stimuli, such as calls played from a loudspeaker. Primatologists pay attention to this in principle because it is easy for us to imagine that monkeys or apes, close to our own species, develop cognitive faculties close to ours. But with our caimans, we should have thought about it: these superpredators, who spend their lives probing what surrounds them with their senses, are quick to realize that experimental stimuli are irrelevant. Plus, they have the memory of an elephant. So we had to go to work in a context where we would have access, for each experiment, to naive individuals. We couldn't work on a lake where all the animals were constantly observing us. We had to look for a river, where the caimans would be spaced a long distance apart. The following year we went to Peter Taylor's house. This time Amélie was on the trip.

Luc was right. Getting to Caiman House, the field station of Peter Taylor and his wife, is part of the expedition: a transatlantic flight from France to Guyana's capital, Georgetown; then several hours in a small bush plane to go deep into the heart of the country, landing on a dirt runway; finally, half a day in a pickup truck before arriving in an indigenous village on the banks of the Rupununi River, where the caimans were waiting for us.

Peter had spent the last few weeks spotting female caimans and their young. He had found a dozen of these small families on the river, about a kilometer apart. It seemed like the perfect situation: we would avoid habituation since the females were far enough apart from each other. "But we'll have to work at night," Peter said. "Caimans have been hunted a lot around here, and they are fearful of humans." The first few nights we were busy recording baby caimans. We decided to do it in two very different contexts. First, a quiet context. The young caimans gather near the shore probably to avoid being swept away by the current. A microphone and a recorder left near them made it possible to record calls of fairly low intensity, which the youngsters frequently emit. Then we moved to a stressful context. Peter grabbed one of the babies by the tail and shook it slightly to mimic a predator attack, and Amélie recorded the calls of the baby. Acoustic analyses showed that the sounds emitted in the quiet context differed from those emitted in the stressful context. The former were short, muffled, and slightly frequency modulated: "Djong! djong!" The stress calls were a little longer, louder, and much more modulated: "Zuh! zuh! zuh!" Even though they sound different, these two calls are built on the same pattern, the same acoustic family in a way: they are frequency-modulated harmonic series, and you can go from the quiet context call to the stressed context call simply by modulating the frequency more—putting more energy into the treble and increasing the volume. Are these calls signals, you may ask? Do they allow individual senders to communicate information to their mothers? To other juveniles? Only acoustic playback experiments could answer these questions. That's what we did.

The loudspeaker is placed on the bank, about 10 meters from the group of young black caimans. The mother is in the middle of the river, motionless. We are in our boat, moored to a tree on the opposite bank, attentive and silent. Judging the moment appropriate, Amélie presses the button on the tape recorder and triggers the emission of calls recorded in a quiet context. The mother does not move an inch. The young ones, however, move toward the speaker, continuing to emit low-intensity calls, and gather around it. Amélie stops the stimulus, and we begin again to wait patiently in the dark. A few minutes later, we play distress calls from the loudspeaker. The mother caiman immediately turns around and rushes in the direction the sounds are coming from. (In follow-up experiments with other individuals, some mothers roar, with a guttural and threatening sound: one of her babies is in danger and she is coming to save it.) The young stop motionless, but carry on calling. The acoustic analysis of their calls will show that they are making distress calls. I'm happy: the Guyana field trip has demonstrated that baby crocodiles have a vocal repertoire that allows them to encode information and target the receiver: contact calls that allow the babies to stay together and distress calls that enrage the mother and alert siblings.[160] I was also delighted for Amélie: her first field trip was a success, despite some problems caused by intolerance to an antimalarial drug. This shows how important it is to observe the animal you are studying in its natural living conditions. This is essential if we want to be able to ask meaningful questions.

Further research showed that all crocodilians share the same acoustic distress code. A few years ago, during fieldwork in the Venezuelan Llanos, Thierry and I tested female spectacled caimans with calls from black caiman, Orinoco crocodile (*Crocodylus intermedius*), and Nile crocodile juveniles. The females responded to all these stimuli.[161] Other experiments using artificial signals that we had created on a computer showed that a simple whistle mimicking the descending frequency of a distress call is enough to provoke a maternal response.[162] Based on these observations, it is likely that crocodile mothers are not able to recognize their own offspring but respond to any newborn call.[163] Moreover, the calls of baby crocodiles do not just inform us about their state of stress. I tell you in chapter 10 how they also code for the size of the emitter. For now, let's go back to the Pantanal, where Nicolas Grimault and Leo Papet, my fellow researchers on this journey, and I have decided to test the ability of an adult crocodilian to locate its young when they vocalize.

"I marked every nest I could find with one of these red flags," says Zairo, smiling broadly. Zairo is Zilca's field assistant. He was born here and is an expert on the wilderness of the Pantanal. To find nesting females, he had to cross thorny bushes surrounding the lakes, well protected by leather clothes. Denis is the team's driver. He is the only one who knows how to start the Toyota engine with a piece of metal when the old truck breaks down. My goal is to test whether caiman mothers can accurately locate a newborn baby based on its vocalizations. Since noise from human activity has become a critical issue for many animals and its impact on crocodilians has never been studied, I wanted to determine if noise could impair their ability to locate a sound source. Nicolas, Leo, and I had hypothesized that, if the noise source weren't placed in the same location as the source of the signal of interest, female caiman crocodilians would locate the signal more easily—a process called spatial unmasking. This is a strategy known in humans and in only a few other animal species, like parakeets, ferrets, frogs, and crickets: the hearing system uses spatial cues to isolate the signal from the noise. Signal detection should be better when noise and signal are emitted at a distance from each other rather than from the same location. When we began our research, knowledge about sound localization in crocodilians was limited. Field data was almost nonexistent, and we had no idea how accurately crocodilians are able to locate a sound source or what acoustic characteristics they use to do so. The first investigations were carried out in our laboratory in France.[164] Leo had trained young Nile crocodiles to go to a loudspeaker emitting an artificial sound to get a food reward. The results showed that these animals locate a sound source very precisely. They quickly turn their heads in the direction of the sound, making almost no directional errors.[165]

To carry out playback experiments in Nhumirim, we first had to record calls from baby jacaré, the local species of caimans. Zairo had spotted a female staying in a large pond where cattle came to drink and where it would be easy to catch young ones. Although the female jumped out of the pond to attack us while we were handling a newborn, we made it out in one piece and got good recordings. Back at the Fauna, our home base during the trip, Nicolas, Leo, and I prepared the stimuli for the next step. The mother's first reaction had already shown us that these calls are emotionally charged for a mother caiman. We were just as moved by her reaction!

In the early morning, Denis, Zairo, and Zilca came to pick us up at the Fauna. "*Nós vamos para a lagoa 42!*" Our companions had decided to make the first attempt on lake "42," a beautiful pond where Zairo had spotted a female and her young a few days before. After a short drive, we parked the truck at a distance and walked to the lake to assess the situation. The caiman family is there, the female immobile and her little ones moving slowly around her. I enter the water without too much apprehension, carrying two speakers. After all, the jacaré caiman is a relatively small species compared to the Nile crocodile or the black caiman, and does not usually attack humans. Still, I'm a bit worried as I walk along, imagining myself walking on a dangerous animal hidden underwater. When I stop, about 20 meters from the shore, the water has already filled my boots. I put the two speakers on their tripods so as to emit the sounds at just above the water level, and we all hide behind the vegetation to observe the female's behavior. Leo turns on the first speaker, which starts to make a continuous background noise. The female

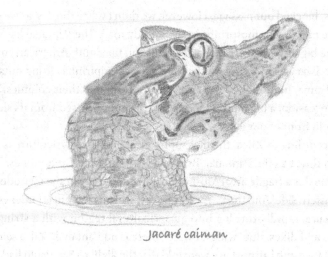

Jacaré caiman

doesn't move. The noise doesn't bother her at all. We then begin to emit a series of distress calls from the same loudspeaker, with an intensity lower than the noise level. Still no response. But when we transfer the distress calls to the second loudspeaker, the mother caiman immediately turns her head, attentively; after a few calls, she starts swimming quickly in its direction. We had just proved that spatial unmasking works in crocodilians. Female caimans are able to detect distress calls when the angle between the loudspeaker emitting the noise and the loudspeaker emitting the calls is just over 4°.[166]

Back at the station, the topics discussed with Zilca began with the behavior of the female caiman we had just tested, and then moved on to the future of the Pantanal. I was mostly just listening. "Poaching is a real problem here," Zilca explained. "Once I met poachers. They thought I was a ranger. I had to convince them I didn't have a gun." I imagined Zilca, with her frail silhouette and hesitant pace, addressing a group of determined poachers. How can we understand each other with such incredibly divergent interests?

One day, we wanted to go to Campo Dora, a large private ranch next to Nhumirim, where Zilca used to work on the jacaré populations. We never made it to the ranch because a branch of the river, too deep to cross, blocked our passage. However, we didn't waste the day, as we saw some nice jaguar footprints on the muddy road. The Pantanal has one of the highest concentrations of wildlife on the South American continent. Every evening, as we shared a glass of caipirinha sitting outside the Fauna, hundreds of ibis and storks returned to their colonies. On one occasion, a big tapir jumped up in the air, surprised from its sleep, right in front of our eyes.

According to Zilca, the development of intensive agriculture is the main threat to the Pantanal. Brazil needs food, water, and money. The Pantanal is a fragile area. Most of its water comes from the Cerrado on its eastern side. And the Cerrado, this vast savanna region, is under great pressure, rapidly turning into an agricultural region with a string of dams and dikes. But "without water, there is no Pantanal," Zilca said.

At Corumbá airport, we were told that the flight to São Paulo had left the day before. So we had to travel a good part of the night by bus to Campo Grande, then wait a few hours before taking a plane to Belo Horizonte, and finally spend the next night waiting for a flight to São Paulo. The Pantanal is far away from everything, a wonderfully remote place in the modern world. It is so threatened, however, that spotting jaguar footprints and surprising a sleeping tapir could soon become impossible.

9

Hear, at all costs

MECHANISMS OF AUDITION

It seems quite commonplace to use sounds to communicate. Most animals that are close to us bark, meow, neigh, sing, whistle, or moo. In short, most of them chatter routinely. And so do we! Spoken language is our preferred means of exchanging information. Yet, in the animal kingdom, acoustic communication is a peculiarity rather than a universally shared characteristic. It is used only by vertebrates (although snakes are not talkative[167]) and arthropods (some crustaceans, some spiders perhaps, and many insects[168]). How is it possible that none of the 100,000 species of mollusks are capable of emitting the slightest sound?[169] Couldn't an animal like the octopus, known for its resourcefulness, make its jaws squeak? Why do dragonflies remain so desperately silent? What about most butterflies? And how about sea anemones? Acoustic signals are very useful for attracting a mate, trying to scare off a predator, or signaling one's presence to others—all things that would be useful to countless species that, despite everything, remain as silent as carp (which we know are not deaf!). The reasons for this silence are certainly varied and difficult to identify. There is no escape, however, from the two following conditions: it is necessary to be able to *hear*, and to *produce sounds*.[170] In this chapter we address the first of these two conditions because, during the evolution of species, hearing has often preceded sound production. In chapter 10, we address the second.

What is hearing? A succession of several stages, from the perception of sound by the ear to the brain's processing of the information contained in the sound signal. The first stage corresponds to what in physics is called a *transduction*—the transformation of a signal of a certain nature into a signal of another nature. When you speak into a voice-activated device, such as your telephone or computer, this device must transform the sound waves into electric waves, which are then processed and lead to the execution of your command. This step of converting sound into an electrical signal is what our ears do when they pick up sound and transform it into a nerve signal that is sent to the brain. However, this is not simple to achieve. The ears have a lot of work to do, and it's a complex task.

All animals' ears (and we will see that there are many different kinds) were formed during evolution from already existing sensory structures whose role was to perceive movements of the animal's body and to inform the nervous system. These structures are called mechanoreceptors. Like all animals, we have many of them inside our body and on the surface of our skin. They allow the brain to know at each instant the position of our limbs and the pressures on our feet, hands, and elsewhere. In short, they are essential sensors for the management of our posture and movements.[171] A mechanoreceptor always works according to the same principle: a vibration, a pressure (that of a finger placed on the skin, for example) causes a part of the body to move in relation to other parts, generating a differential motion that is perceived by the mechanoreceptor. The production of a nervous signal is the task of the sensory cells, which are neurons specialized in the transformation of mechanical vibrations into electrical waves (the nervous impulse). These cells are in fact sensitive to deformation. In a vertebrate ear, these cells have extensions—in the form of cilia, or ciliary tufts if you like. When these cilia move, curve, or bend, the auditory cells feel it and produce the electrical signal in response. Depending on how these cells and their cilia are organized and how the sound vibrations reach them, we have two very different types of ears: short-distance ears and long-distance ears.

I don't want to bother you too much with physics, but there's an important notion to understand before going any further, and it's simple. It's the notion of near and far sound fields. If you stand very close to a speaker and the music is loud and rhythmic, you can feel the vibrations with your body, and even a sort of blast every time there is a "boom" of drumming, for example. In fact, you really feel the air molecules that come and go with the transmission of sound waves. Their vibrations are so strong that if you dangle a piece of string between your fingers, you will see it swing. You are in the *near sound field*. Now move a few meters away. You can still hear the sound. But you no longer feel that physical sensation of vibration. The piece of string hangs motionless between your fingers. Only your ears can hear it. The air molecules still come and go, of course, transmitting the sound pressure waves, but with much less amplitude. You have moved into the *far sound field*. What does this mean for our animal ears? That to perceive sounds in a near sound field, a hair or a filament suffices.[172] It is enough for this hair or filament to be directly connected to the sensory cell to allow the perception of sound to take place. However, in a distant sound field, it is essential to stretch a membrane (an eardrum) to receive pressure variations and to connect this eardrum, more or less directly, to the deformable cilia of the sensory cells.[173] One last little notion of acoustical physics before continuing: since sounds propagate better in water than in air, the near field extends further away from the sound source in this medium. Of course, the extent of the near field also depends on the power at which the sound is emitted and on its wavelength. Here are some numbers to illustrate this: the near field of a cricket extends up to 4 centimeters from the animal; that of a blackbird, up to 10 centimeters; and that of a codfish, up to 5 meters.[174] With these three examples, you can guess that short-distance ears are mainly observed in animals living in the water (crustaceans such as lobster or shrimp,[175] cephalopods,[176] and most fish[177]) and in insects in the air that communicate very closely, while long-distance ears are the prerogative of the majority of animals in the aerial environment (most insects and vertebrates) or aquatic species communicating at long distances (certain fish and marine mammals).[178] Before we talk about how vertebrates hear, let's take a look at insect ears as an example, since both short-distance and long-distance ears are found there.[179]

Insects! What a world. At least a million species have been described, and the total number of species is probably close to six million. Insects are one of the most diverse groups of animals—from species where individuals live in perfect solitude to the densest animal societies on the planet. Insects conquered the aerial environment some 400 million years ago. As their ancestors were already equipped with mechano-receptors, the first insects could probably perceive the vibrations of the substrate, such as a stem or a leaf. The acquisition of eardrums, the membranes needed to hear from afar, took place in a disorderly fashion. Note that hearing is rather uncommon: most insects cannot hear any-thing![180] Nonetheless, over time, the ability to hear sounds developed independently in different groups of insects. As a result, the location of long-distance eardrums on the bodies of insects differs between groups—they can be in as many as 15 different places.[181] In locusts, the eardrums are on each side of the first segment of the abdomen, i.e., on the sides of the animal behind the wings;[182] in grasshoppers, they're on their front legs; and some moths of the Sphingidae family have their ears on the appendages surrounding their mouths.[183] From an anatomi-cal point of view, the tympanum of insects corresponds in principle to a slimming of the cuticle (the external envelope that serves as a skeleton for insects) leaning against a small cavity filled with air derived from the respiratory system[184] and connected to sensory neurons (scolopidia[185]). The sensory neurons end with cilia that the vibrations of the eardrum distort, causing the neurons to become excited and generate the nerve message.

One of the essential evolutionary drivers behind the development of insect eardrums is the bat. Throughout the history of insects, the ear-drum has appeared independently more than 20 times. And in 14 of those cases, the eardrum was (and still is) adapted to perceive ultra-sounds, the high-pitched frequencies produced by bats when they search for their prey. Fossils, molecular analyses, and biogeographical data indicate that these ultrasonic ears were in place very late in the his-tory of insects, only 65 million years ago. The first bat fossils date back to that same moment in time.[186] As my colleague Michael Greenfield says, night flying became far too dangerous for insects at that time, unless they were properly equipped to detect and avoid these terrible new predators.[187] Thus, there was probably considerable selection pressure favoring individuals with anatomical and physiological changes that would allow them to hear a bat coming. Nocturnal or deaf individuals were in danger of rushing headlong into the jaws of death. They had to hear at all costs to survive in the dark. Moths provide the best evidence that there is a parallel evolution (coevolution) between insect hearing and bat predation. In insects, the ultrasonic ear has appeared independently in a dozen different lineages. Butterflies without ultrasonic ears live in areas where bats are absent, or fly at times of the day or year when bats are not active. It should be noted that insects with ultrasonic ears react to hearing bats by turning or diving toward the ground, or even by emitting ultrasonic sounds themselves that interfere with the bat's signal. Clever little buggers.

Bats have not been alone in promoting the evolution of long-distance ears in insects. Sexual selection has also played its part. Insects are small in size, and often low in density. Imagine a female cricket in a meadow, surrounded by tall grass. The probability that she will find a male by moving randomly through the meadow is not very high. In any case, it would take time, a lot of determination, and a lot of energy. But if she hears a male chirping, she has a much greater chance of detecting and locating him. Acoustic signals, therefore, are an efficient means of finding a partner: they can be used day and night, cutting through vegetation; their infinite modulations can easily encode information, such as the identity of the species; the emitting individual has total control over the beginning and end of its emission; and the origin of the message is generally easy to locate (unlike chemical signals, which can stagnate in the air and whose source is more difficult to identify).[188] You are likely familiar with the praying mantis, this magnificent predatory insect with large prehensile forelegs. There are many species of them all over the world. Some have a single long-distance, ultrasound-sensitive ear on the ventral side of their thorax. Well, this ear was already present in mantises 120 million years ago, long before the arrival of bats. Perhaps it was already used for communication, or to detect prey, or to spot predators we don't know about.[189]

Let's move on to the short-distance ears, and stay with the insects. There are many different kinds of these ears in insects, but perhaps the most impressive are those of flies and mosquitoes.[190] These animals have short-distance ears on their antennae.[191] I guess you've never observed a mosquito antenna; here's a project to pencil in your diary. It's kind of like a pretty, fluffy feather duster on the front of the animal's head, filled with sensors that allow the bug to navigate its environment, find its food, locate partners, etc. One of those sensors, known as the Johnston's organ, serves as the mosquito's short-distance ear. Located in the second segment of the antenna, the Johnston's organ has no less than 15,000 sensory neurons, a number comparable to the neurons found in the human ear. This mosquito ear is incredibly sensitive.[192] In the *Drosophila* fly, there are only 500 neurons in the Johnston's organ, but a second system—the arista, positioned on the third segment of the antenna—completes the array. The *Drosophila* fly is therefore equipped with two types of short-distance ears. While the tympanic ears can be sensitive to a wide range of sounds, up to very high-pitched signals whose frequency can exceed 100 kHz, the antennal ears perceive mainly low-pitched sounds (frequencies below 1 kHz). And, as you can see, these ears no longer hear anything beyond a few centimeters, i.e., as soon as you leave the near sound field. So what are these short-distance ears for? Essentially for love conversations.[193] Talking about love requires intimacy, we know that! In flies, these ears allow the female to hear the male's song during the courtship parade. The two individuals are then very close to each other, resting on a leaf or fruit, and the male vibrates his wings to produce his songs. Singing is not the prerogative of male flies; researchers have recently reported that females sing during copulation. Singing with pleasure? We don't know, but their singing starts with the arrival of seminal fluid, modulates the amount of sperm transfer from the male, and lasts for 15 to 20 minutes in an apparent feedback loop coordinating female singing and male sperm transfer. Does the male respond due to the female's song?[194]

Fanny Rybak, an associate professor at the University of Orsay in Paris, and her colleagues have shown that these acoustic signals act in concert with chemical signals during the male courtship.[195] These very short-range communications are also common in bees. During the dance on the hive combs, the worker vibrates her wings at a frequency of 200 to 300 Hz. These sound waves are perceived by her fellow bees through their antennae.

In mosquitoes, the ears hear the sound produced by the beating of the wings of other mosquitoes. We also hear it, by the way, and we know this song that announces the bite only too well, don't we? Male mosquitoes use it to detect, locate, and pursue females. The sensitivity of their Johnston's organ (its *frequency selectivity*, as it is called) is even tuned to the frequency of the beating of their sweetheart's wings.[196] An experiment has shown that male and female mosquitoes can tune the frequency produced by the beating of their wings so that they emit the same sound.[197] The next time you hear the particularly annoying song of a mosquito in flight, think of it as a love song . . . maybe it will help you endure it!

All vertebrates (fish, amphibians, reptiles, birds, and mammals) hear thanks to a particular anatomical structure: the *inner ear*.[198] The inner ear has two distinct parts. The first, the vestibular system, is mainly dedicated to the detection of the movements and positions of the body.[199] It is fundamental for the control of body balance but does not intervene in hearing, so we do not discuss it here. The second part is the place of sound detection—the ciliated sensory cells (or hair cells), which are sensitive to vibrations. As in insect ears, these cells transform sound vibrations into nerve signals sent to the brain. The inner ears of all vertebrates are variations on this theme. The main difference lies in the way in which pressure variations (from the far sound field) or particle oscillations (from the near sound field) are transmitted to the inner ear. Let's clarify. We are no longer dealing with insects here, but with vertebrates, and listening a few centimeters away with hairs attached to antennae is no longer relevant. For the vast majority of vertebrates, their ears are therefore long-distance ears, and it is an eardrum that receives the variations in sound pressure. These variations are then transmitted to the inner ear by a middle ear . . . except in fish! Remember: in the air, the near sound field is not very large; and vertebrates, which are rather large animals compared to insects, are generally too far away from their fellow creatures to perceive their sound emissions.[200] In the water, however, this field can extend for a few meters.

But these poor fish have another problem: they live in the water. (I hope you knew that!) And the density of the water is the same as their bodies. This is annoying for the sounds, because it means that a sound wave that hits a fish will make its whole body oscillate evenly, just like the water around it. It is therefore impossible for the sensory cells in the ear to detect the slightest movement of their cilia if they also oscillate in concert. This is a difficult problem to solve. How can it work? In practice, the cilia of the inner ear cells are stuck in an otolith, a kind of small crystal of calcium carbonate that is much denser than water. When the body of the fish, including the sensory cells, vibrates under the effect of the oscillations of the near sound field, the otolith reacts with a delay. This deforms the cilia, and voilà! Nerve cells are excited by this deformation and inform the fish's brain.[201] This ear is not very efficient, limited to low frequencies of a few hundred hertz. But we haven't reached the end of our surprises: in about a third of fish, the ear can also detect pressure variations and thus perceive sounds in a far sound field. How does the ear do this? By associating a small air pocket with the inner ear. Yes, you read correctly: an air pocket in the body of a fish. This pocket of air changes in volume due to pressure variations, and these movements are transmitted to the nerve cells. More precisely, the wall of this pocket vibrates, acting like the membrane of a long-distance ear. So these fish do have a kind of internal eardrum. Great, isn't it? For some species (more than 8000), the air pocket is simply the swim bladder, the structure that allows the fish to regulate its water depth, which finds here a new function.[202] Definitely a multitask organ. A succession of small ossicles (the *Weberian apparatus*) transmits the oscillations of the bladder to the inner ear. Such structures considerably increase the capacity of the fish's ear: it becomes more sensitive and able to hear frequencies of several thousand hertz. There are even species related to the herring that hear ultrasound up to 180 kHz, which is considerable . . . and perhaps useful in trying to escape the killer whale that is looking to devour them.[203]

What about other vertebrates—amphibians (frogs, newts, salamanders), reptiles (snakes, lizards, crocodiles, tortoises), birds, and mammals?

The great majority have a tympanic membrane (one for each ear) in contact with the external environment, which collects pressure variations due to sound waves. In most frogs, this eardrum is clearly visible, placed just behind the eye. Behind it a small bone transmits the vibrations from the eardrum to the inner ear. Frogs' eardrums work in water as well as in air. An anatomical feature that frogs share with birds, crocodiles, and lizards is that the eardrums of the right and left ears are connected by an air-filled canal.[204] As a result, each eardrum not only vibrates in response to the arrival of sound waves but also transmits its vibrations to the other eardrum via this channel that connects them. An eardrum thus receives vibrations on each of its sides, external and internal. The connecting channel therefore makes it much easier to locate a sound source.[205] Known as *pressure-differential ears*, they are especially beneficial for small animals because the clues usually used for sound localization (the difference in sound amplitude and the difference in sound arrival time between the two ears)[206] are not sufficient. This is because the sound arrives at practically the same time and with practically the same intensity in both ears when the ears are very close together, whether it comes from the left or from the right. It is therefore difficult to use these indices of time and intensity to locate the source of the sound. The pressure-differential ear makes it possible to perceive them more finely. In fact, it is found in many insects.[207]

A quirk of frogs and other amphibians is the relationship between their ears . . . and their lungs. When the frog's lungs are inflated with air, they are set in vibration by sound waves. These lung vibrations are transmitted to the inner surface of the eardrum and oppose the tympanic vibrations due to the sound waves hitting the outer surface of the eardrum. The lung vibrations thus attenuate the intensity of the sounds perceived by the frog. Where the matter becomes subtle is that only certain sound frequencies can vibrate the lungs and are therefore attenuated. Which ones, you may ask? Well, those caused by the vocalizations of other frog species. The lungs act as an attenuator of ambient noise, and by listening less to the croaking of other species of frogs, individuals can better hear the croaking of their own species. This is useful when vocalizing in ponds where several species of frogs live together.[208]

But all frogs don't hear that way. There are frogs that do not even have an eardrum—only the inner ear is present. Renaud Boistel, a French researcher specializing in the bioacoustics of frogs, Thierry Aubin, and their colleagues have studied the *Sechellophryne gardineri*, a little frog which, as its name indicates, lives in the Seychelles.[209] Even if its inner ear, well hidden inside the head, is not connected to any eardrum, the Seychelles frog hears and communicates through vocalizations, as shown by Renaud and Thierry through playback experiments. They used sophisticated methods to measure the densities of various parts of the frog's body and found that the conduction of sound waves to the inner ear involved the bones of the head . . . and the mouth! By simulating the structure of the mouth through mathematical calculations and computer modeling, they demonstrated that the mouth was a *resonator*—a sounding board, like a drum—for the characteristic frequencies of the frog's call (around 5710 Hz). In other words, when sound waves from a call reach the frog, they enter the mouth, which amplifies them through resonance. The wall separating the mouth from the inner ear is extremely thin (about 80 micrometers), making it easier for the amplified sound waves to be transmitted to the sensory cells. Complicated but effective.[210]

In birds, the eardrum is indeed there, at the bottom of a small pipe opening through a hole at the back of the eye, where a system of ossicles connects it to the inner ear.[211] Sometimes the feathers around the auditory hole form a kind of auricle—a sort of outer ear—concentrating the sound waves. This device is clearly visible in owls. Generally speaking, birds have quite good hearing: their range of audible frequencies easily extends from a few hundred hertz to more than 6000. Aquatic birds—especially those that dive deep to pursue prey—have developed specific adaptations to cope with the acoustic properties of the water and the high pressures during deep dives.[212]

In mammals, we find the same configuration as in birds, except that the right and left eardrums are not connected. The mammalian eardrum is a simple pressure receiver. Most species of mammals living in the air have a true external ear, with a clearly visible, sometimes mobile pinna, which allows more efficient collection of sound wave energy. Look at your cat or your dog listening: it can direct the pinna of its ear toward the sound source. In marine mammals (whales and dolphins), there is no outer ear.[213] This would not be very practical for swimming. In addition, they have the same problem as fish: their body vibrates in phase with the water in which it is immersed. The ear canal is narrow, is full of cellular debris, and has lost its role. It has become a useless relic. So how are the auditory sensory nerve cells excited? In toothed whales (odontocetes, dolphins, sperm whales), sound vibrations are transmitted to the ear through a blubber-filled canal in the lower jaw. Well sheltered in the skull, the ear consists of two small bones suspended by ligaments in an air-filled cavity, a bit like in the fish we were talking about. One of these bones, the tympanic bone, vibrates in response to incoming sound vibrations. It is connected by a short chain of ossicles to the inner ear, where the sensory cells are located. It is at this level that the vibrations generate nervous action potentials, which are transmitted to the auditory areas of the brain. This system (which is quite complex[214]) is very efficient, and dolphins are known to hear a wide range of sounds, from medium frequencies (a few kHz) to very high frequencies (200 kHz).[215] In baleen whales (mysticetes), although the ear appears to have roughly the same characteristics as in dolphins, things are much more mysterious. It must be said that it is not very easy to study these massive animals. We don't really understand how sound waves are transmitted to their eardrums—probably through the bones. In any case, although it has never really been measured, their hearing range seems much less extended than that of dolphins—from a few hundred hertz to a few kilohertz.[216] But perhaps we underestimate them.

Perceiving sound waves and producing a nerve signal in response—or signal transduction, discussed earlier—is only the first step toward hearing sounds. The nerve signal must then be processed by the brain. We would have to enter the world of neuroscience to explain these neurophysiological mechanisms, which is not the purpose of this book. However, let's look at a process that allows a receiver to extract useful information from the sound signals it receives: *categorical perception*. The sensory world of animals, including humans, is indeed overloaded with nonessential information. Critical information—an incoming predator, the presence of a fellow animal—can be drowned out. Processing all this information and extracting useful information is a real challenge. Categorical perception is an important mechanism that allows the individual receiver to sort the stimuli into distinct categories. We humans practice categorical perception every day. The most common example is the perception of the two syllables /ba/ and /pa/. For /ba/, the delay between the beginning of the lip movement that produces /b/ and the production of the vowel /a/ by the vocal cords is almost zero. On the other hand, for /pa/, the lips move before the vocal cords vibrate. We can construct by computer synthesis intermediate syllables between /ba/ and /pa/ by which we progressively vary the duration of this delay. When listening to the progression of these variants, human adults identify each variant as either /ba/ or /pa/—they do not perceive a continuum between the two. For them, the signals they hear do not gradually shift from /ba/ to /pa/; the boundary is perfectly clear. They are two distinct categories.

In animals, surprisingly, categorical perception has essentially been studied by looking at their ability to distinguish . . . syllables from human language. Just think where anthropocentrism leads! A few rare studies have considered natural situations, such as the swamp sparrow *Melospiza georgiana*, which perceives the variation between the acoustic elements that make up its song in a categorical manner;[217] the Japanese macaque *Macaca fuscata*, which does the same for two calls in its vocal repertoire;[218] or the túngara frog *Engystomops pustulosus*, when it has to distinguish the parade call of its fellow frogs from that of another frog species.[219] My colleague Nicolas, who traveled with me to the Pantanal, and I decided to investigate this issue in crocodiles. These elite predators spend their lives picking up information from their environment, through sight, smell, and hearing. And they are constantly making important decisions. Am I or am I not going to that calling baby? Could the noise I hear over there be potential prey? We had to find an experimental context to test the crocodiles' ability to categorize sounds. The idea came to Nicolas during one of our stays at *Crocoparc*, the zoological park of Luc Fougeirol (who was also the founder of *La Ferme aux Crocodiles*, discussed in the previous chapter) and his family, in Morocco.

In the ponds of the park, the North African green frog (*Pelophylax saharica*) is very common. It provides the park's soundscape with its croaks. A frog croak has an acoustic structure that is not far removed from a small crocodile's contact call: a short sound, with a fundamental frequency and small harmonic series. On the other hand, it is weakly modulated, sounding like "yhah! Yhah!" instead of the " djong! Djong!" of the crocodile. How does a small crocodile distinguish the croaking of a frog that forms the background of its environment from the call of one of its fellow crocodiles? With Julie Thévenet, our PhD student, we conducted an experiment to find out.

Imagine a little crocodile, a few months old, a good 20 centimeters long, in a basin several meters in diameter. On the edges of the basin, four loudspeakers re-create the background sound of the frogs: "yhah! Yhah!" The crocodile is accustomed to them, ignoring them and going about its business. Suddenly, one of the loudspeakers emits a crocodile contact call: "djong! Djong!" The little one reacts by moving toward what it believes to be a fellow crocodile. It can tell the difference between the two signals, the frog and the crocodile. To evaluate its ability to distinguish between the two categories, we constructed chimeric acoustic signals, such as 10% frog/90% crocodile, 20% frog/80% crocodile, and so on up to 90% frog/10% crocodile: "yhaong," if you like. We presented them to our little crocodile in his pond. The results were clear. As soon as the chimeric signal included more than 20% of a crocodile call, the little one turned its head toward the loudspeaker, sometimes moving a bit. It had noticed that this signal was no longer "pure frog." We observed these same reactions—visible but nevertheless quite weak—each time for all the signals until the call was 80% that of a crocodile. There was no progressive increase in the intensity of the reaction. However, when the chimeric signal was 80% crocodile, suddenly our young crocodile showed a more sensitive interest in it, moving decisively toward the loudspeaker. With this 20% frog/80% crocodile signal, we reached a plateau. With more than 20% frog in the call, the little one classified the signal as "environmental noise that is not too worthy of interest." With more than 80% crocodile, the response became, "It's a fellow crocodile call; I'm going!" Being able to categorize is a great decision aid.

What will the next step in our research be? Identifying the acoustic parameters that allow the crocodile to categorize the two types of signals: Is it the pitch of the calls? The difference in modulation? Or some other characteristic? Julie has already begun to explore this question by constructing chimeric signals in which only the frequency or modulation varies. I must confess that I'm looking forward to the results. Exploring the sensory world of crocodiles in this way is fascinating.

Let's stop there for now. Of course, other physiological and psychological mechanisms deserve to be detailed, such as the *precedence effect*, which characterizes the fact that insects or frogs caught in a chorus of acoustic signals pay attention to the first signal received, ignoring the next.[220] It would also be interesting to explore recent developments combining behavioral and computational approaches used to reveal the acoustic features that receivers rely on for signal recognition and discrimination.[221] But you already know a lot about how animals hear. Now let's see how they make sounds. The Okavango expedition is waiting for us.

10

Tell me what you look like

PRODUCTION OF SOUND SIGNALS

Panhandle, Okavango Delta, Botswana. The motorboat suddenly swerves and we zigzag quickly on the narrow river. "Hippos!" Sven shouts excitedly. At the helm, however, Vince frowns, a shadow of anxiety passing over his face. Hippos are the most dangerous animals in Africa, and they could easily overturn our boat. That was a close call![222] The afternoon is over, the horizon is on fire, and we have been sailing up one of the arms of the Okavango for almost an hour. Now the GPS indicates the place we are looking for. Vince throttles the engine. Sven ties us to a bunch of papyrus and we stop. Silence . . . binoculars pointed . . . "She's here," says Vince with his finger pointed. "Her nest is on the bank, right here; I remember." Of the female Nile crocodile we see only a ridge line, nostrils, eyes, and ears. The rest of her body is submerged. Vince starts the engine again, and we are going to attach our loudspeaker in the vegetation, near the entrance of the nest. Back at our waiting position, we suffer in silence. The mosquitoes arrive in squadrons, tightly packed and buzzing. Thierry is smacking his cheeks, grumbling. "*Le terrain* forever," he had written to me one day. . . . I wouldn't give up my place for anything in the world. We came to the Okavango to test whether Nile females react differently to the distress calls of newborns and older hatchlings. Through our acoustic analyses, we had already learned that the vocalizations of baby crocodiles change as they grow. But do mothers pay attention to this information? The Okavango expedition had to answer that question.

Years before, with Amélie, we had tested the hypothesis that young crocodiles have a vocal signature that allows their mothers to recognize them. Unfortunately, that hypothesis had fallen through. However, analysis of the acoustic characteristics of the calls in the days following hatching showed that the calls change with age.[223] Each day calls become more low-pitched, accompanying the growth of the small crocodile. At the time, we did not test whether this correlation between body size and voice provided information that might be of interest to the mother during the weeks when she stays with her young to protect them.

However, over the years, I had been accumulating recordings of young crocodilians of various species and sizes. Ruth Elsey had welcomed me warmly to the Rockefeller Wildlife Refuge in Louisiana. She had studied the reproductive biology of the American alligator *Alligator mississippiensis* for years, and thanks to her, I was able to record many newborns of various sizes. Thierry and I had also recorded many small spectacled caimans and Orinoco crocodiles during our stays in Hato Masaguaral, Venezuela. Black caiman, American alligator, Nile crocodile, Morelet's crocodile *Crocodylus moreletii*, Orinoco crocodile—for all these species we had found that the acoustic structure of the call is correlated to the size of the individual. We needed to know if this apparently solid information had a biological role. For that, we had to go into the field. Going back to Peter Taylor's house in Guyana was an interesting possibility, but I didn't have a large enough sample of black caiman calls. The most numerous recordings—those with the greatest diversity in hatchling sizes—were Nile crocodile calls. Thanks to Luc Fougeirol, I had been able to access many broods of this species. But in order to experiment on the mothers, I had to go and meet them, obviously in the wild. Where the Niles live. In Africa.

Thierry and I had discovered Vince Shacks while watching a documentary on crocodiles. He was immediately interested in our project. "I know the area very well. We have to go to the Panhandle. That's where we can most easily find females with their young." With his colleague Sven Bourquin, Vince had just spent several years studying the biology of the

Nile crocodile in the Delta. The two partners were a great pair: Vince, thoughtful, verging on anxious, but perfectly organized; Sven, more whimsical and playful, confident—nothing was ever a problem for that Crocodile Dundee. Both men were familiar with the Nile crocodile and its habits. Both professional and friendly, they soon inspired confidence in us. When we arrived, they had organized our expedition down to the last detail. The two all-terrain cars, well equipped, were ready to go. From Maun in the south of the Delta where we had landed, it took us a few hours on a fairly good road to reach the trail leading to the camp. That's when things got serious. Botswana's sandy trails live up to their reputation. The next few weeks would be spent cut off from the world.

The camp is surprisingly luxurious. Thierry and I are staying in a cabin on stilts. It has a small terrace with papyrus as far as the eye can see; bathroom in the open air; comfortable beds. "Not much to fear in the area. The big cats aren't here this time of year," Vince reassures us. I ask nervously, "And what about the snakes?" Vince smiles: "Gentlemen, you are in the region of Botswana with the highest density of black mambas. We found one on the kitchen table one morning a year or two ago!" The mamba is a very nice snake, one whose sudden acquaintance you'd rather not make: a good 3 meters long, agile, of the efficient kind that kills you unceremoniously in about 30 minutes. "And if you get bitten?" "No problem, we'll give you a piece of paper and a pencil to write your last words on," Sven laughed. "But don't worry too much; he's discreet and you're unlikely to meet him . . . and if that happens, back off quietly. No need to run; he'll be much faster than you." The last crocodile hunter in the area was found on the trail dead and still in the driver's seat of his Land Rover. Bitten by a mamba, he had attempted to reach the village, but the powerful venom had quickly paralyzed him. This was an exciting start, but we are full of courage, make no mistake!

The aim of the field trip was to compare the behavioral response of Nile crocodile mothers to calls from very young, and therefore small (30–40 cm, including tail), crocodiles with that of slightly older, and therefore larger (60–90 cm), crocodiles. A few weeks before our arrival, Sven and Vince had made a first visit to locate the nests. "Do you see these GPS points?" Vince asked, pointing to his map. "Each one is a nest. We have about 15 of them. Taking into account the losses due to predators, we should still be able to count on about 10 females with their young." In a period of two weeks, we managed to test 9 females. The females responded to the playback by either moving toward or away from the loudspeaker. Sometimes the approach was fast and furious, sometimes just a swim of a few meters toward the loudspeaker. It was always in response to a call from a very small individual being emitted. But if the call came from a larger individual, in principle the female would not move, or would move backward. Only one female approached the loudspeaker when she heard the vocalizations of a bigger youngster. These individual variations did not mask the general rule: Nile crocodile mothers are attracted to smaller juveniles and ignore larger ones. They therefore pay attention to the size information encoded in their calls.[224]

Do you remember the hypothesis of honest communication, which I first introduced in chapter 5 about the cries of baby fur seals and warbler chicks? I mentioned to you that one of the reasons a communication signal could carry reliable information about the transmitter was the cost it could represent; the energy expenditure or the risk of predation, for example. With this correlation between the body size of the young crocodile and the characteristics of its call, we discover another reason for honesty: a signal can be reliable if the transmitter is unable to lie! This is indeed the case with our crocodile. As its body grows, the vibrating membranes at the back of the crocodile's throat, which produce the sounds, grow with it. Its call becomes lower. It cannot lie; it has no choice; it could only do so if it did not grow. To understand this relationship between anatomy and the coding of information in acoustic signals, let's leave the crocodiles behind and go to the world of mammals.[225] The mechanisms of sound production have been very well studied there.

In all mammals, vocalizations are the result of two independent processes: the production of a source signal by the vocal organ—the larynx—and its modification by the cavities of the vocal tract. This process has been formalized in the *source-filter theory*.[226] The larynx is a very complex organ whose presence is noticeable in men, thanks to their Adam's apple.[227] At the level of the larynx are the vocal cords, which are cords in name only. They are actually a kind of membrane that can be made to vibrate when air comes out of the lungs.[228] These vibrations produce the source signal—sort of a primary sound wave, if you will. Since the vibration of the vocal cords is complex, this primary wave is not a simple hissing sound. It is a complex sound, consisting of a frequency called the fundamental frequency, as well as harmonic frequencies, which are multiples (×2, ×3, etc.) of the fundamental.

The value of the fundamental frequency is primarily determined by the length of the vocal cords of the individual transmitter. The longer the cords, the slower they vibrate, producing a lower fundamental frequency. This fundamental frequency is therefore the main factor responsible for the pitch of the voice. When it is high, the voice is perceived as high-pitched. When it is low, a low voice is heard. As a first approximation, larger mammals produce the lowest fundamental frequencies.[229] In the 6-ton African elephant *Loxodonta africana*, the fundamental frequency measures about 20 hertz, compared to tens of thousands of hertz in a bat weighing only a few grams. But be careful! The bodily dimensions of a species and the length of the vocal cords are not always so related. In monkeys, for example, some species of macaques will have higher-pitched voices than others of the same size.[230] Within the same animal species, more subtle fundamental frequency differences can be observed depending on the relative size of the individuals, and also sometimes their age and sex in the case of sexual dimorphism (if female and male are of different sizes). However, some research has shown that the fundamental frequency is an unreliable indication of an individual's size. Indeed, this frequency results from a complex interaction between the density of the tissue constituting the vibrating membranes of the larynx, their degree of stretching, and the length of the actual vibrating part of the membrane. An individual sender can to some extent modulate the fundamental frequency by controlling the power of the air flow coming out of its lungs and the tension of its vocal cords. When the sender varies the fundamental frequency, modulations can be heard in the intonation of its vocalizations.

What about harmonic frequencies, you might ask? Harmonic frequencies also contribute to the characteristics of vocalization. But they are modified as the vocalization passes through the vocal tract—the conduit that goes from the larynx to the exit of the mouth or nostrils. The first sound wave produced by the larynx will indeed propagate, and during its journey it is modified. Certain frequencies are reinforced; others will be filtered. Let's draw a picture to visualize things. You probably know that the light that comes to us from the sun is made up of several colors. You can easily see this when raindrops break the light down into a rainbow. Let's say that sunlight is the first wave produced by the larynx and that the different colors are the frequencies that make up the first wave. Place a sheet of paper in the sunlight. It is white because sunlight, with all its colors mixed together, is white. Now place a blue tinted glass over the sheet. The sheet is no longer white but blue. The blue glass plays the role of the vocal tract between the larynx and the mouth. It acts as a filter, allowing only part of the light to pass through it. The light that comes out of the glass is different from sunlight because some of its components (red, green, etc.) have been caught in the glass and are no longer there. When we vocalize, the first wave produced by the larynx undergoes fairly similar treatment as it passes through the vocal tract. Certain sound frequencies, called formants, are reinforced. What determines which particular frequencies are reinforced (in other words, what determines the formants) are the resonance properties of the vocal tract. This tract is in fact made up of a succession of cavities: the pharynx, the oral cavity, and the nose. Each of these cavities can resonate with the sound waves that pass through it, and reinforce one or the other formant. The way in which the formants are distributed allows the characteristics of vocalization to change (for example, in humans, to pronounce the vowel /o/ rather than /a/) and affects the timbre of the voice, so that a voice will be more or less soft, nasal, or metallic.[231] Let's take a specific example with the human voice. Our fundamental frequency can vary between 60 Hz (very low voice) and 300 Hz (high voice). On average, it is around 210 Hz for women and 120 Hz for men, who have slightly longer vocal cords. If the fundamental frequency is 100 Hz, then the associated harmonics are 200, 300, 400, 500 Hz, etc., since they are multiples of the fundamental. This series of frequencies (100 and its multiples) constitutes the first wave, which is then filtered by the vocal tract. In the human species, the vocal tract has four main formants, distributed between 500 and 3500 Hz. Let us consider the first two formants, which are centered, respectively, on 500 and 1500 Hz. The 500 Hz formant reinforces this frequency in the voice, while the lower frequencies become less audible. The 1500 Hz formant reinforces this frequency, which means that frequencies between 500 and 1500 (say, around 1000 Hz) are attenuated. When we change the position of our tongue or move our jaw, we can change these formants and give them more or less importance, which allows us to modulate the timbre of our voice and even to pronounce this or that vowel. Don't worry if you've lost your way a bit. Just remember that the sound produced by the vocal cords is modified as it passes through the throat, mouth, and nose, and that we have the ability to modulate these changes by moving our tongue and lips and changing the opening of our mouth.

This ability to substantially and precisely modulate the physical structure of our vocal tract, and thus the properties of the formants, makes the human species a little special. Most nonhuman mammals do not possess this ability: their formants are somewhat fixed and in principle even quite easy to predict simply by measuring the length of the vocal tract. The vocal tract can be considered to resemble a tube from the larynx to the lips. The longer the tract, the lower the frequency of the formants; therefore (and logically), the bigger an animal is and the longer its vocal tract, the lower its formants will be. This type of information directly related to the physical characteristics of an individual is called a *quality index*. This reliable relationship can be observed not only between species but also between individuals of the same species.[232] Such is the case in our crocodilians, where the membranes at the back of the throat play the same role as the larynx of mammals and the oral cavity plays the same role as the vocal tract, and where small individuals have higher-pitched voices than large ones.[233]

The source-filter theory predicts that the biomechanical characteristics of the vocal cords and of the vocal tract translate into quality indexes in vocalizations. Body size, and sometimes sex, physical condition, or age, is static information imposed by the process of producing vocalizations.

This information can be very useful in the context of mate choice (intersexual selection) or competition between individuals of the same sex (intrasexual selection). Moreover, these selection mechanisms can lead to additional effects. It has been shown, for example, that male terrestrial mammals produce vocal signals whose formants are even lower when the size dimorphism between females and males is pronounced.[234] In these animal species, sexual selection favors males whose voices are lower and therefore sound bigger and scarier.

My colleague David Reby has studied acoustic communications in the Cervidae—the deer family—for many years, and has become a specialist in identifying the mechanisms by which these animals produce their vocalizations and the information they encode.[235] Maybe you've had a chance to hear the deer *Cervus elaphus* roar. It's absolutely breathtaking. It happens in the evening or early in the morning, in the woods; it's almost dark, and the bellow begins, something like a long, powerful, low call, "RoooaarrrhhhHHHHHH!," which the deer repeats over and over again. Let's take a close look at the animal as it bellows. It extends its head and stretches it upward, tilting its antlers toward its back. When it does this, we can clearly see a bump moving on its throat—its Adam's apple, or larynx. The larynx descends toward the sternum as the neck stretches, away from the animal's mouth. At the same time, the bellow becomes deeper. By lengthening its tract, the deer lowers its formants. In fact . . . it is cheating. It exaggerates its vocal size. Like Jean de La Fontaine's frog, it tries to pass itself off as bigger than it is. Of course, the maximum length of the tract, obtained when the larynx is at its lowest point, depends ultimately on the size of the individual. Since all deer lower their larynges when they bellow, eventually their signals reliably reflect the differences in size between individuals. The signal therefore remains an index of quality.[236] We can suppose that this behavior of lowering the larynx, now shared by all males, is the final stage of a lie that has gradually become widespread over the course of evolution. When the ancestors of today's deer began to lower their larynges, males probably had a serious advantage in terms of reproductive success. Like today's deer, those with the deepest voices were preferred by the hinds;[237] moreover, it was easier for them to frighten off an opponent with their voice. By promoting the success of these males, sexual selection has done its job in order to achieve what we now call a *stable evolutionary equilibrium*. It's difficult for the male deer to lower his larynx any further as it goes down to touch the sternum. But who knows? The deer of the future may have developed other tricks to cheat vocally about their size. However, females don't rely solely on voice quality to judge a male. They also pay attention to the number of bellows per minute.[238] Contrary to what we might believe, bellowing is a very tiring exercise, and only males in good shape are able to do it continuously and win the prize. This situation was explained in the *handicap theory*, formalized by Amotz Zahavi a long time ago. This theory states that the cost of sending the signal (in all its aspects, which we have already discussed) represents a handicap that only certain transmitters are able to bear. In the case of the bull deer, the "bellow" signal becomes reliable because of its high energy cost. The deer's voice says so much that David and his colleagues have demonstrated with playback experiments that a female can tell the difference between a male that has already won a harem and one that hasn't . . . just by listening to it.[239]

Relying on vocalizations to choose one's partner is very common. In the animal kingdom, many acoustic signals are produced in the hope of attracting a soul mate. When sexual selection is intense, which is the case in polygynous mammalian species (where males have harems and sexual dimorphism is usually pronounced), anatomical adaptations giving males low-pitched voices have often developed. Being as large as possible is often an advantage when competing with rivals; but size is not always part of the equation. A very original example is provided by the koala *Phascolarctos cinereus*. You probably know this Australian teddy bear that feeds on eucalyptus leaves.[240] It's not very large, about 15 kilograms at the most. Yet its territorial call, which is used to repel male intruders and attract females, is incredibly low-pitched, a succession of frightening breaths in and out: "Dro-he dro-he dro-he"—a Kawasaki engine noise. The call's fundamental frequency averages 27 Hz, which is 20 times less than one would expect given the size of the animal. In fact, this frequency is very close to the one emitted by . . . an elephant. Ben Charlton, David Reby, and their colleagues have shown that the larynx of the male koala has no particular ability to produce a low fundamental frequency. But, to their surprise, they discovered a second pair of vibrating membranes higher up in the vocal tract, at the junction between the nasal cavity and the mouth. Koalas have a second vocal organ! And these membranes are long enough to produce frequencies as low as 10 Hz.[241]

Other mammals have adaptations that allow them to produce fundamental frequencies that are abnormally low for their body size. For example, the male hammer-headed fruit bat *Hypsignathus monstrosus*, from the African tropical forest, has a disproportionate larynx that fills the entire thoracic cavity, i.e., more than half of its total body volume (the flying larynx, as some people call it).[242] Its loud "Honk! Honk!" is emitted when males gather and compete for females.[243] In other mammalian species, laryngeal hypertrophy is a little more modest but real.[244] It is sometimes accompanied by vocal pouches (in howler monkeys, for example) or nose expansions (as in the saiga antelope *Saiga tatarica*), which increase resonance and lower formants. Sometimes it is the structure of the vocal cords themselves that is peculiar, with an increase in their mass that lowers their frequency of vibration. This type of device is found in large cats, such as the lion *Panthera leo*, the tiger *P. tigris*, the leopard *P. pardus*, and the jaguar *P. onca*. If you are lucky enough to have heard a lion roar at night, you will agree that it bears little resemblance to the mewing of its cousin, the domestic cat. And in the human species, you might wonder? While our species does not present a very pronounced sexual dimorphism on many traits,[245] the octave difference between men and women is one of the largest among mammals, with only howler monkeys being more extreme.[246] Besides, having a low-pitched voice is not always favored in animals. In bonobos (*Pan paniscus*), the apes closest to the human species, both females and males have very high-pitched voices. Their vocal cords are about twice as short as those of their chimpanzee cousins (*Pan troglodytes*), which are of comparable body size. But unlike chimpanzees, the bonobo society is known for its codominance between females and males. Equality between the sexes? Maybe, maybe not; but in any case there is no selection pressure on them in favor of low-pitched vocalizing in males.[247]

The fact that communication signals can carry reliable information about the sender is of major importance to the receiving individual. Without a minimum of reliability in the transmission of information, the communication would lose all its value and would probably not have been maintained over the evolutionary history of the species.[248] Imagine if every time someone spoke, what they said had nothing to do with reality, was totally unpredictable, and contained no reliable information.[249] After a while, that individual would probably stop being listened to, especially if listening and paying attention represented a cost to their listeners or a disadvantage compared to not knowing what they were saying. This is a simplified picture of what happens in the evolution of species and how they communicate. As we have seen, there are many attempts at cheating. In the green frog, for example, when a small male meets a large male, it starts to sing at a lower pitch.[250] The fact that senders have the opportunity to cheat suggests that communication signals should not be considered completely honest or totally deceptive. As usual in biology, a black-and-white perspective would be wrong, and remember that the balance between honesty and deception in communications is a subtle one. In principle, it results from what the senders can produce and what the receivers are willing to accept. Do you remember the concept of the arms race I mentioned in chapter 5 when I was talking about communication between parents and their young? The principle is simple: the sender exaggerates its signal in order to get something from the receiver, while in return the receiver becomes increasingly reluctant to respond to the received signal. We have seen two processes that ensure the relative reliability of the information encoded in acoustic signals: 1) the cost of the signals, which means that not everyone can afford to do as they please ("a signal has to be expensive to be honest");[251] and 2) the fact that the quality indexes are inseparable from who the sender really is, making it difficult for them to alter the information sent.[252]

My colleagues Kasia Pisanski and David Reby at our ENES Bioacoustics Research Laboratory in Saint-Etienne, France, had the idea to explore these processes in humans. Their starting point was that, while vocal exaggeration is widespread in the animal kingdom, it is difficult to establish its actual capacity to confuse listeners. Kasia and David used the fact that the depth of the human voice is correlated with the size of the individual: taller people have on average a deeper voice than smaller people. However, everyone is able to modify their voice somewhat by forcing it. Kasia and David asked women and men to pronounce the vowels A-E-I-O-U in three different ways: first, in an "honest" way, speaking naturally (first condition), then as if they wanted to sound taller (second condition), and, finally, as if they wished to sound smaller (third condition). Kasia and David then played these recordings to 200 adults, asking them to estimate the real size of the speaker and to guess whether he or she had tried to deceive them. Acoustic analysis of the vocalizations showed that, while it is possible to modify one's voice to appear larger or smaller, the level of cheating is still limited. The body size of an individual imposes an inescapable anatomical constraint on the vocalizations that he or she is able to produce. The playback experiments also showed two interesting things. First, listeners were able to guess quite accurately when the individuals had voluntarily modified their voices: it is not so easy to cheat. However, strangely enough, listeners were still fooled by the altered voices. People's heights were indeed overestimated by about 3 centimeters when they tried to sound taller, and underestimated by 4 centimeters when they tried to sound smaller. How to interpret these results? Well, we have here an example of the arms race that leads to a balance between honesty and cheating. In the human species, the depth of the voice plays a role in the choice of a partner. Thus, for example, women are, on average, more attracted to tall, deep-voiced men (which does not mean, of course, that all women are attracted to tall, deep-voiced men—individual tastes differ, and these are certainly not the only parameters of choice). Therefore, a man may have an interest in forcing his voice lower to be perceived as taller, but this strategy will quickly reach its limits since it will be detected by the listener. These results have been published in the journal *Nature Communications*, and if you are curious about the details, I invite you to read the scientific publication in its entirety. Isn't it exciting to see once again that the human species does not escape the evolutionary processes that run through the living world?[253]

Body size is certainly not the only static information useful to a receiving individual when choosing a partner, or when evaluating an opponent on the basis of its vocalizations. Let's consider birdsong. It provides information on many aspects of the sender. In monogamous bird species where both parents are involved in brooding and rearing the young, it is important for the success of the brood that the male invest time and energy. In short, and without anthropomorphism, it is in the females' interest to pay special attention to the quality of the male they choose. Let's take the example of the canary (*Serinus canaria*). In this species, females are solicited by a particular song of the male, and they may respond favorably by adopting a very explicit posture to invite mating. This song of invitation is the accelerated repetition of a syllable composed of two notes, both very strongly modulated in frequency: "Zee-oop! zee-oop! zee-oop! zee-oop!" This extremely fast trill was named the A phrase by Eric Vallet and Michel Kreutzer, from the University of Paris Nanterre.[254] Through playback experiments, they showed that females prefer the fastest A phrases, suggesting that these versions demonstrate superior vocal performance in males.[255]

Although physiological experiments have shown that singing probably does not cost birds much energy (since oxygen consumption is only slightly increased),[256] it is likely that being able to perform such trills at full speed can be a marker for a female of some interesting qualities in a male. This can be supported by the *developmental stress hypothesis*, which suggests that the complexity of the song in adulthood is partly determined by the development of the embryo in the egg and then of the young bird—a period that coincides with the establishment of the brain structures involved in sound production.[257] If during this period the chick is under particular stress, such as insufficient or unbalanced nutrition, the development of its singing abilities will be disturbed.[258] For a female canary, mating with a male who masters the production of fast A phrases means having a partner who has not had any particular stress during the crucial phases of its development and is therefore in "good working condition," a well-qualified father-to-be. Singing is therefore an easily assessable indicator of the male's history and life characteristics, such as his motor or cognitive abilities. It should be noted that stresses with long-term consequences on an animal's vocalizations are not limited to the perinatal period. An individual may experience dietary restrictions, heat stress, parasites, and other stresses throughout its life. Any unpredictable change in the environment is likely to be stressful to an organism, and the response to any stressor results in physiological changes (such as an increase in blood levels of certain hormones or a decrease in the performance of the immune system).[259] An example? The repertoire size (i.e., the number of different notes in the song) of the sedge warbler *Acrocephalus schoenobaenus* decreases by about 20% when the bird has parasites.[260] By attesting to the physical condition of the sender, communication signals are valuable informants for the receivers. Be careful, though! I wouldn't want you to think that receivers make reasoned choices, such as, "If Pierre sings like that, it's because he's in great shape. . . . Quick, let me take him as a partner!" Receivers are, of course, unaware of the value and meaning of these signals; their choice is the result of sexual selection that has gradually established itself over time.[261]

Appearing to be the biggest or the most beautiful is not always what succeeds in seducing a partner. The males of the superb lyrebird *Menura novaehollandiae* have another strategy: when they try to mate with a female, they play the illusionist. Their vocalizations give the impression that a group of birds is harassing a nearby predator. Everything is there: imitations of calls of various species accompanied by wing noises for good measure. The bird switches from one signal to another at an astonishing speed, superimposing calls of several species. It's a real tour de force. Playback experiments demonstrate the quality of this imitation: when listening to it, individuals of the imitated species start to nag the loudspeaker as they would in response to natural calls. This false danger signal is produced only during the 45 seconds of copulation (which is a long time for a bird). During this moment, the male is perched on top of the female and flaps his wings in front of her, blocking her view. The female can only remain motionless, trapped in this polyphonic sensory trap.[262]

Let's stop for a moment to review. From the previous examples, an important element stands out: sexual selection is one of the essential evolutionary drivers of the mechanisms that produce communication signals. In chapter 5, we saw that parent-offspring interactions are also a driving force in this evolution. Whether it is potential partners, parents, or opponents who judge the qualities of the sender by its voice, in all cases it is always the receiving individual who ultimately determines whether the vocalization produced is a valid communication signal and is worth keeping or needs to be altered. (Let us specify once again that this decision is the result of selection; it is not a reasoned judgment.) A communication signal is a biological characteristic peculiar to an animal species (in the same way as the shape of its head, the color of its fur, or any other characteristic) whose evolution over the history of the species has made it possible to influence the behavior of the receiving individuals. Senders and receivers can therefore be seen as engaged in an evolutionary game, i.e., a relationship where each moves its pawns according to what the other does.[263] And this game can be more or less cooperative or conflictual.

I can feel a question coming. . . . How on earth, at the very beginning, does a signal appear? How can the ability to produce information-carrying sounds be acquired during the evolution of a species? If you remember chapter 1, you know that we are dealing with the most formidable of the questions of ethology formulated by Nikolaas Tinbergen: the fourth question—the question of origins. Behavior unfortunately leaves little or no fossil record;[264] vocalizations even less. We must therefore turn to tools other than those of archaeological excavations to try to understand the beginning of the history of acoustic communication.

Let us first bear in mind that the evolution of biological systems is a gradual process. An animal species is defined as a set of individuals sharing many common characteristics, in principle capable of reproducing among themselves, but all slightly different. This individual variability, partly due to chance, is the raw material of evolution. From one generation of individuals to the next, more complex features may emerge from less complex features in small steps. This is cumulative selection. Evolution is therefore a contingent process, which means that new features do not appear out of thin air. They are always preexisting traits that are gradually transformed over generations through complicated interactions between chance (so-called neutral drift) and natural selection (the fact that the communication signal increases the probability that the individual producing it will have offspring). The evolution of acoustic communication signals is not immune to these mechanisms.

Two main processes are identified that lead to the initiation of new communication. The first is the *precursor sender model*, in which the new signal results from the transformation of an already existing behavior, which produced sound and was already carrying information but almost inadvertently. The second is the *model of sensory bias*, in which the sender exploits sensory capacities already present in the receiver, but which were used for something other than communication. These two models are not always in opposition; they can be complementary. Of course, if nobody ever emits anything or nobody ever responds, there will never be any sound communication. I repeat: evolution always works from elements that already exist. Let's look at it together.

The precursor sender model usually involves *ritualization*.[265] When an animal moves or makes a movement, associated sounds are often produced. Think of the sound of the wings of a bird flying away, or the "knock-knock" of the spotted woodpecker looking for larvae in a tree trunk. These sounds are not real communication signals, but they can still carry information for other birds. "We're flying away! Danger on the horizon? Be careful!" or "Woody Woodpecker's up in the tree, stuffing his face." If individual receivers react by modifying their behavior and this interaction results in a benefit for everyone, senders and receivers alike, then the signals gradually become stronger by exaggerating primitive noises. Their components will often be repeated. They will become more stereotyped and easier to detect, more efficient. The "knock-knock" in the random rhythm of the woodpecker extracting larvae from the tree trunk becomes a characteristic drumroll, heard at a long distance: "TRTRtrtrtrtrtrtrr." Sometimes the very anatomy of the structure that emitted the original sound changes.[266] When frightened, the crested pigeon *Ochyphaps lophotes* will fly away, producing with its wings a series of two notes very quickly repeated: "D-rrrr-d-rrrrr-d-rrrrr." In response, the other pigeons fly away too. These are real communication signals informing the group of the presence of danger. They are produced by a feather, the eighth flight feather, considerably thinner than the others.[267] Over time, the shape of this feather has changed, diminishing its role in flight but transforming it into a musical instrument. Evolutionary biologists call this an *exaptation*—when a structure that initially devolved and adapted to a certain function (in this case, flight) is more or less reconverted into another function by undergoing transformations. Nature is full of examples.[268] Many other birds have modified feathers that enable them to produce acoustic signals, and many other animals have sound organs whose primary function was quite different. In fact, we may say without too much risk of overstating that everything *always* starts with exaptation.[269] As biologists often say, nothing is created; everything is transformed.

Let's take the case of the bugs. Have you ever observed a cricket singing? Take a good look at it. It raises its two elytra and rubs them together, producing the well-known "tchip-tchip-tchip-tchip." This mechanism of sound production is called stridulation. In concrete terms, the underside of each elytron has a row of small teeth, known as the rasp, while its edge, called the scraper, is thickened. Stridulation is produced when the rasp of the elytron rubs against the scraper. The surface of the elytron is thinned in two places, the mirror and the harp, which resonate with the vibrations, and something akin to ribs plays a role in their conduction. If you still have some memories of physics, the wings of the cricket function as *coupled oscillators* that finely control the sound produced.[270] In short, the elytra of crickets, which were originally wings, have become a sophisticated musical instrument, producing communication signals. Generally speaking, exaptations are commonplace in insects: modifications of the wings, legs, parts of the thorax, abdomen,[271] or even elements of the reproductive apparatus—all of which have led to the establishment of structures producing acoustic communication signals. It must be said that the rigid external skeleton of insects is an excellent raw material for the creation of musical instruments![272]

Fish provide other interesting examples of exaptation for acoustic communication.[273] We should not be surprised, since we already know that some species of fish can hear. However, it is not enough to be able to hear; there must also be sound production—and vice versa. Acoustic communication requires both hearing ability and sound emission. In fish, therefore, a significant number of species produce sounds and use them to communicate information to other fish.[274] Here there are no larynx or syrinx, but gnashing or snapping jaws, swim bladder vibrations, chirping, and fin movements. Eric Parmentier, a great specialist of fish sound productions and professor at the University of Liège in Belgium, has shown with his team that the clownfish *Amphiprion clarkii* produces a series of a few clicks (between one and eight) by rapid oscillations of its head and clapping of its jaws. More precisely, the clownfish has a particular ligament (which Eric calls the sonic ligament) responsible for the rapid elevation of the lower jaw, which in turn creates the click by forcing the jaws to snap.[275] Another way to produce sounds is to make the membrane of the swim bladder vibrate through particular muscles, whereby the bladder acts as a sound box. Eric and his team have shown that the piranha fish *Pygocentrus nattereri* uses this mechanism to produce a varied repertoire of sounds that males use when competing against each other.[276] The barking sound, called type 1, is emitted when two males face each other. It is probably some sort of warning accompanying an attempt at intimidation. The type 2 sound is usually associated with a competition for food. The type 3 sound is only emitted when one individual chases another and tries to bite him. A piranha bite! Now you know what you have to do if you ever hear a type 3 sound while swimming in an Amazonian river!

Finally, there's nothing very original about the acoustic communication functions of fish. Attracting a partner and scaring away uninvited guests are the main goals.[277] The information conveyed by the acoustic signals therefore depends first of all on the identity of the species: a piranha does not produce the same sound as a clownfish. However, there is more subtle information, such as quality indexes like body size.[278]

It is not easy to demonstrate that the information encoded in the acoustic signals produced by fish is actually perceived and used by the receiving individuals. A few years ago, with my colleagues Marilyn Beauchaud and Joël Attia, both associate professors in our ENES laboratory, we decided to tackle this question, with the primary objective of transposing to fish the experimental approach classically used in mammals and birds; that is, to perform playback experiments with signals

Metriaclima zebra

whose structure could be perfectly controlled. At the time, very few people had tried to make fish listen to sound signals. It has to be said that it is quite complicated technically. We advertised for a student who wanted to prepare a thesis on this subject. Almost immediately, I received a response from a particularly motivated student, Frédéric Bertucci. His enthusiasm and seriousness made it possible to take up the technical challenge. Frédéric succeeded in setting up playback experiments in aquariums, which is not easy.

The animal chosen was a pretty little fish with blue stripes, from the cichlid family—the *Metriaclima zebra*. Frédéric showed that the sounds produced by this fish (a succession of clicks) have the value of territorial signals ("Beware! This is my home!") and bear an individual signature that depends on the length of the body.[279] Only a fish already established in a territory responded to the playback signals—"I'm here, I'm staying put." Frédéric also showed that the receiving fish remained very tolerant to changes in the artificial signal. The rate of clicks could be considerably slowed or accelerated without changing the response of the test individuals.[280] A few other studies have shown that the information carried by the fish sound signals was indeed taken into account by the receiving individuals. For example, Clara Amorim, Paulo Fonseca, and their colleagues at the University of Lisbon in Portugal found that the female painted goby, a small fish you may have seen in tidepools at low tide on the Atlantic coast, prefers the more talkative males but pays little attention to the size-encoded information in the signals. The amount of time spent talking is correlated with the amount of fat reserves in the animal. Females are therefore sensitive to the physical condition of the male and assess it during courtship. To be honest, however, it should be pointed out that acoustic signals are not the only issue here, as the female will only pay attention to sounds if she *sees* the male.[281]

Let's get back to business, so to speak! Let me remind you of the question that guides us: understanding how acoustic communication takes place during the history of an animal species. Bird feathers that become whistles, a woodpecker that drums instead of pecking on a tree trunk, insect wings that turn into musical instruments, fish jaws that snap, swim bladders that become a bass drum: All these exaptations are associated with the establishment of ritualized behaviors that produce well-structured sound signals—behaviors that have been favored throughout the history of each animal species by their ability to transport information between senders and receivers.

However, as we mentioned before, this model of the sender as a precursor is not the only one that explains the implementation of acoustic communication. Let us now consider the model of sensory bias, where the sender exploits a sensory capacity already present in the receiver. You'll see that it's very easy to understand. Let's take the crickets as an example. In the field cricket *Teleogryllus oceanicus*, females are attracted to sounds with frequencies below 16 kHz, whereas they are repelled if the frequency is above 16 kHz (higher-pitched).[282] In chapter 9, we saw that these categorization phenomena are widespread in animals and that they facilitate rapid decision making. Our female crickets choose to go toward a singing male (sound of about 5 kHz), whereas they flee from bat calls (of very high frequency). Keep this information in mind: females process sounds in two categories—one attracts, the other repels. Now let's look at a group of cousins of the field cricket, the *Lebinthini*. In their home, the males' song is very high-pitched, up to 20 kHz.[283] This is quite surprising already. Moreover, when you play this song to the females . . . they don't move. On the contrary, they stop suddenly, in a trembling motion, exactly like when they hear a bat call. It's a behavior often observed in insects because it allows them to escape from these predators. So how do *Lebinthini* females meet their partners? Well, the females shiver, and the vibrations generated are transmitted to the plant on which they are standing. The male perceives these vibrations and he comes right away![284] As a result, the high-frequency songs of the males derive from sensory exploitation. The males of this group of crickets have developed a sound signal that uses a previously existing behavioral response of the females. Perhaps they have simply reinforced the higher harmonic frequencies of their original signal? Anyway, the result is there, and they take advantage of a sensory capacity of the females that had been selected for reasons other than communication between the sexes.

Note that this example of the *Lebinthini* crickets is not isolated.[285] For example, the same story can be observed in moths. In several species, the males produce sound signals during courtship that the females cannot distinguish from bat calls. Male calls and bat calls again provoke the same response in females: they freeze. The male will take immediate advantage of the situation. As the female stands still, it's easier for him to approach her and convince her to mate.[286] Here again, the sender exploits a sensory disposition of the receiver that was there for other reasons.

Notice the difference between this model and the precursor sender model that we were talking about earlier. In the precursor sender model, the sensory capabilities of the receiver play a critical role, since it is the receiver that either does or does not perceive the signal carrying the information and then "decides" whether or not a communication can start. Here, in the model of sensory bias, the receiver is ready to hear and react to sounds that the sender does not yet produce. The receiver's sensory system has evolved to detect predators, as in *Lebinthini* crickets and moths, for example. For a variety of reasons, some of which are probably due to chance, the sender one day produces a new signal that exploits the sensory predispositions of the receiver, initiating a new communication process.

You now have a good idea of how a new acoustic communication can occur, i.e., how new signals are set up. Let's tackle the next question: How can we explain the diversity of acoustic signals observed in nature?[287]

Look at birds, for example. There are tens of thousands of species, and almost as many vocal repertoires, even if we exclude the few birds that use little or no acoustic communication. Within each species, there are populations that often have their own dialects and individuals whose vocalizations often differ. How is this possible? Why this extravagant divergence between animals that are, after all, quite similar? It's true that the common ancestor of today's bird species dates back 65 million years,[288] and a lot has happened since then.[289] We've already evoked some of the mechanisms driving this diversification.

We've just seen that sexual selection can be a powerful driver of signal evolution. The senders producing the most convincing vocalizations (for example, the fastest trill of the song or the most elaborate repertoire) are chosen in a privileged way, directing the evolution of communication for this species or that population in this or that direction.[290] When we look at animal species forming social groups in chapter 14, we will see that communication signals can be subject to social selection by kin groups: one alarm call, more effective, will be favored over another. When we were walking in the tropical forest with the white-browed warbler and its cortege of Brazilian species, I told you that some birds emitted vocalizations whose acoustic structure was adapted to the conditions of propagation of sound waves in the environment (for example, songs traveling in a particularly effective way despite the density of vegetation). Acoustic communication signals can thus be under the influence of ecological selection. Those that are most effective in a given environment will be favored. As bird species inhabit a variety of environments, this ecological selection can lead to a diversification of the signals used from one species to another. Moreover, the propagation conditions of sound waves are not the only element that can influence the evolution of signals. It is the entire set of environmental conditions that must be taken into account. In particular, we sometimes observe a real acoustic competition between different animal species for the sound space. It has been suggested that animals can share this space, for example, by emitting lower-pitched or higher-pitched sounds so as not to interfere with each other.[291] You can appreciate that things are very complex—many different factors can come into play.[292]

Sometimes the same signals can be preserved throughout the history of the species, but the role of the signal changes. About 20,000 years ago, astyanax fish living in rivers in Mexico became trapped in underground rivers and lakes, in caves. Since then, this fish has morphed into two forms—a form living in surface rivers with well-developed eyesight and a subterranean form that has lost the use of its eyes. Sylvie Rétaux, a researcher at the CNRS and a specialist in genetics and development, is trying to understand the mechanisms that allowed *Astyanax mexicanus* to change into its cave form. With my colleague Joël Attia and their postdoc Carole Hyacinthe, Sylvie has recently been interested in the acoustic signals produced by astyanax, because this fish talks, of course, like everyone else. Its repertoire is even quite varied, since it can emit at least six different sounds. The two astyanax forms differ in their use of acoustic signals. It appeared to Sylvie and her team that sharp clicking was produced during aggressive interactions by surface fish, whereas cave fish use the clicking signal when feeding.[293] In caves, food is scarce, and recycling an initially aggressive signal into a way of informing each other is probably a beneficial adaptation.

To understand the history of communication signals, evolutionary biologists construct phylogenetic trees, based on the genetic heritage and acoustic characteristics of current signals. The basic idea is to reconstruct all the historical changes from a common ancestor that have led to the diversity observed today. I recently conducted a study with a team of colleagues on the evolution of drumming behavior in woodpeckers.[294, 295] As I mentioned earlier, drumming is an exaptation from foraging behavior, as most woodpeckers spend their time vigorously beating the bark of logs in search of food. It's also how they dig the cavities where they're going to make their nests. Some woodpeckers drum ("DrDrDrDrDr . . .") very regularly; others speed up ("Drrr-Drrr-Drrr-Drr-Dr-Dr-Dr"); others slow down; and some are content with a short "TOK-TOK" that can be heard far away. Like birdsongs, these signals are used to attract a mate or deter an intruder. We first analyzed the acoustic structure of the drumming of 92 species of woodpeckers from around the world, then tried to reconstruct the history of these drummings over the history of the woodpecker family, and then attempted to explain

Great spotted woodpecker

their diversity. Our results showed that drumming is a very ancient technique, which was already mastered by the ancestor of all present woodpecker species 22.5 million years ago. We have also shown that the diversity of drumming types is essentially explained, quite simply, by chance genetic mutations. The closer two species are genetically related, the more similar their drumming patterns are. The drumming of sister species is difficult to distinguish. Those of first cousin species are a little easier to tell apart. As for those of second cousin species, they are quite different. Through playback experiments, Maxime Garcia, then a postdoc on my team, showed that the great spotted woodpecker *Dendrocopos major* cannot distinguish between the drumming of its species and that of a sister species. On the other hand, it does discriminate well between those of a second cousin species (sister or cousin—these are analogies, of course, to illustrate a close or distant relationship). But then, how do woodpeckers of different species living in the same forest manage to distinguish between each other? Fortunately, sister or first cousin species rarely live in the same places. We studied five woodpecker communities (in Switzerland, Guatemala, the US, Malaysia, and Guyana), and in each case the six to eight woodpecker species likely to live together in the same forest were almost always distant cousins. Their drumming patterns are therefore very different from each other, thus limiting the risk of confusion. When two sister species live in the same place, which does happen from time to time, then a very well-known phenomenon in evolutionary biology occurs: *character displacement*. The drumming of each species changes and becomes less and less similar to that of the neighboring species. To say that the character is displaced means that the drumming pattern changes, differs, and moves away from the drumming of its sister species. It can take thousands, even millions of years.[296]

I won't close this chapter without presenting you with one last mechanism explaining the divergence of communication signals between species: the magic trait! You have certainly heard of Darwin's finches. These small, somewhat dull birds inhabit the Galápagos Islands and provided Charles Darwin with a prime example when he established the theory of evolution. There are several species of Darwin's finches. They all come from the same original species and have gradually specialized in the different food sources available on the islands. Darwin's finches are an example of adaptive radiation, as it is called in the jargon of biologists: *radiation* because the new species went in different directions during their evolution (they radiated around the starting area); *adaptive* because the trajectory of each species corresponds to a food specialization for which the bird's morphology, especially its bill, has adapted to pick up food. In short, big beaks, big seeds; small beaks, small seeds. Studies of the songs of the different species show that there are correlations between the shape of the beak and the acoustic properties of the songs. Large-billed finch species have simpler songs (less difficult to produce) than small-beaked finch species. These correlations can be explained by mechanical stresses related to the type of bill and its associated musculature. A large, powerful bill cannot be opened and closed extremely quickly, making it more difficult to produce songs with rapid modulations. Changes in bill morphology during the diversification of finch species are partly responsible for the song diversification.[297] The diversity of the songs, which seems to appear by magic, is in fact nothing magical. It is the consequence of another phenomenon that has nothing to do with acoustic communication.[298]

Do you find all these mechanisms complicated? You're right, because they are! I warned you; it's difficult to reconstruct the history of communication behavior. Now let's get some air in our brains. Let me take you to the Amazon rainforest.

11

Networking addiction

ACOUSTIC COMMUNICATION NETWORKS

Amazonian rainforest, near Belèm, Brazil. Invisible in the high branches, a screaming piha blares out its song, one of the most powerful in the Amazon. First, some very soft trills that make the singer seem far away: "Wuuu—wuuu—wuuu." Then a brutal burst: "Weee-weee-YUUU!!" Right behind me, another one: "Wuuu—wuuu—wuuu . . . weee-weee-YUUU!!!" Then a third on the right, and several others join the sound demonstration. The forest fills up for several minutes with the calls of about 10 individuals. The chorus of *Lipaugus vociferans* submerges all the other sounds. Suddenly, for no apparent reason, all fall silent. The cicada concert can be heard again. Half an hour later, the *Lipaugus* resume possession of the sound space. Why do these birds congregate like this to sing? Does their apparently chaotic chorus obey a certain logic?

It was Thierry who suggested that we should take an interest in the *Lipaugus*. After our disappointing experiment with black caimans in the Kaw marshes (Do you remember? The female caimans had stopped responding to our signal broadcasts on the second day of the experiments . . .), we decided to leave our floating platform as soon as the opportunity arose. After a week, a refueling helicopter brought us back to the Camp Caïman hostel, and we got our rental car back, determined to explore French Guiana as tourists during the remaining week before our flight back to France. French Guiana is surprising: you drive on a perfectly maintained road, passing a post office or police cars similar in

every way to those encountered in metropolitan France, then you stop
at a small parking lot, put on your walking shoes, and after a few meters
on a trail, you find yourself transported into the middle of a primary
rainforest. It's a brutal, magical jump—and it's magnificent. Gigantic
trees. Here, a tree trunk lying across the path being devoured by hundreds
of huge beetle larvae—an impressive sound of jaws chomping. There,
up in the branches, quiet, a lazy sloth swaying. Everywhere, termite
nests, strange insects, big metallic-blue butterflies, colossal caterpillars,
plants with huge leaves, furtive birds, and Julida—an order of enormous,
astonishing millipedes with shiny black bodies. I was in heaven, chatting
away with Thierry in the wilderness. Then, suddenly, "Wuuu—wuuu—
wuuu . . . weee-weee-YUUU!!" We were to spend a long time looking
for it in the branches, our singer, perched as it was about 10 meters high.
However, it is a rather large bird. About 30 centimeters long, its grayish
color camouflages it perfectly. "A bird that calls terribly loudly and that
we can't see . . . a perfect model for a bioacoustic study!" Thierry has
always been skilled at identifying research subjects. A little look in our
ornithological guides would increase our curiosity. The *Lipaugus* form
exploded leks. In other words, the males gather regularly to sing in particular places in the forest, always the same males but remaining a good
distance from each other, between 40 and 60 meters. They stay there
for a while, then go back to their other occupations and return a little
later. No one understands why they do this. Our microphones, computers, and loudspeakers were not going to stay in our suitcases. That same
evening, while we were drinking a little rum at the restaurant and watching the leatherback turtles laying eggs on the beach below, we decided
to start the *Lipaugus* project.

Our first day was dedicated to the observation and the first recordings of the vocalizations. As good ethologists, we had to know a little
more about the behavior of this animal before trying to ask it any questions. It was quite easy to spot several leks: the Guianese forest is easily
accessible, at least at its edge. The pihas' chorus always started in the
same way: first a few powerful, isolated calls ("TSSIOO!! . . .
TSSIOO!! . . . TSIIOO!!") coming from several individuals; then a series of "wuuu—wuuu—wuuu," muffled sounds whose origin could not

Screaming piha

be located; and, finally, the complete songs, followed by a culminating explosion exceeding 110 decibels, a sound level between a rock concert and a jackhammer. Imagine the bird opening its beak wide, retracting its head, its whole body behaving like a high-fidelity loudspeaker projecting the sound waves as far as possible: "Weee-weee-YUUU!" These calls were first emitted by single individuals and then taken up by the whole chorus: Did the birds alternate among themselves? Was there a kind of organization in time, in space? A chorus conductor, perhaps? And why the different calls? These were the questions we had in mind.

After a few days spent recording these vocalizations, we decided to do some initial playback tests. Our first goal was to test if the order of emission that we had observed—first the "TSSIOO," then the "wuuu—wuuu," and finally the "weee-weee-YUUU!"—had a functional meaning. In other words, was it imperative that the birds first heard "TSSIOO" to start the lek chorus? Our second goal was to determine what happens if birds hear vocalizations from another lek. Do the individuals singing in a lek know each other? Do they react differently when they hear strangers? At that time, we were not sure if the birds in a given lek were always the same, but it seemed likely. So the principle of the experiment was as follows. Arriving at the place where we had spotted a lek the day before, we tied our loudspeaker to a tree; then we moved about 30 meters away, our tape recorder connected to the loudspeaker by a long cable. We waited until the pihas had been silent for at least half an hour before playing a first signal—one of the three types of calls—and then observed the birds' response. After a few days on this regime, we gained our first insight: the "TSSIOO" as well as the "weee-weee-YUUU!" quickly triggers strong behavioral responses from our singers. To our "TSSIOO," individuals responded with their own "TSSIOO." When we played the "weee-weee-YUUU!," the chorus would start. In both cases, it was not uncommon to see a bird silently flying close to our loudspeaker in response to our sound stimuli, then flying away again immediately, before it made its first calls. On the other hand, the "wuuu—wuuu" emission alone did not seem to provoke any particular excitement: the birds generally remained silent. The problem with the pihas was that we could not see them, so we could not tell who was singing. To understand the structure of this communication network, we had to be able to identify who was who in the lek and who was where. We were not equipped for that. In addition, our return flight to metropolitan France was scheduled, so we left French Guiana with no feather in our caps. The calls of the pihas resounded in our heads, and we had only one desire: to get back into the field.

A few months later, we met up with our friend Jacques Vielliard at the International Bioacoustics Conference in Italy. Do you remember Jacques, the professor at the University of Campinas in Brazil, with whom we had studied the white-browed warbler? Malu, Jacques' wife, had just obtained a professorship at the University of Belèm in the Amazon, and Jacques had decided to spend most of his time there. "Folks, I'm building a field station on the edge of a beautiful primary forest . . . you must come and see us," he told us, a glass of beer in hand. We immediately asked him, "Do you have any *Lipaugus* there?" In his ever slow and modulated voice, he answered, "Guys . . . think a little! . . . I'm talking about the Amazonian forest! . . . Of course, we have some. The *cricrió*, as they say in Portuguese . . . *A voz de Amazônia*." The voice of the Amazon! Indeed, its song is frequently used in films to reproduce the ambience of the rainforest. The screaming piha is widespread throughout the Amazon basin. Some even call it the captain of the forest. That's how, a few months later, we joined Jacques in the Amazon.

Belèm is a large city, bordered by a wide arm of the Amazon River. Although the modern part looks like any other, with its large, aesthetically challenged buildings, the old town is charming, a real postcard— with its small port where urubu vultures nonchalantly walk in search of some fishy leftovers, its colorful *Ver-o-Peso* market, and its old colonial houses, whose walls, roofs, and balconies are overrun with epiphytes plants. In the tropics, nature creeps in everywhere. An hour's drive away is the field station built by Jacques. He doesn't do things halfway: a large, beautiful house, with wide terraces where hammocks can be hung, immediately facing the forest. In the evening, we can see colorful flights of parrots and toucans. However, you should not get too idyllic an idea of the situation: mosquitoes, ticks, poisonous spiders, and scorpions are all part of daily life. The scorpion, a sneaky wee beastie, loves to slip into your shoes at night. One morning, Thierry shook his shoe a little by chance and discovered an intruder who hastily scuttled off. He had already put on the other shoe without taking this precaution, and had been lucky, and he never forgot to empty both shoes again before putting them on. Not to mention the antimalaria medication that had to be taken every day, never to be forgotten, and the compulsory vaccination against yellow fever, that very nice disease that makes you die from vomiting streams of black blood. The tropics are teeming with invisible life. Every small wound can quickly take on worrying proportions, and Jacques recommended that we carefully disinfect the slightest scratch.

As our friend had promised, the screaming pihas were on the scene. We had access to several leks on foot from the house. The objective of this first mission was simple: to see if we could identify and locate the individual singers of a lek, to follow the temporal dynamics of their choral group, and to write the score of the chorus. How to locate the position of invisible birds? It's not so difficult when using the principle of *sound triangulation*. Imagine six microphones positioned at different places and at different heights in the forest. When a piha sings, the microphone closest to the bird is the first to receive the sound waves; the furthest away is the last. Since the sound propagates at 340 meters per second, the delay in reception between two microphones corresponds to the difference in the distance to the bird. By using six microphones, whose exact position is known, we can then calculate the position of the sender bird quite accurately. The microphones were wireless, and we could centralize all the recordings on a multitrack tape recorder, which received the signals from each microphone via radio waves. Chloe Huetz, the engineer in Thierry's research team, was in charge of developing the computer program to process all this data. At the same time, it was necessary to establish the vocal signature of each individual participating in the lek. This meant recording each bird, then analyzing the acoustic structure of its song to identify its individual characteristics. It soon became clear that each individual was recognizable by its voice. A comparison with the vocalizations of birds from other leks in the forest further showed that, like the existence of a regional language, birds participating in the same orchestra share a lek signature. And by comparing the Brazilian recordings with those we had obtained in French Guiana, it appeared that these two populations, the Brazilian and the Guianese, did not have quite the same way of singing, even if the difference remained minimal (small variations in the modulation of the "weee-weee-YUUU!"). Despite the presence of these local and individual signatures, the singing of the piha was remarkably similar everywhere. As we will see in chapter 12, some birds learn their songs by copying other individuals, and this cultural transmission is likely to result in significant variations in the acoustic structure of vocalizations between geographically distant populations. On the other hand, species that produce their sound signals without the need for learning by imitation have songs that normally show little variation, which is directly related to the genetic distance between populations. The screaming piha probably belongs to this second category of birds.

Once the multimicrophone recording and triangulation system had been developed, and the individual signatures well characterized, we conducted playback experiments inspired by our first tests in French Guiana. The results confirmed our first impressions: the different calls produced different responses. The "wuuu—wuuu" was rarely followed by a vocal response. On the other hand, the birds reacted strongly to "TSSIOO" and "weee-weee-YUUU!" and even to an isolated "YUUU!" More precisely, when we broadcast this type of call, one individual would fly over the loudspeaker and then call in response. Only the "weee-weeee-YUUU!" seemed to systematically trigger the chorus. I should make one essential point: we were only able to spot the singing males, so, obviously, we didn't know—and still don't know—what the females were doing. Yet, of course, they had to be there too. Otherwise, why else would the guys gather together? Not for a rugby match, that's for sure. One hypothesis is that the "wuuu—wuuu" is addressed to the females, like a drumroll signaling that the males are gathering in the lek and are available. The "weee-weee-YUUU!," higher-pitched and thus allowing a precise location of each individual, would be addressed to both females and males. No discrimination. As for the "TSSIOO," they could be alarm calls, signaling the arrival of a possible danger. To find out more, it would be necessary to spend some time on the spot—a lot of time. Neither Thierry nor I could abandon our duties and our families for that long, so we hired Frédéric Sèbe on a postdoctoral contract. Frédéric, now a researcher on my team, was a former PhD student of Thierry's, a field man par excellence—afraid of nothing, and certainly not of living in the Amazonian forest for long periods. It was the perfect opportunity: Frédéric left to spend a year studying the pihas.

It was a big challenge, and despite extensive work, Frédéric would not, alas, unravel all their secrets. But he did come back with data that made it all a little clearer. Frédéric had been able to follow the activity of several leks throughout the year. By combining the system of locating individual singers by triangulation and their individual identification through vocal signatures, he was able to describe their dynamics precisely. A few surprises awaited us. We thought that the pihas gathered to sing during special seasons, but the recordings showed that the leks were active year-round and that they always had more or less the same number of individuals. Going every day to sing in the lek seemed to be part of the daily routine of the *Lipaugus*. On closer inspection, however, it became clear that the individuals singing together in a given lek were not always the same. At intervals of a few months, a good one-third of the males had been replaced by other individuals we had never recorded before. Let's be precise: this turnover varied between 0% (the same individuals were all still present in the lek) and 75% (three-quarters of the males had been replaced by others). In a way, the leks are like theaters where males enter the performance in turn. Some of these theaters have the same troupe performing year-round, while others see most of their actors change over the months. Frédéric's data also made it possible to describe the performance of these actors. It begins a little more than two hours after dawn, with a fanfare opening where all the singers vocalize at the top of their voices. Then, regularly, new explosions of calls occur, albeit more modest. Finally, two hours before twilight, comes a closing fanfare. Remarkably, each singer has its place on the stage and rarely moves from it. Here again, let's be precise since the data allows it: 90% of the birds found from one month to the next in the same lek remained within a radius of 30 meters around their initial position. An individual can have up to ten singing posts—i.e., trees on which it perches to sing—all very close to each other, but it spends most of its time on two or three of them. Just imagine: the same male will be singing for months from nearly the same perch! Now let's get into the figures. While the surface area occupied by all the individuals of a lek can reach 50,000 square meters (which corresponds to a circle 250 meters in diameter), each individual is confined to a maximum of 700 square meters. Moreover, males are not all equal when it comes to the distribution of roles. The most active singers are always in the center of the lek and can call their "weee-weee-YUUU!" up to six or seven times a minute. Those who are on the edges, less vocal, will be satisfied with making the call once over the same length of time. You can bet that there is no such thing as equality among the pihas.

What happens when a new male, unknown to the others, appears one day in a lek? To find out, Frédéric took up the playback experiments that we had initiated in French Guiana. This time, he was able to follow the behavior of each of the males in the lek. Frédéric mimed the arrival of a new individual by playing the calls of an individual recorded in another lek during the dawn chorus. The loudspeaker was placed in the singing place of an already present individual. The disturbed individual would first react sharply by nervously flying over the loudspeaker, accompanied by a clear increase in the number of calls he made per minute. Then he would move a few dozen meters to perch on an unoccupied tree and resume his vocal activity—no more disturbed than that. The other males of the lek obviously couldn't care less, continuing their vocal routine from their singing places. The next day, if Frédéric stopped the playback, the disturbed individual would return to its initial position. The *Lipaugus* were therefore ready to accept newcomers. When the number of males increases, they simply enlarge the surface area of the lek. In short, what have we learned from the *Lipaugus*? That this dull-colored bird spends a lot of its energy singing a very repetitive repertoire; and that the lek, this aggregate of males, is organized both in time and space. We will probably never know if and how the females make their choice among all these suitors, but it is reasonable to think that not all males are the same for them. Moreover, perhaps their preferences are variable: if we can assume that the stars— the males in the center of the lek—are more attractive to the majority of them, we can also imagine some females preferring the more discreet ones. Loudmouths versus wallflowers.

The *Lipaugus* project had whetted my appetite for studying *communication networks*. It was Torben Dabelsteen, my colleague from the University of Copenhagen, and Peter McGregor, then also a professor at the same university, who had formalized this concept a few years earlier.[299] Starting from the observation that most bioacoustic studies to date have considered any communication process as a simple sender-receiver duo, they had written several articles in scientific journals about the idea that, in nature, individuals are usually in the middle of a network, in which each individual can be alternately a sender and a receiver (not forgetting some who will in fact only be receivers, simply listening to what others say to each other). Communicating in a network, with the possibility that the intended listener is not the only one who hears the communication, can have consequences. We know this—we who modify our speech and way of speaking according to who might be listening in.

I decided to launch a new project to explore this issue, this time in my laboratory, with a species that is easy to breed in captivity: the zebra finch. With a size comparable to that of a sparrow (about 10 centimeters from beak to tail and weighing about 15 grams), this bird, native to the

Zebra finches

Australian semidesert zones, flies in large groups that can number in the hundreds of individuals in search of food or water.[300] The pairs nest in the same bushes, forming breeding colonies. But beware! This species has a very strict moral code: male and female are faithful for life. Let us note, nevertheless, that the life of a zebra finch is short compared to ours—a few years at the most. Females and males are easy to tell apart: the female appears to be in half-mourning, with a grayish plumage and a clearly visible black tear falling from her eye, while the male has an orange cheek, a zebra-striped chest, and a red flank peppered with white dots. Their vocalizations are much more complex than in the piha. A song emitted only by the males, lasting a few seconds, is a succession of squeaky notes, which are not very pleasant to hear: "Tzeek tzeeek tzeek tzeeek didiguezic dziduck." Many short calls, each consisting of only one note, are emitted by both sexes. Julie Elie, one of our students who was passionate about observing animals and who has since become a renowned specialist on the vocal repertoire of the zebra finch, has analyzed the acoustic structure of thousands of vocalizations of adult individuals and has categorized the calls into eight types: the whine, the nest, the tet, the distance call, the wsst, the distress call, the thuk, and the tuck.[301] As you might guess, each of these calls is made in a more or less precise circumstance. For example, the whine and the nest calls are produced when the male and female exchange nest duty. The wsst accompanies aggressive behavior. The thuk and the tuck are alarm calls, the first being intended for the young and the partner, the second for the whole troop. The one that caught my interest was the distance call, which allows the couple to find their partner when they lose sight of each other. This is the easiest to record because, to encourage the birds to produce it, all you have to do is separate the pair by isolating the male in one room and the female in another.

Initially, inspired by Thierry's work on penguins that I told you about in chapter 4, I wanted to test the ability of males and females to recognize each other through the distance call in the chaotic context of a breeding colony. I also imagined that I could identify the neurophysiological bases of this recognition: Are there regions of the brain dedicated to this task? A few observations suggested that this call allowed the couple to identify each other after separation, but experimental evidence was lacking. I could not carry out this project alone and hoped to convince a student to share this adventure. At the end of a course I was giving at the *École normale supérieure de Lyon* to prepare students for the prestigious French exam, the *Agrégation*, a young woman stood in front of me—smiling, determined, and confident: "I want to work with you," she said. I asked her about her motivations. Why would a student of the *École normale supérieure*, who must have had all the research laboratories of this famous school making eyes at her, want to join my team? "Because you do things that nobody else does. I didn't even know you could get paid for it! I'm studying biology because I'm interested in living things, but I find myself in front of pipettes and molecular formulas. Getting to listen to birdsongs—that is what I'm passionate about." Clementine Vignal really was a godsend. We didn't know that we were going to spend more than 10 years exploring the world of the zebra finch together.

We started with the resources we had on hand: a small aviary built by the technical team at the University of Saint-Etienne; a few cages; and some birds bought at the nearest pet shop. In a short time, we had our colony of zebra finches, with our first mating pairs and their nests.

First experiment, first disappointment. When we isolated a male from its female and made it listen to distant calls from females, including those of its partner, the poor male usually remained motionless, seemingly bewildered. We started by checking that the females' calls had an individual signature. The result of acoustic analysis and statistical comparisons between individuals was clear: each female had her own voice. Even we were able to distinguish between them. Our males had no excuse not to recognize their sweethearts. We had to look elsewhere. We knew that the zebra finch is a very social bird and not used to being alone for a long time. So we came up with the idea of placing a few individuals in another cage, next to the cage containing our male and the subject of our experiment—some companions to relax him; some friends from the bar, so to speak. That's when we made what, for us, was the discovery of the century!

"Look," said Clementine. "The results of the experiments are really quite odd. I get the feeling that the male's response to his female's calls varies depending on his companion birds." And so we found ourselves comparing the male's behavior in response to his partner's calls with that of other familiar females—all in different social contexts—all the while changing his companions. Next to some males, we placed a cage with a couple, a female and her male. For others, the two companions were an unmated female and male put in two separate cages. Finally, others were placed in the company of two other males. The first impression was confirmed: the male's response to his female's calls varied according to his audience. If the male was in the presence of a single female and a single male, or two single males, he would respond to any female call played by the loudspeaker, regardless of whether it was his own beloved or another, with no preferential reaction to those of his own female. But when he was accompanied by a well-established female-male couple, he reacted like a madman to his partner's calls (the number of calls emitted was multiplied by five!) and barely reacted to the calls of a familiar female who was not his own. Male zebra finches were therefore perfectly capable of recognizing their female partner by her voice,[302] but they were sensitive to the social context—to their surroundings. There was an effect of the social audience on the behavior of the males.[303] More importantly for us, it was the first time that an experiment had demonstrated a bird's ability to grasp the nature of social bonds between other individuals. In order to explain the results of our experiment, we had to admit that the male, who was made to listen to female calls, was able to distinguish between a couple of single birds and two birds in a couple. This social intelligence had so far only been shown in monkeys.[304] Excited by this discovery, we wrote an article for the famous magazine *Nature*, which accepted it. Champagne!

Champagne, yes. For two reasons. First, because we were extremely happy that we had made a scientific discovery. Second, because we hoped it would enable us to obtain funds to continue our work. In order to explore the world of the zebra finch, we needed money, and the article in *Nature* would be a great asset in finding it. We wanted to build larger aviaries, monitor the behavior of each bird in the colony from a distance, record each individual in its nest . . . and Clementine dreamed of going to work in the field, in Australia. Money is something I haven't told you about yet. Being a researcher is a bit like having the only job where, once you've been hired for life, you're told, "Now it's up to you to find the money for doing the work; no one else will do it for you!" What you don't realize right away is that you're going to be looking for money and then looking for more money for the rest of your life—which means writing more and more research proposals to persuade those who read them that the research work you propose is exciting and new, that it will revolutionize our knowledge, and that it has the flavor of excellence; basically, that you will save the world. For Clementine and me, the publication in *Nature* was to act as a real magic wand: it suddenly transformed a research-team-ignored-by-everyone into a lab-of-potential-international-notoriety. Not being in the habit of shooting myself in the foot, I won't say that it's easy to publish in this kind of journal, but luck has a lot to do with it. As Peter Marler, the high priest of bioacoustics, once told me, "It's almost impossible to predict what will interest them." At this stage, it didn't matter to us. Our little male zebra finch refusing to respond on command to the voice of his female had helped us reach the grail of scientific journals. And the National Research Agency, the main backer of French public research, agreed to finance the continuation of our work.

We were not at the end of our surprises with these birds. We had recruited Julie Elie, who would spend hundreds of hours observing them. She would notice that the monogamous couple—a female and a male "married and faithful"—is the foundation of their social organization. Female and male spend most of their time side by side, touching beaks, grooming each other's plumage, and conversing softly. Nice life as a couple, isn't it? Moreover, it is easy to distinguish a group of single birds from one made up of paired couples: the single birds are considerably noisier, with sudden explosions of sound where everyone is vocalizing at the same time; while the group of couples is calmer, each pair of birds whispering sweet nothings.[305] Can we go so far as to talk of wedded bliss? Moreover, in the case of the zebra finch, forming a couple for life does not seem to be an option but rather an obligation: whatever the circumstances, you have to pair up. When we had isolated single females in one cage and single males in another, we discovered a surprising thing indeed: in the absence of an individual of the opposite sex, the zebra finches soon developed same-sex pair bonds.[306] The males paired up with their brothers, and the females did the same with their sisters. Let's be clear: true single-sex couples are formed where both partners go so far as to mimic copulations and build their nests. And they're faithful to boot. When these birds were put back into mixed aviaries, each one stayed with its partner. With it I am, and with it I stay. As Clementine says, "The zebra finch pair is a true social partnership." To associate with a fellow bird to form a couple is a vital need for the zebra finch. How to explain it? It's possible that the intense constraints to which this bird is subjected in the wild—little water and food in a semidesert environment—favor a rapid pairing, with very strong bonds between partners: a solid and ready-for-anything couple will be better able to reproduce and raise young.

A recent study on a small African bird, the blue-capped cordon-bleu *Uraeginthus cyanocephalus*, has shown that what we observed in the zebra finch is not anecdotal. In the cordon-bleu, the couple performs very special courtship rituals in which both male and female tap-dance with their legs.[307] You read that right: tap dancing! The two birds jump very quickly on their perch by clapping their legs. For a cordon-bleu, that sound is super sexy, as long as the rhythm is perfectly controlled. To complete the scene, male and female sing. In short, the cordon-bleu's courtship is akin to Fred Astaire and Ginger Rogers crossed with Sonny and Cher.

Manfred Gahr, researcher at the Max Planck Institute for Ornithology in Germany; Masayo Soma, professor at the University of Hokkaido in Japan; and their student Nao Ota tested the influence of a social audience on the cordon-bleu's dance. The Max Planck Research Institute is located next to the small village of Seewiesen in Bavaria. This is where Konrad Lorenz studied goose imprinting. Do you remember that? We talked about it in chapter 1. The geese that had hatched in Lorenz's presence followed him everywhere, even when he swam in the lake next to the research station. Today's lake is not Lorenz's lake—it was filled in and dug anew—but the spirit of the man who was one of the founding fathers of ethology still permeates the place.

Manfred, Masayo, and Nao showed that male and female cordon-bleus adjust the modalities of their courtship if they are in the presence of other birds.[308] In particular, they increase the number of dances, combining singing and tap dancing. What is the reason for this? Perhaps to make it clear to the audience that they are courting each other and thus to keep away other possible suitors. Furthermore, during their dance, the partners point their tails toward their lover: while you never know for sure who a song is intended for, in this case there is no room for doubt. Another hypothesis is that the dancers do not want to put all their eggs in one basket and may be trying to charm members of the audience rather than their partner. We don't know. Perhaps marital morality is more elastic in cordon-bleus than in zebra finches.

The *audience effect* in acoustic communication networks hasn't only been studied in these two species of birds.[309] Let's give Caesar back what belongs to him: the first time an audience effect was suggested, it was for . . . chickens. The rooster makes calls to indicate the presence of food only when hens are there. Otherwise, it's silent. The same behavior is true of his alarm calls when a predator approaches.[310] It is now well established that many animals modulate their communication behavior according to the individuals around them. In primates, examples abound. For instance, when a chimpanzee calls out in response to an attack from a fellow chimpanzee, it exaggerates the intensity and duration of its calls if there is another individual in the vicinity that is known to be stronger than the attacker and therefore likely to drive it away.[311] As another example, rhesus macaque mothers (*Macaca mulatta*) respond more quickly to their babies' requests when they are close to adult individuals known to be aggressive toward the young.[312] When threatened by a predator, Thomas's langurs (*Presbytis thomasi*) emit alarm calls until each member of the group has responded with an alarm call, suggesting that they are able to identify who has responded and who has not.[313] In short, it is clear that we are far from animals vocalizing by simple reflex. This is very similar to what we observe in our own species.

Moreover, *social intelligence*, that remarkable ability to analyze the relationships between individuals in the group, as identified in the zebra finch, has since been found in many animals. These discoveries are not very astonishing if we think about it; obtaining an acute awareness of the social relations between the individuals that we are close to makes it possible to limit errors in judgment that can have serious consequences— for example, when you have to form alliances with other individuals. Take the chacma baboons (*Papio ursinus*), large African monkeys living in organized packs—males with canines that never end, and an ironclad hierarchy between individuals. When two males fight, the dominant one grunts while the dominated one squeals at the top of its lungs. With playback experiments, the famous primatologist Dorothy Cheney and her colleagues have shown that the other members of the group are attentive to these vocal exchanges and learn from them. See for yourself: when the researcher used a loudspeaker to emit sequences mimicking a reversal of hierarchy between two individuals (the dominant squealed while the dominated grunted), she observed that the other monkeys were clearly astonished: they immediately turned their heads toward the loudspeaker and stared intensely at it.[314] This behavior shows that chacmas track the evolution of the hierarchical situation within their social group by listening to the vocal exchanges of their fellow primates. They seemed very disturbed by what they were hearing. You might object,

"Yes, but social intelligence—this ability to analyze the relationships between fellow creatures—is probably reserved for animals with big brains." Well, it's not. Evidence of social intelligence can be found in . . . fish.[315] The first studies on this subject were conducted more than 20 years ago by Claire Doutrelant when she was still a student (she is now a researcher at the CNRS). Claire and fellow researcher Peter McGregor had placed a female Siamese fighting fish *Betta splendens* in an aquarium near another aquarium containing two males. In this species, the males fight each other with impressive demonstrations. No acoustics here, but visual signals: The males spread their fins and gill protectors (opercula) apart and twisted their bodies, all to appear as big as possible and impress the opponent. When the female was put in the position of choosing which male to approach, she systematically chose the winning male.[316] She had therefore observed the interaction between the males and had been able to deduce, and memorize, which one was the strongest—the best candidate. As another example, a male fighting fish exaggerates the visual signals he deploys all the more in front of an opponent that he has previously seen surrender to another competitor.[317]

In birds, too, females may listen to discussions between males. Eavesdropping on conversations between males influences the decisions females make when choosing who to breed with. A study of the black-capped chickadee *Poecile atricapilla* has demonstrated this experimentally. Paternity testing has shown that an average of one-third of the chickadee's chicks in a nest have not been sired by the male who looks after them, but by one or more of the males in neighboring territories. Morals are flexible in these birds. The females do not hesitate to look elsewhere—but why? Scientists have tested whether they base their reproductive decisions on information gained by spying on male vocal competitions. The first step in the study was to observe pairs of neighbors to find out which of them showed more ability to dominate the other. This was easy to see: a high-ranking bird did not wait until its neighbor had finished singing before it began singing. It did not think twice about butting in. Then, from their loudspeaker, the scientists produced various situations by playing chickadee songs for six minutes. In the control situation, they mimicked situations where the natural order was preserved: the neighbor entered the territory singing submissively, without barging in, while the owner of the territory sang aggressively. In a first experimental situation, the researchers reversed the established order, this time simulating an intruder singing aggressively in front of a landowner previously identified as being of high rank. In a second situation, they simulated an intruder singing submissively, i.e., without ever interrupting the owner of the territory, even though the latter had been identified as submissive himself (submissive perhaps, but never interrupted).

Some time after the hatchings, scientists took blood samples from the chicks and compared their DNA with that of the local males. The high-ranking males that had been confronted with the control situation (nonaggressive playback of the neighbor) could claim paternity of 90% of the chicks present in their nests. On the other hand, in the nests of the high-ranking males shaken by an aggressive playback of a usually submissive neighbor, 50% of the chicks were fathered by the neighbor! However, the individuals identified as submissive during the observation phase did not gain anything from having, once in their lifetime and for six minutes, passed for dominant near their female.[318] As easy as it is for a male to lose his rank, it's no mean feat to win the favor of female black-capped chickadees.

This ability to listen to signals that are not intended for the "receiver" (a sort of eavesdropping) sometimes extends beyond the subject's own species. It has been extensively studied in the heterospecific groups (i.e., groups with several different species) that birds often form. The red-breasted nuthatch *Sitta canadensis*, an elegant little Canadian bird, has a habit of joining groups of black-capped chickadees and takes advantage of their alarm calls when a predator approaches. But the nuthatch is cautious when it comes to interpreting these calls. Let's take a closer look. The structure of the black-capped chickadee's alarm call varies with the size of the predator. If a large owl arrives, the chickadees that spotted it make a short call ("chick-a-dee-dee-dee"); if it's a small owl, the call is long ("chick-a-dee-dee-dee-dee-dee"). In short, the call codes for the size of the predator.[319] It has even been shown that there is a proportional relationship between the size and the number of "dee's" in the call. Subtle, isn't it? Let's talk a bit about the nuthatches' reaction. If one of them sees the predator, it also modulates its calls: short, high-pitched calls faced with a small owl, longer and lower-pitched calls faced with a big owl. But if it does not see the predator and hears only the chickadee calls, it starts to make alarm calls of intermediate duration, regardless of the duration of the chickadee calls,[320] as if it didn't completely trust that information. You never know: What if the chickadees are wrong? Information gleaned by listening to others has never been worth strict adherence. As one of my grandmothers used to say, it's better to be safe than sorry. The nuthatch won't risk sending the wrong information to its fellow nuthatches![321]

To sum up, the next time you hear a singing bird, you have to imagine it as a sender inserted in a communication network, listened to by many receivers and listening to other senders.[322] There are many other things to say about communication networks.[323] For instance, an emitter individual can vary the size of the network likely to listen to it by modulating its vocalizations. Private conversations can exist, but then you have to be very discreet and produce signals that are inaudible from a distance. We talk about this later, in the story of how some animals living in groups can use acoustic signals to cooperate and form alliances. For now, let's return to another issue mentioned above—that of learning. You now know that some species of birds have to learn to produce their songs by copying others. How does this work? More generally, how does communication behavior develop over the course of an individual's life? This is, you remember, the third question asked by Nikolaas Tinbergen. It's also the focus of the next chapter.

12

Learning to talk

VOCAL LEARNING IN BIRDS AND MAMMALS

Marin County, California. In spite of the intense sunshine, the air is crisp and the wind strong. The ocean can be heard roaring in the distance. Perched on the top of a bush, a small bird with the appearance of a finch is singing its "yeeeee . . . peee peee peee pew pew pew pew pew pew" ritornello. It has a pretty head with black and white stripes, a yellow beak, and elegant marbled wings, brown and beige. "Have you seen this bird? What is it?" I said, turning to my colleague Frédéric Theunissen. "Come on, Nic! It's the white-crowned sparrow. Peter Marler's bird." Peter Marler! The high priest of bioacoustics, the one who unraveled many mysteries concerning the learning of song by imitation in songbirds. In his famous study, published in the prestigious journal *Science* in 1964, he recorded white-crowned sparrows in various locations around the San Francisco Bay Area: in Berkeley to the east, the home of the famous university where Marler once worked and where Frédéric is now; in the Sunset Beach area to the south; and in Marin County to the north.[324] By comparing their songs, Marler showed that the sparrows had dialects, and it was easy to distinguish the songs of the Berkeley residents from those of Sunset Beach and Marin County—a bit like how one can tell the French apart from their accents, be they Marseillais, Parisians, or inhabitants of Saint-Etienne. Later, Marler raised young sparrows in his laboratory and showed that these birds learned their song by imitating the one broadcast from a loudspeaker; it was possible to teach the Berkeley dialect to a sparrow born in Marin County, provided the bird was very young and had not yet begun to sing. So at this moment I am amazed, thinking that the individual we are watching may be a great-grandson of the sparrows recorded by Marler!

Peter Marler was not the first to work on how birds learn to sing. His studies were inspired by some illustrious predecessors, some of whom go back a long way. If 2000 years ago Pliny the Elder had already noted in his *Natural History* that the parrot was a good imitator, it is to the Austrian Ferdinand Pernau that we owe the first documented observations on the learning of song by birds. That was in 1720! Some 50 years later, in 1773, a man named Daines Barrington raised goldfinches (*Carduelis carduelis*) in the presence of adults of other species and observed that the little goldfinches imitated their guardians. Barrington also noted that, when given a choice, goldfinches preferred to imitate adult goldfinches rather than the adults of another species.[325]

However, it was not until the twentieth century that our knowledge in this area began to skyrocket. This was mainly due to technical progress with recorders (tape recorders) and the sonograph, an instrument that makes it possible to represent sounds graphically, much like musical notes. Two scientists got the ball rolling: the British William Thorpe and the Danish Holger Poulsen.[326] Then, by combining recordings of dialects in the field with experiments in his laboratory, Marler provided compelling evidence that an animal could learn to produce its vocalizations by imitating adults. It must be realized that, at the time, many still firmly believed that this ability was reserved for the human species and that animal songs and calls were innate reflexes, engraved in genetics. Proving through scientific experiment that an animal learned to produce its vocalizations by imitating adults was a real revolution.[327]

From there, songbirds became the preferred model of study for understanding human language learning. Since then, thousands of scientific papers have been published on the subject, and our knowledge of how birds learn to sing has become considerable. Frédéric Theunissen, with whom I was observing Marler's sparrow, is one of the scientists seeking to understand the mechanisms of song learning; that is to say, understanding how birds hear, how they memorize, and how they produce sounds. Let's set the stage. At the risk of repeating myself, we are right in the middle of the third question formulated by Nikolaas Tinbergen: the question of the *ontogeny* (i.e., the establishment) of a behavior. With Marler and Tinbergen by our side, we are making solid progress.

It's important to remember that there is no equality when it comes to vocal learning in birds.[328] Only three groups of birds learn to sing by imitation: a category of passerines (oscines, or singing passerines, such as the sparrow, American robin, wren, or zebra finch), parakeets and parrots, and hummingbirds.[329] To give you an idea of numbers, these learning birds represent more than half of the bird species currently living.[330] There are almost 10,000 species of birds on the planet, including just over 4700 species of oscine passerines, about 350 species of parakeets and parrots, and more than 330 species of hummingbirds.[331] You can see that the number of bird species practicing vocal learning is far from anecdotal. Other birds seem to develop their vocalizations innately.[332] In other words, nonoscine passerines (such as the screaming piha, whose leks I told you about in chapter 11) and all other birds do not learn to sing.[333] Their voice comes to them as they grow up, automatically as it were. A nonoscine passerine, the eastern phoebe *Sayornis phoebe*, if artificially deafened at a very young age (we are sometimes cruel, but rarely) produces a completely normal song as an adult.[334] Admittedly, there are a few examples that suggest that these two categories, capable and not capable of learning, are not totally watertight.[335] For example, the three-wattled bellbird *Procnias tricarunculatus*, a large Central American bird of the same family as the piha, emits versions of a nasal song, "aaaiiar!," which differs from region to region. So there are dialects! It has also been said that a young bellbird in captivity learned to imitate the vocalizations of a Brazilian bird, the chopi blackbird *Gnorimopsar chopi*.[336] Thus, vocal learning in nonoscine birds may be more widespread than previously thought.

How does vocal learning in oscines take place? Classically, it is done in two stages: the chick first memorizes the sounds it will have to produce, and then it has to practice and gradually match what it produces to what it has memorized. In the wild, a young chick is traditionally fed in the nest by its parents and therefore hears its father singing, and sometimes its mother, depending on the species—and sometimes adults in the neighboring nest, for species in a colony, for example. At about three weeks of age, the chick becomes able to memorize the songs it hears. At about five weeks, it is independent and goes off to explore other horizons. It is then exposed to the songs of other individuals. Some time later, sometimes even the following year, on its return from winter migration, the young adult bird will establish a territory by singing in turn. This pattern is, of course, variable—ranging from the length of stay in the nest and the age of passage to adulthood to the tendency to be sedentary or migratory—and many factors may differ from one species to another.[337] But the question remains the same: When, where, and from whom did the bird learn its song?

Studies that have explored these issues abound. Much of the laboratory work, inspired by that conducted by Peter Marler, has provided convincing evidence that young songbirds learn to sing by imitating one or more adult song masters, whether real or simulated by a loudspeaker. Our little zebra finch is an ideal species for these studies, a so-called biological model. The development of singing has been particularly well studied in this species, and we have been able to draw from it some fairly complete knowledge. Let's take a look.

The song of the zebra finch is a short sequence of syllables: "tzeek tzeeek tzeek tzeeek didiguezic dziduck." The song of each adult male is a version of this theme. Very standardized, the song does not change throughout the life of the individual. This bird lives, as we discussed in chapter 11, in the semidesert areas of Australia, which do not experience marked seasonal variations, as is the case in temperate zones. Like the climatic conditions, the song of the zebra finch remains immutable. The pairs are faithful for life. The zebra finch thus offers a remarkable model of social stability. In the male zebra finch, the song appears gradually, between 20 and 90 days after hatching. The process begins with a *sensitive period* (or *sensory period*) during which the young bird needs to hear a song from an adult male. This phase occurs roughly between the twentieth and sixtieth day after hatching. It is during this period that the young bird imprints in its memory the song pattern it will try to copy.[338] At the age of one month, the young chick begins to babble; this is the *sensorimotor phase*. At the beginning, its vocal productions are not a real song but a series of sounds that have neither head nor tail, known as a subsong. During the days and weeks that follow, the young bird will sing and sing again, thousands and thousands of times. At the end of the 90 days, the song will be impeccably mastered. Three months to become a professional singer is not too bad! Basically, this learning process is a lot like the way we humans learn to speak. As babies, we're immersed in a bath of words. Day after day, we memorize and memorize again syllables and words. We keep repeating "ba-ba-ba-pa-pa-pa" and other babblings, and then we get better and better at it. The major differences between birds and humans are the length of learning (a few weeks for the zebra finch, many months for humans) and the complexity of the vocalizations learned.[339]

So baby birds learn to sing by imitating adults. Humans are not the only ones with this ability. However, there's a flaw, isn't there? I can hear you from here saying, "That's all well and good, these zebra finch observations are very interesting, but weren't they made in captivity in the laboratory, in very special conditions, far from what birds experience in the wild?" Now you've touched on a real question. The laboratory is certainly a *controlled* environment, as biologists say. It allows us to ask specific questions and to isolate the role of this or that factor. But captive birds are not aware of the wealth of social interactions that their free counterparts experience on a daily basis. Nor are they aware of all their problems, such as searching for food or escaping from predators. So how does vocal learning take place in real conditions? Let's take a tour to Daniel Mennill's home at the University of Windsor in Canada. Mennill and his team were the first to conduct an experiment on song learning in a natural environment, with wild birds, in absolute freedom. Their work, published in 2018 in the journal *Current Biology*, is a landmark study that confirmed that observations made in the lab are consistent with what happens in the birds' real-life environment.[340] Mennill chose the Savannah sparrow *Passerculus sandwichensis* as his subject of investigation. As its name suggests, it is a species quite similar to Marler's white-crowned sparrow. The song of the Savannah sparrow sounds like this: "zit zit zit ZEEEE zaay." Mennill worked in eastern Canada on Kent Island, a strip of land about 3 kilometers long and 800 meters wide that the sparrows are fond of. Researchers installed 40 loudspeakers on the island, which played songs during the nesting season for six consecutive years; in other words—and this was a brilliant idea—six generations of birds bathed in an artificially modified sound environment. Mennill already knew that a young Savannah sparrow learns its song by mixing what it hears from several adult males. He decided that the loudspeakers would emit two types of songs. One had been recorded from adult males living on Kent Island and thus reproduced the local dialect, and the other came from a remote population of sparrows singing a different dialect. A different foreign dialect was used each year. The bet was risky because laboratory studies had shown that young birds generally learn to sing much better when their tutor is physically present, which is why researchers like Sébastien Derégnaucourt of the University of Paris Nanterre use small, birdlike robots to enhance the impact of the songs produced by the loudspeaker. The probability of Mennill's sparrows copying songs simply emitted by loudspeakers was therefore minimal—especially since the young birds were simultaneously in contact with real adults, which they could both hear and see. A real challenge! At the end of six years of observation, however, the results were clear. Not only had the young birds copied songs from the loudspeakers, but dialects previously unknown to this population were passed on to the next generations. Mennill had created a new sound culture on the island of Kent . . . and won his bet![341]

Now that we know vocal learning is not a laboratory artifact, let's examine the role of social influences. A bird does not learn very well if it is isolated in a cage with only a loudspeaker repeating songs. In this situation, it mostly learns in a somewhat automatic way, without any richness of expression, so to speak—a bit like trying to learn to play the piano with a tutorial on the internet, in the absence of a real music teacher. In a remarkable study, David Mets and Michael Brainard of the University of California, San Francisco, experimentally demonstrated how important the presence of a flesh-and-blood tutor is when learning to sing in the Bengalese finch *Lonchura striata domestica*.[342]

By comparing different strains of finches with different singing rhythms (some families of Bengalese finches in Japan sing slowly and others sing quickly), Mets and Brainard found that computer-based learning resulted in strong heritability: a bird sings with the same rhythm as its father and grandfather, whatever the rhythm of the song produced by the computer. In other words, genetics ruled. Inversely, when the sparrows were trained by adult birds—real teachers, if you like—each one adopted a new rhythm inspired by that of the teacher. This time, it was not genetics.

Another study shows that the presence of a female, and the way she reacts to the young bird's vocalization attempts, also greatly influences the learning process. This was an experiment in which young zebra finches saw an adult female on a video screen. These young zebra finches more accurately copied the song of the tutor if they saw the adult female react to their song, and they made more copying errors if the female's behavior appeared disconnected to them.[343] When you see that the audience is satisfied, you learn better.[344]

This leads me to point out that songbirds show great diversity in the way songs are learned. For example, the length of time a bird may learn varies from a few weeks (as in the zebra finch), to the first year (as in the common chaffinch, for example), to a lifetime (as in the canary, starling, and many others). Species that only learn early in their lives are called "closed-ended learners." Species that are capable of learning beyond their first year of life—those that do not have a critical period for learning to sing—are known as "open-ended learners."[345] It is among the latter that we find the most virtuosos.

The variation does not stop at the temporal sequence of learning; it also involves the size of the vocal repertoire. A male zebra finch stubbornly sings only one version of the species' song. He produces the same song all his life, the one he learned once and for all, without changing it one iota. In many other birds, the individuals sing several songs. In 80% of these species, this repertoire remains modest (for example, a great tit *Parus major* sings less than five different songs) or moderate (about ten songs). But in some, the repertoire of a single individual can be quite large (more than a hundred different songs in the nightingale

Luscinia megarhynchos) or even absolutely extravagant: the brown thrasher *Toxostoma rufum* from North America can sing more than a thousand songs![346] Well, it had to memorize them. Do I dare speak of an elephant's memory?

Species of birds also differ in the degree of fidelity in imitating their song. If the copy is never perfectly like the original, it is still very successfully replicated by the zebra finch. By rendering almost exactly what it hears, it's safe to say this is a bird that doesn't go in for originality. Depending on the species, the quality of the imitation can be more or less . . . mediocre. Some birds even riff on the theme of the song they hear.

Finally, there are real composers who simply invent their songs and seem to take no notice of what they hear. Moreover, while some only copy the songs of their own species, there are some that are open to diversity and will not hesitate to imitate the songs of another species. This is the case, for example, with the song sparrow *Melospiza melodia* and the swamp sparrow *M. georgiana*, two North American cousins that are very similar. While the song sparrow only imitates the songs of other song sparrows, the swamp sparrow does not think twice about incorporating the songs of its cousin into its repertoire. The highest level of open-mindedness is reached in certain species. The brown thrasher is a remarkable case, an outstanding imitator with a fabulous repertoire. With more than a thousand songs, it must really like musical diversity. Some can copy almost any sound from their environment, such as the northern mockingbird *Mimus polyglottos* and the famous common hill myna *Gracula religiosa*[347]—a kind of black starling with a yellow adornment on the back of its neck, which I once heard whistling to perfection the tune of the French national anthem "La Marseillaise." However, the prize probably goes to the superb lyrebird *Menura novaehollandiae*, an Australian species. It is a large bird with a splendid tail whose ability to imitate and incorporate environmental sounds into its vocal repertoire apparently knows no limits. It is estimated that 70%–80% of its vocalizations are imitations of other bird species. But it's not just birds that it imitates. I remember an international bioacoustic conference where one of the presenters played a recording of a lyrebird that faithfully reproduced the sound environment . . . of a building construction site! Hammer noises, saw squeaks, orders issued by the workers, nothing was missing. It was so faithfully reproduced that it sounded like the soundtrack of a movie.[348]

How can we explain the diversity of these learning programs? The question is still unresolved.[349] It has been suggested that one of the interests of birds that build their songs by imitating others could fall under the principle of good neighborliness. The sharing of songs between neighbors, i.e., the fact that individuals with territories close to each other sing in the same way, is a principle common to many species. This makes it easy to differentiate a neighbor from a newcomer. Do you remember? We have already talked about the dear-enemy effect in chapter 3 with the black redstart, the favorite model of our friend Tudor Draganoiu. Birds forming leks—like the piha—also apply this principle of sharing songs between neighbors. It could explain the prevalence of species with small repertoires: it is easier to sing the same thing if the number of songs is not too large. On the other hand, the more songs you learn, the easier it is to adapt to a new neighborhood, for example, when you return from migration. In that case, it's a bit like arriving unexpectedly at a party where people are singing: if your song repertoire is large, you'll be integrated more easily. We should therefore see more variety in the songs of species where individuals regularly change neighbors. This is the hypothesis that Don Kroodsma defends.[350]

Kroodsma is a seasoned maestro in the field of birdsong. Now professor emeritus at the University of Massachusetts, he has spent several decades exploring how and why birds sing. As an extremely rigorous scientist, he is known for his extensive field studies and for writing books on birdsong of major importance to both the scientific community and the general public. The sedge wren, one of Kroodsma's subjects of study, provides the impetus for his hypothesis.[351] Let's take a closer look. Sedge wrens (*Cistothorus stellaris*) living in North America are migratory and seminomadic during the breeding season, so they change neighbors often. When one of these birds is trained to imitate a song, it tends to improvise or even invent. In contrast, its cousin, the marsh wren *Cistothorus palustris*, which is particularly sedentary and never changes neighbors, faithfully copies the proposed model. In the wild, marsh wren neighbors do indeed sing similar songs. To complete the picture, the sedge wrens fortunate enough to live in the tropics are sedentary there, so they always have the same neighbors, and this little world sings the same songs. The constant proximity probably leads them to copy each other.

Let's not forget an important element that has hardly been mentioned so far. The females of many bird species sing, especially those in the tropics. The analysis of the evolution of birds shows that in the ancestral species common to all current oscines, both females and males sang. It was only during the evolutionary history of birds that female song seems to have been lost.[352] In this ancestral species, both sexes may have had a vast repertoire of songs that allowed them to defend their territories more effectively as a pair than alone. Sexual selection is also likely an important driver of this evolution, and many argue that the manner in which a mate is chosen plays a decisive role. Here, a partner capable of producing a wide variety of songs will be chosen; elsewhere in other species, more attention will be paid to the stability and quality of its sound production. Moreover, some researchers have ventured to develop the same kind of hypothesis for the evolution of human language: having a complex language could have been favored either by sexual selection (one prefers a partner mastering a vast repertoire, for various reasons), or by so-called kinship selection, with the sharing of information between members of the community (the more one is capable of exchanging complex information within one's family group, the more everyone's survival increases). There are many other hypotheses attempting to explain the genesis of human language. We talk about them later. So let's be careful—especially since, even if one or the other of these explanations is valid in birds, none of them explains all the observations and the diversity of situations, whether it is the importance of the neighborhood, sexual selection, or information sharing. It must be acknowledged that we do not yet know why in one species individuals learn many songs while in another the vocal repertoire remains limited. In short, it is still a matter of artistic blurring, and there is room for new generations of researchers. All are welcome!

Now, let's go back to the very process of vocal learning—in other words, what happens in the brain when the bird learns its song and produces it.[353] The neurophysiological basis of these mechanisms is beginning to be well understood.[354] In the brains of birds, there are groups of neurons (the cells that transmit and analyze nerve information) that are specialized in learning and song production. These groups of neurons, called "song nuclei," are interconnected by nerve fibers. They have names such as HVC, aire X, lMAN, etc. I will not detail their respective roles here, but you should know that they are the subject of many studies.[355] It's not every day that you find a model to study the brain structures responsible for memory and learning, so researchers seized on the birds. The first nuclei were discovered in the canary by Fernando Nottebohm, a former student of Marler and now a professor at Rockefeller University in New York.[356] This "song system" was then identified in all the oscine passerines that were studied. In species where only males sing, it is present only in males. In those species where females also make themselves heard, song nuclei are also found. Among hummingbirds, it's the same story: a study published in Nature, in which Jacques Vielliard took part, reports that hummingbirds have regions in their brains that specialize in vocal control, very similar to those of the oscines.[357] We are probably facing a phenomenon of evolutionary convergence: phylogenetic reconstructions (those studies that make it possible to follow the history of a species) strongly suggest that the common ancestor of oscines and hummingbirds did not learn its songs. Vocal learning would therefore have appeared independently several times in birds. Showing similar brain organizations underscores a structural and functional convergence of brains. This is not uncommon in biological systems: the same problem (learning and producing a song); the same solution (specialized brain nuclei for these functions).[358, 359]

What is interesting and unsettling at the same time is that this convergence also exists between the brains of songbirds and the human brain.[360] Yet our brains are very different in their general organization. Nevertheless, the neural circuit of the songbird brain shows important similarities to the areas that are dedicated to language in the human brain. Like our brain, the bird brain has groups of neurons that control the production of vocalizations stored in the brain's memory, which is called motor control of sound production. You may also know that our brain's production of language is lateralized. To oversimplify, the left hemisphere is more specialized in the production of words and sentences, while the right hemisphere mainly manages intonation. In birds, there is also a dominance of the left hemisphere for the control of vocal production. In any case, it has been found in two well-studied species, the canary and the Bengalese finch. But things are complex. Remember that the syrinx, the sound-producing organ of birds, is a double structure in songbirds—a kind of double whistle, if you like. In the northern cardinal *Cardinalis cardinalis*, the low-pitched part of the notes of the song is produced by the left syrinx, while the high-pitched part is produced by the right syrinx.[361] Let's take a note from the cardinal's song—for example, "piyou!"—which starts high-pitched and ends low-pitched, all in less than a second. Well, "pi" is whistled from the right while "you!" is whistled from the left. Can you imagine the necessary motor coordination? And, of course, this vocal lateralization corresponds to a neural lateralization: each hemisphere controls the half-syrinx on the opposite side. In other species, one of the half-syringes and its hemisphere produce most of the notes of the song; the other side is mute. In the zebra finch, the two hemispheres alternate rapidly, controlling each of the syringes and song production in turn. Hardly simple, that's for sure.

On the hearing side, there are also similarities between birds and humans. In both groups, the processing of sound signals by the brain takes place differently between the right and left hemispheres. Let's skip the details; just remember that the right brain hemisphere of birds seems to mainly process the spectral properties of sound (Is it high-pitched or low-pitched?), while the left hemisphere would rather deal with information encoded in the sound signal, such as the identity of the individual whose song is heard.[362] These results, obtained in a small number of species, such as the canary or the starling, are to be taken with a grain of salt . . . nobody knows if they are valid in other species. However, in the case of humans, we observe similar characteristics: the left hemisphere is more concerned with understanding the meaning of words and sentences, while the right hemisphere deals with context. What I am saying here is, of course, extremely oversimplified, and I will probably be pilloried by my colleagues who are specialists in the neurobiology of language and its perception. The main point is that one can draw both anatomical (brain structure) and functional (process) parallels between the bird brain and the human brain when it comes to studying the production and perception of vocalizations.

In terms of neurons, some amazing things are observed in birds. For example, some neurons activate only in response to certain sounds, certain syllables, or certain combinations of syllables in the song of the species. Other neurons only respond if the bird is listening to the entire song. Frédéric Theunissen has shown that this selectivity appears during the learning process.[363] Birds' brains change when they memorize. In addition, many factors play a role in modulating the activity and development of the song system. Hormones are involved, of course, which in many species will make the number of neurons, and therefore the size of the song nuclei, vary according to the season: more neurons in spring, fewer neurons in winter, for example. Another factor that seems to be essential for song learning is sleep. It is often said that, to memorize something well, you have to sleep on it. It seems that birds follow this rule to the letter. It was in the early 2000s that Ofer Tchernichovski, a professor at Hunter College in New York, and Sébastien Derégnaucourt, then a postdoctoral student in Ofer's laboratory, made this strange discovery—a winning ticket that would earn them a publication in the journal *Nature*.[364]

What had tipped them off was the fact that certain neurons involved in the control of song production in adults were particularly active during the chick's sleep. To be more precise, the neurons of the sleeping chick were as active as the same neurons in an adult who was awake and singing. Would we go so far as to say that the baby bird dreams that it is singing like a grown-up bird? Well, why not? In any case, this result was quite puzzling. Sébastien and Ofer decided to examine in detail how vocal learning progressed during sleep-wake cycles in the young zebra finch. They trained 12 zebra finches to imitate a song from a loudspeaker, and they recorded their vocalizations day after day. One might have expected the chick to progress steadily throughout the learning period. But the whole affair turned out to be more subtle. During the day, as the nestling practiced singing at the top of its voice, its singing became more and more structured; that is to say, it had an increasingly regular and standardized organization of the notes. But every morning, after a good night's sleep, the song produced by the chick had lost much of its organization from the previous day. It's not always best to sleep on it. It is only after two to three hours of morning training that it finds the tune again, and improves on it in the afternoon. And so on every day. But here's the remarkable thing: the birds showing the greatest morning loss of structure were also the ones that, in the end, would best imitate the singing pattern broadcast by the loudspeaker. The morning's vocal doodling that followed sleep was correlated with better learning. It is better to sleep on it! Sébastien and Ofer proposed an interpretation of these results: starting the day with a more variable song could give the bird the opportunity to explore its vocal abilities and improve imitation. Using melatonin, the famous sleep hormone, the researchers put birds to sleep for a few hours during the day. After this forced siesta, the birds had a destructured song. The researchers also found that the singing of young birds that had been made deaf after a few days of learning experienced the same degradation as that caused by sleep. In summary, during the day, the chick trains and improves its performance; in the morning, its highly variable singing allows it to adjust things by realizing how well it is able to control its sound production; at the end of the day, it falls asleep satisfied, having improved its vocalizations; at night, it obviously does not sing and is not likely to hear itself, but it dreams about it! When it wakes up, it is not in top singing form—it does not immediately regain its singing abilities. But its memory has worked well during its sleep. At the end of this new day, by dint of training, it will have perfected its vocal imitation a little more. Sleep helps learning—in birds and humans alike.[365]

Oscine birds are experts at vocal learning. A recent study estimated that the types of syllables sung by the swamp sparrow *Melospiza georgiana* could persist for over 500 years, so efficient is the copying accuracy.[366] The existence of dialects, however, suggests copying errors, and imperfect imitation leads to small changes in the typical song of a population—changes that accumulate, are culturally transmitted from one generation to the next, and may lead to divergence between isolated populations.[367, 368]

An interesting study on this issue has focused on the greenish warbler *Phylloscopus trochiloides*. The range of this small forest bird begins in eastern Europe, extends over western Asia, descends to the Himalayas, skirts the Himalayan massif in the south, rises on its eastern side, and ends in China to come into contact with western Asian populations. I hope you can visualize this kind of loop, circling the Tibetan Plateau, which is too high for forests. It is not for nothing that we speak of this bird as being a ringed complex of species and subspecies.

The song of the greenish warbler gradually changes as you move along the ring. The further you move away from eastern Europe, the more different the song becomes from the typical song of that region. Once the ring is crossed, in the contact zone between the eastern European and western Chinese populations, the dialects are so different that the birds do not recognize each other. Why? Because the colonization of the ring by the warbler populations has taken place gradually from Europe to China via the southern Himalayas. This colonization was accompanied by a progressive variation in song. The acoustic difference in the contact zone is such that the warblers on either side no longer consider themselves to be of the same species: females and males do not pair up.[369]

Copy errors from one generation to the next are probably the main reason for this song drift. Does this mean that nonoscine birds—you know, those that don't learn their song—don't experience any drift in their vocalizations? Of course not. It is genetic drift—the fact that distant populations accumulate small mutations—that causes changes in their vocalizations. Thierry and I took part in a study to compare different populations of the woodcock *Scolopax rusticola*, a very pretty little wader that has left the marshes for forests and moors. We found that, among these nonoscine birds (therefore a priori unable to learn to sing), there were dialects nevertheless: woodcocks from the forests of the Paris region do not sing with the same accent as those from the Azores. It is likely that these populations do not mix and that small genetic differences accumulated over the generations explain their small acoustic differences.

Like human language, the songs of oscine birds are transmitted by imitation from generation to generation, forming true *cultures*. Within

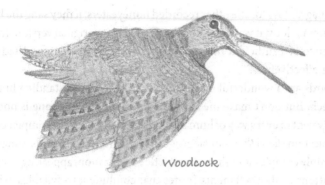

Woodcock

the same bird species, the various dialects that differ between isolated populations are *cultural traditions*, a bit like our different human languages. The parallel with humans does not end there. Just like what happens to human languages, vocal cultures observed in a bird species can disappear. A recent study has shown that an Australian bird, the regent honeyeater *Anthochaera phrygia*, is forgetting the songs of its ancestors.[370] The cause is the decline of regent honeyeater populations and the resulting low density of individuals. Until the middle of the twentieth century, hundreds of honeyeaters used to travel hundreds of kilometers during migrations across all of southeastern Australia. The destruction of their habitat means that there are now only between 200 and 400 individuals left, spread over an area of 300,000 square kilometers! In the regent honeyeater, the young do not learn the songs of their fathers because the latter do not sing during the breeding period. In order to learn the typical songs of the species, the newly emancipated young must join other adult males. And with only about 10 conspecifics in every 10 square kilometers, it is not easy to find company. A recent study of the songs of 146 male regent honeyeaters showed that 18 of them were no longer singing the regent honeyeater song. Instead, they sang the melodies of other bird species, such as those of the little waterbird *Anthochaera chrysoptera*, the noisy friarbird *Philemon corniculatus*, and several others. Deprived of listening to their fellow birds, the honeyeaters learn the language of other bird species—a bit like the character Tarzan who, raised in the fictional *Mangani* ape tribe, had learned their language.[371] As for the other recorded honeyeaters, if they sang the honeyeater melody, it was simplified compared to its original version. In the absence of a sufficient number of tutors, the singing lessons had lost their effectiveness.

Birds are a wonderful model for the study of understanding human speech. But don't make me say what I did not say: birdsong is not the equivalent in every way of human language. Song is a much simpler communication signal than our language. In certain species of birds, songs are certainly complex and show a syntactic organization, appearing as combinations of short elements (notes that combine into syllables, which themselves combine to form sentences whose combination forms songs). However, bird vocalizations show little compositional syntax, i.e., they are not combinations of elementary structures that have meaning (words) that themselves generate meaning (sentences). However, as we will see in chapter 14, birds sometimes combine calls to generate meaning, particularly to designate the identity or size of a predator.

Do you remember the birds that parasitize other species' nests by depositing their eggs in them? Some of these birds are oscine passerines, and the young are supposed to learn to sing by imitating adults. How do they do this when they have been raised by adoptive parents of another species? By studying the brown-headed cowbird *Molothrus ater*, Mark Hauber, a professor at the University of Illinois, showed that these birds have an incredible solution to this problem: the young birds rely on a password to choose who to imitate. As soon as they leave the nest, the young birds recognize the adult cowbirds, without any prior learning, when the latter emit a call (the *chatter*).[372] Then they learn by imitating the entire vocal repertoire.[373]

I can hear you from here: "That's enough for the birds! What about the other animals? The bats? Whales? Dogs? Monkeys? Bugs? What about the others?" It must be said that vocal learning seems to be rare in the animal kingdom.[374] Apart from the categories of birds that I mentioned, learning would seem to be strictly reserved for a few mammals: cetaceans (dolphins and whales), pinnipeds, elephants, bats, and humans.[375] It *would* seem so, considering that for many of these groups the data is still fragile. As a recent paper points out, deciding whether an animal species belongs to the "very select club" of vocal learners is still often a matter of debate.[376] Probably because being capable of vocal learning is a matter of degree: if some species or group demonstrates it without any ambiguity, this capacity is more or less developed in others. Let's take a look at this.[377]

The animal for which we have the most certainty is the humpback whale, which is already familiar to you. You already know that this animal, which is part of the baleen whale group, produces absolutely fabulous, very complex songs. If you've never heard them, listen to *Songs of the Humpback Whale* on the internet. These are the first recordings of humpback whales made by Roger Payne in the 1960s.[378] The humpback whale is a cosmopolitan animal, present in all the oceans of the planet. While the males of a given population all share the same song, each humpback whale population sings in a slightly different way. Moreover, the song of a population is not unalterable; it tends to change. It is a cultural transmission.[379] The whales copy each other; they imitate the songs produced by other whale populations.[380] One population will switch to another's dialect in just a few years' time. How is this possible? It's not that complicated. Whales make colossal migrations, over thousands and thousands of kilometers, which gives them the opportunity to meet individuals from different populations and to borrow some vocalizations that are particularly pleasant to their ears. Also, remember that sound travels well in the water. Even without moving, whales can hear one another singing several hundred kilometers away. It's easy, then, to change their tune![381]

Even though it is one of the most popular and best-studied whales, the humpback whale is not the only one to practice vocal learning. The bowhead whale, to name but one, also produces complex songs that vary from year to year. This species, like many other baleen whales, is most likely a vocal learner. On the other hand, if we look at other cetaceans, such as dolphins, things are a little less clear. However, several observations and experiments have shown that dolphins learn to imitate whistles— artificially produced by computer or emitted by other dolphins—that they did not produce before.[382] There is also the beluga, the famous white whale, which more or less succeeds in imitating human sounds.[383] But we still lack studies that offer results that I believe to be wholly convincing.

However, it is quite likely that dolphins and other toothed cetaceans are capable of learning to produce sounds that are new to them.[384, 385]

Now let's take the pinnipeds and the story of a seal known as Hoover.[386, 387] Having lost his mother, Hoover had been raised from a very young age as a pet, without contact with other seals. When Hoover came in contact with humans, he had transformed his seal bark into humanoid expressions, such as "Hello!" or "Come over here!" While Hoover himself showed some serious vocal plasticity by modulating his vocalizations to "talk," from my point of view he didn't really learn how to vocalize. Instead, this was simply a matter of distorting sounds that the seal was already producing instinctively. You will agree with me, I hope, that in this case we are quite far from the songbird that must imperatively hear an adult to become able to sing—and just as far from the human child who learns to produce thousands of words. In fact, experimental evidence that would demonstrate vocal learning in pinnipeds remains limited. To my knowledge, only one study, conducted on gray seals (*Halichoerus grypus*), has shown that these animals can learn to change the intonation of their voices, to copy human vowels, and even to imitate sequences of sounds of different frequencies.[388] But again, while there is no doubt about vocal plasticity in these animals, vocal learning seems to remain very limited.[389]

In elephants, the situation is somewhat comparable.[390] Since these creatures have brains as big as a large watermelon (5 kilograms, compared to the kilogram and a half we carry with us) and a very complex social life, it is reasonable to think that they have some ability to learn their vocalizations by imitation.[391] But, you see, there is poor evidence of this. Only three anecdotes have been reported, all concerning individuals in captivity. The first was a 10-year-old female African elephant mimicking the sound of a truck. The second was a 23-year-old bull of the same species who had spent his life in the company of Asian elephants—the only other surviving species of elephants—and who reproduced their trumpeting, which was much higher-pitched than that of African elephants. Our bull had taken on the Asian accent, so to speak.[392] The third was a male Asian elephant named Koshik, born in captivity in a Korean zoo in 1990. Koshik had spent the first five years of his life in the presence of two adult females. Then he found himself alone, without an elephant nearby. His keepers and the public who visited him were the only living beings around him. He had been trained to obey several vocal commands, such as "Sit!" and "Down!" When Koshik reached the age of 14, his caretakers noticed an astonishing thing: the elephant was talking! To be exact, he was producing five or six "words." When bioacoustics researchers recorded the elephant, they showed that Koshik mimicked human vocalizations very well. The researchers played the recordings to people who were unfamiliar with the elephant and their work, and the recorded words were perfectly understandable to the listeners. What did he say? Oh, of course, nothing very complex. Koshik simply repeated the injunctions of his caretakers: "hello," "sit," "no," "down," "good." To produce the words, he used the resources available to him: he bent his trunk back and pressed the end to his mouth.[393] The story does not say whether the elephant selected the appropriate word for the situation, but it is likely that by imitating the only people around him, he was trying to establish the social contacts without which every elephant gets depressed. Hoover the seal may have been a similar case. Parrots do the same, as they only learn to imitate us if they are deprived of parrot company. As we say in France, "If there are no thrushes, we will eat blackbirds"; we make do with what we have.

Among bats, the picture is not very clear either.[394] Most of these animals are very social and very talkative. In the Egyptian fruit bat *Rousettus aegyptiacus*, a large bat that you may have seen in a zoo, the vocalizations carry a variety of information: the identity of the sender, the context of the call (aggression, distress, etc.), and even who the vocalization is addressed to. Young fruit bats raised alone with their mothers develop their calls later than when they are in a social group with many individuals.[395] If they listen to calls of their species that have been artificially altered in frequency, they modify their own calls by trying to imitate the artificial sounds. And if you raise young fruit bats in the absence of adults, they take a very long time to develop the vocal repertoire typical of their species. Exposure to playback of adult calls makes their task much easier.[396] Besides, we know that the pale spear-nosed bat *Phyllostomus discolor* requires auditory feedback for normal vocal development and that the baby of the bat species *Saccopteryx bilineata* babbles like human babies.[397, 398] Different dialects between populations of bats of the same species have also been identified. It is not known whether these variations are imposed by genetic differences or whether they emerge from a cultural drift, as in songbirds.[399] Perhaps we are on the eve of great discoveries about bats. Indeed, only 2% of studies on vocal learning concern bats. No one has yet tested whether they are capable of mimicking truly unfamiliar sounds. A while ago, I was visiting my former student Julie Elie, an expert on the calls of the zebra finch, in her new lab at the University of California, Berkeley. Julie is now working with bats and is investigating whether they are capable of vocal learning. During my visit, I was impressed by the experimental device she had set up. She records the individual vocalizations of bats living in groups while filming the animals. She knows who is who, who talks to whom, who talks before or after whom, etc. Her expertise in vocal repertoire analysis makes me feel that, if there's anything interesting to find out about bat vocal learning, she'll find it. Other amazing scientists are also working on the subject.[400] When the results of all of their research converge, we will be able to measure the extent of bats' ability to learn to produce a vocal signal . . . or not. Research is also about confirming that a hypothesis is not validated, that nothing can be found.

Lately, it is the naked mole rat *Heterocephalus glaber* that has been in the spotlight. This small hairless rodent, living in a colony underground, was featured in the magazine *Science* because researchers discovered it has vocal dialects that individuals copy from each other. At last, had we found a mammal that ensures a true cultural transmission of vocalizations like the birds? When we look at the details, the reality is a little less exciting. Mole rats are eusocial mammals, which means that their groups are dominated by a queen and a king, the only ones to reproduce, while the other individuals cooperate in the service of their majesties. These animals are very talkative, constantly peeping, chirruping, and grunting. It must be said that, in the darkness of their galleries, communicating by acoustics is rather a good idea. No less than 17 types of vocalizations have been identified in this animal. The most common is referred to as the soft chirp, which is emitted when two individuals meet at the bend of a meander. It is the mole rat's "hello" in a way, letting the animals know who they are dealing with. In the study, scientists recorded 36,190 soft chirps from 166 animals, belonging to seven different colonies. The analyses

showed that it is possible to guess the identity of the colony of an individual mole rat from its calls alone, suggesting that mole rat families
speak with the same voice. In addition, playback experiments showed
that individuals respond with more calls when they hear a soft chirp
produced by a member of their own colony than when they hear a call
from an individual from another colony; in other words, mole rats are
more likely to say hello to family members than to strangers. Scientists
then found that young individuals placed in a colony other than the one
where they were born adopt the accent of their new family. This final
discovery was made by chance. During the course of the study, one of
the colonies observed by the researchers lost its queen twice. In the ensu-
ing periods of anarchy, the individuals in the colony lost their common
accent—everyone started to call in their own way. In conclusion, the
mole rat does not seem to do better than other mammals: it modifies its
vocalizations according to the ambience of the colony, by copying its
mates—just to please the queen, moreover. This social conformism provides us with a good example of vocal plasticity, but here again, we are
far from the vocal learning capacities observed in birds.[401]

So while most animals don't really learn to talk, many show great
vocal plasticity.[402] In fact, it is quite possible that the number of species
in this category is vastly underestimated, especially in mammals. David
Reby, who knows a lot about the mechanisms of vocal production and
who is also a very observational person—an essential quality for an
ethologist—recently played me some vocalizations of his own dog. He
had recorded the dog's lively barking at the television in response to TV
series theme songs—proof that dogs can be distracted, like restless
children, by putting them in front of a screen. "Listen carefully," he said.
"This is funny. My dog barks along with the melodic line of the theme
music!" And it was true. It was quite perceptible to the ear. Computer
analysis of the acoustic structure of the music and his dog's barking
confirmed David's first impression: a dog singing the theme tune of his
favorite show. Try it with yours, to see.

And what about monkeys and apes, you may ask? Nonhuman primates
are very close to us, aren't they? Our common ancestor with apes is some
six million years old, which is not much by geological standards. On top of

that, nonhuman primates are physically similar to us—comparatively speaking, of course. Almost all of them vocalize, and often they have complicated social lives that require complex exchanges of information between individuals. In short, they have everything to make their language resemble ours, don't they? Well, as surprising as it may seem, and according to the current state of our knowledge, we have to admit the reality that we are the only primates alive today who learn to speak by copying individuals who have already mastered language. Human supremacy at last. The gibbon shows no willingness to learn its screams by imitation.[403] Young squirrel monkeys raised in isolation produce the same calls as if they were raised in the presence of other monkeys. Young macaques raised by monkeys of another species do not learn the calls of their adoptive parents. Finally, superhuman efforts were made during several decades of the twentieth century to try to teach chimpanzees—and even an orangutan—to talk. Some of the people involved in this work reported that "their" monkey had learned to master words.[404] For example, a chimpanzee named Vicki is said to have uttered "daddy" and "mug." I have listened to one of her recordings, and let's just say . . . her pronunciation leaves something to be desired. In short, the results were very disappointing.

However, we must nuance things. A human being raised by elephants or orangutans might try to imitate them, but chances are that this would also be without much success. Wait, I can hear you from here taking me to task and quoting the famous "wild children" and "wolf children" who have been the subject of much discussion and have made great movie subjects. About 50 cases of children "raised" by wolves or bears have been recorded over the centuries. None of them spoke, some were characterized as barking, others growling. All of them reportedly had enormous difficulty in acquiring human language. A little bit of anything and everything has been said about them, to support theories about the innate or acquired nature of human language. In any case, the information has always been insufficient to draw the slightest scientifically valid conclusion. We do not know, for example, at what age they were abandoned or how long they lived with animals. What can be said, beyond fantasy, is that for a human to acquire human language, contact with human speech is necessary from a very young age and in a constant manner.

I don't want to leave you with the impression that monkeys are less capable than birds or whales when it comes to their sound communications. First of all, the vocal repertoire of many primates is very complex, using combinations of vocalizations whose secrets we are just beginning to unravel (and I'll come back to that soon). Their vocalizations are not just set-in-stone genetics. In recent years, evidence has been accumulating that certain species of monkeys modify their vocalizations through learning. In particular, studies of marmosets in captivity have shown the importance of hearing adult vocalizations in order for young animals to acquire a normal vocal repertoire.[405]

One of the first in-depth experimental studies on this vocal plasticity was conducted by two field researchers: Marie Charpentier, director of research at the CNRS in Montpellier, France, and my colleague Florence Levréro. Energetic and determined, Florence spent many years in various African countries studying the lives of great apes, gorillas, chimpanzees, and bonobos. Equally energetic and determined, Marie is leading exciting research on the biology of the mandrill *Mandrillus sphinx*, a sturdy African monkey, whose males proudly display an impressive red and blue snout complemented by a short, yellowish beard. Marie invited Florence to set up an experiment on mandrill acoustic communication. I didn't take part in it—we can't do everything—but I regret it a bit because it must be something to work with this animal. The two researchers and their team were about to make a major discovery. The voice of the mandrills is certainly partly genetically determined, since closely related monkeys have more similar voices than individuals who are distant kin. But it is also shaped by experience: the mandrills take on the accent of the social group in which they live. It was the first time that such vocal plasticity based on the imitation of social peers was demonstrated in a monkey. Furthermore, playback experiments showed that mandrills can recognize by voice related individuals they have never seen before: even if the accent is not the same, they can still recognize their family's song. It's the genetic background that speaks. These intriguing results earned Florence and Marie the publication of their work in the journal *Nature Communications* (which is no easy task).[406] Other similar examples exist in monkeys and apes.[407] With Florence and our doctoral student Sumir

Bonobo

Keenan, we went on to show a few years later that cultural influence also permeates the voice of bonobos, the closest cousin of the human species. Besides, a study carried out on several natural sites in Borneo and Sumatra shows that orangutan populations develop "vocal personalities" that depend on the number of individuals: vocal repertoires are more stable and more complex in populations with a low density of individuals.[408] Other observations, carried out in other monkeys, suggest that the dual influence of genetics and learning on vocal signals is found in many, if not all, primates. The vocalizations of monkeys and apes should no longer be seen as innate and totally fixed.[409, 410]

13

Inaudible speech

ULTRASOUNDS, INFRASOUNDS, AND VIBRATIONS

Goegap Nature Reserve, Northern Cape Province, South Africa. The sun is vanishing behind the rocky mountain. "Follow me!" Céline says, "we're going to set the traps." We walk quickly across the small, dry plain dotted with bushes, placing small rectangular cages here and there. A few minutes later, a sudden slam: The first trap closes on an unfortunate little mouse, too greedy, attracted by the smell of the bait. Céline then opens the trap, delicately takes the little animal between her skillful fingers, notes the number on the tag attached to its ear, and releases it. "This one is a regular. I've been catching her for the past two years." Farther on, a student holds up a large antenna, trying to receive the radio signal sent by the transmitter attached a few days earlier to the neck of another mouse, to locate where it is now. And here's one of them running through the bushes! The African striped mice are scurrying about, trying to find a bite to eat. Sometimes two of them argue over a few seeds. I can hear some sporadic, very high-pitched calls. Most of their conversation is ultrasonic and out of my hearing range. The silence of the karoo weighs on the whole space. The celestial vault is a festival of stars. It's getting seriously cold. A few meters away, a large oryx antelope with swordlike horns calmly grazes on the steppe grass. I feel strangely at home. This is just the place to come to work, I think to myself.

A few days before, I had received an enthusiastic email: "I am Céline, postdoc at the Succulent Karoo Research Station in Goegap Nature

Reserve. If I remember correctly, you'll be at the reserve next week. If you want to see what we are doing in the field, at the moment, we are conducting a mouse trapping session from 8:10 in the morning to 5:00 in the evening. You can come with us if you wish. Our field is right in front of the research station. Or you can visit us at another time during the day if you prefer. It would be super nice to put a microphone near a mouse's nest; if possible we should do this at sunset (17h 45). Hope to see you there!"[411]

I was on a family holiday, and we were on our way across southern Africa from Cape Town, at the southern tip of the continent, to Kasane, in northeastern Botswana—several thousand kilometers of travel—where we would experience the Kalahari nights with the roaring of lions and being awakened by the trumpeting of an elephant bursting into the quiet darkness of our camp. On the itinerary, I had planned to sleep two nights at the Goegap Nature Reserve. Intrigued by the place, I had previously contacted Carsten Schradin, director of research at the CNRS, who runs a field station there to study the life of small mammals in extreme conditions. It rarely rains in Goegap, and only in the summer months. The winter is terribly dry. Carsten is trying to understand how the striped mouse's physiology and behavior enable it to adapt to this desert environment. His research has shown that a drop in corticosterone hormone levels allows the mice to reduce their energy expenditure during the dry season, when food resources are reduced.[412] Energy savings everywhere, even in mice. I asked him if he had ever been interested in the acoustic communications of his little animals. "Apart from making a few recordings, we've never studied their vocalizations," he replied. "Come and see us! See for yourself if there is anything you can do." As usual, guided by my curiosity, I responded favorably to his invitation.

According to Carsten, the striped mouse *Rhabdomys pumilio* has one of the most complex and interesting social systems described in rodents. In many ways, this behavior resembles that of some monkeys. Consider this: these mice live in groups of up to 30 individuals of both sexes sharing a nest in a bush. A real community. At night, they sleep together. In the early morning they separate, each one going off to find food for itself.

Striped mouse

At the end of the day, they return to the nest and seem to enjoy each other's company, saying good night and sniffing each other. Mice from two different nests do not like each other very much: when they meet by chance during the day, they chase each other away. Males are aggressive toward foreign males, while females are aggressive toward both sexes. If an outside mouse approaches the nest, it is immediately attacked.

At the Goegap Nature Reserve, a mouse family consists of several males and females living together in the same nest. With the use of cameras, Carsten has seen that the males actively participate in the care of the young. They keep them warm and maintain and care for their coats. So do the other members of the family. Raising children here is community based. In other regions of Africa where the same species of striped mouse is present, the social structure can be quite different: a monogamous nuclear family (only one female and one male in a pair), a polygynous harem (only one male, several females) . . . all cases exist.[413] The striped mouse is a very good example of social flexibility. In this respect, it is very much like us—families can be nuclear or extended, with many variations in between. Our social organizations can also change over time. None of our other primate cousins show such flexibility. Although the common ancestor we share with the striped mouse is older than the one we have with nonhuman primates, this small rodent represents an excellent animal model for understanding the factors that modulate social organization.

I didn't record anything on my first visit to Goegap because I didn't bring the necessary equipment. It was only a few months later, in our ENES Bioacoustics Research Lab in Saint-Etienne, that the first vocalizations were recorded, using striped mice that Carsten had entrusted to us. Nicolas Boyer, our technical manager, always ready to take up the challenge, and Aurélie Pradeau, the animal keeper, had set up a five-star hotel for the striped mice. My colleague Florence Levréro immediately had an interest in studying this species. As a specialist in acoustic communication in great apes, she was more aware than anyone of the challenge of this study: she had just returned from a field mission in the Democratic Republic of Congo that she had organized to work with bonobos. The mission had been difficult: bonobos are constantly on the move and hide very easily in the virgin forest. Recording them and especially recognizing which individual produces the vocalizations is a real challenge. So you can understand that studying an animal with a complex social system while freeing yourself from the immense difficulties encountered with the great apes was a fantastic opportunity. We decided to tackle the striped mouse adventure together.

The main problem was technical: mice produce ultrasound, and we had neither the equipment nor the know-how to deal with this constraint. Fortunately, Michael Greenfield had joined ENES a few months earlier as a research associate. He had been a professor in France and in the US and had spent a good part of his career working on species communicating by ultrasound, particularly insects. He was the right man for the job and enthusiastically agreed to participate in the project. I am telling you this to show you that science is almost always a collective adventure. Find people who are passionate, competent, and driven by great curiosity, and you have the ingredients; then put them together. That's the recipe.

What do mouse vocalizations sound like? There are two main categories, similar to those we find in birds: calls and songs. The calls are short, emitted when the animal is startled, for example. The songs are a series of very fast, very high-pitched notes (I remind you that mice mainly produce ultrasounds, whose frequencies exceed those perceived by our hearing system): "Twup-twid-twup-twidup. . . ." Different notes almost always follow one another, in a kind of continuous babbling.[414] This may disappoint you, but I must confess that, for the moment, we still have no clear idea what striped mice are saying to each other. Deciphering their language is one of my goals for the next few years. The next mission is already planned: I'm leaving soon for Goegap with Florence, Michael, and our new PhD student Leo Perrier to put ultrasound recorders on the mice domain. Our first hope is to map their communications throughout the day, with the hypothesis that the nests and their immediate surroundings are privileged spaces for exchanging information about food-rich places. You may find it strange how little we know about the acoustic communications of a rodent that is quite common in the wild. I see two explanations. First, since we humans do not hear ultrasound, we are not very interested in it. Second, they are more complicated to study than audible sounds. It was not until the late 1930s that Donald Griffin and Robert Galambos discovered that bats emit ultrasound.[415]

Why use ultrasound to communicate? Scientists currently agree on several reasons: to allow precise localization of the origin of a sound, to avoid being heard by predators, to avoid ambient noise (as in the case of ultrasonic frogs living near streams, which we talk about later), and because the transmission of high-frequency sounds is more effective for small animals.

To locate the source of a sound (e.g., a baby mouse who has strayed from its nest), whether it's an ultrasound or a sound audible to our ears, the receiving individual (the mother mouse) compares the sound information coming from each of its ears. If the baby is on the mother's left, she will hear its calls louder with her left ear than with her right ear. The mother's head acts as a filter that absorbs some of the sound waves. Conversely, if the baby is on the right, the mother's right ear will receive more sound energy. And, of course, if the baby is in front of the mother, both ears will receive sound waves of the same intensity. This *interaural difference in intensity* between the two ears provides valuable information about the location of the sender. These differences in intensity are particularly noticeable if the wavelength of the sound vibrations is small compared to the size of the animal's head. In other words, for a small animal such as a mouse, very high-pitched sounds (of very short wavelength) will be useful for perceiving interaural differences in intensity. Ultrasound, therefore, is perfect here.

Sound intensity is not the only parameter that differs between the two ears if the sender is to one side. Indeed, if the sound wave comes from the left ear, it will arrive earlier in the left ear than in the right ear. The time difference between the two ears is simple to calculate. Remember what I told you in chapter 2: sound waves travel at 340 meters per second. Since the two ears of a human head are 22 centimeters apart on average (or 0.22 meters), a sound wave coming from the side that hits the first ear will hit the other ear $0.22/340 = 0.0006$ seconds later. This *interaural difference in time* of a little less than one-thousandth of a second is certainly small, but it is enough to induce a shift between sound waves. This shift is a second cue analyzed by the brain to locate the position of a sound source. Remember this: a sound is a pressure wave that travels. Air pressure rises, then falls, then rises again very quickly. The rhythm of these pressure oscillations is the frequency of the sound. The faster they are, the higher the frequency and the higher the sound. If you have understood correctly, you will realize that the shift between the two ears will be easier to perceive if the frequency is lower. In fact, imagine a sound wave coming from the left. It hits the left ear first and then reaches the right. If the oscillations of the wave are fast (high-pitched sound), in other words, if the pressure changes very quickly, the left and right ears will have difficulty feeling pressure differences. On the other hand, if the oscillations are slow (low-pitched sound), the left ear will feel an increase in pressure before the right ear, and when this increase reaches the right ear, the left ear will already feel a decrease in pressure. And so on. At any time, the difference in pressure between the two ears will be more noticeable with a low sound than with a high sound. So, in mice? Well, the explanation is simple: mice have very small heads. Moreover, the smaller the head of the animal, the smaller the time difference will be. In a mouse whose two ears are only 9 millimeters apart, the time difference is $0.009/340 \sim 0.00000002$ seconds for a sound coming from the side. Therefore, interaural differences in time are not a very useful parameter for mice. It is better for them to rely on interaural differences in intensity.

This is the first reason for using ultrasound: to locate easily and to be easily locatable. The mother mouse will thus find her straying baby very quickly. Another property of ultrasounds, as I mentioned earlier, is that their energy is progressively absorbed as they are propagated in the environment. The amount of energy loss depends on the frequency of the sound; high-pitched sounds lose energy very quickly. This makes them much harder to propagate than low-pitched sounds. In other words, they are attenuated much faster and become inaudible more quickly. Ultrasound is by definition a hyper high-frequency sound, so it does not propagate very far into the environment. To give you an idea, calls from baby mice are no longer recordable beyond a few tens of centimeters. The ultrasonic signals emitted by the babies therefore will not be perceived by a possible predator, unless it has already discovered the nest. This is reinforced by the fact that rodent predators, such as birds and snakes, are not able to hear in the ultrasonic range. Even when the predator is able to hear ultrasounds—which is the case with some carnivora, such as foxes—the signal will only be perceived by listeners close to the youngster calling out: the mother and sometimes other individuals taking care of the young, such as the father. Using ultrasound to send private messages—that's pretty smart. This may explain the use of ultrasound by many rodents.

We do not yet know what the African striped mice say to each other, but we do have some ideas. Our knowledge of rodent ultrasound communication is becoming substantial, although over the past 20 years, research on rodent ultrasound communication has focused mainly on captive species: white mice and laboratory rats.[416] Let's take the latter first. In the rat, there are three main types of vocalizations: calls emitted by the young when they are isolated (around 40 kHz) and to which the females respond by picking up the young; calls emitted by adults in contexts of fear—in the presence of a predator, for example—or aggression (around 22 kHz); and attraction calls in the context of play in the young or interactions between sexual partners in adults (with a very variable frequency, between 30 and 90 kHz, centered around 50 kHz).[417] In the white mouse, things are very similar. You also find calls of isolation in the young, calls of fright, and other calls made in contexts of social interaction.[418] The males emit long sequences of calls, all different. By analogy with birds, we even talk about the "song" of the mouse! Moreover, for a few years, scientists thought they had found in the white mouse a model for language learning that would have advantageously replaced songbirds: mice are mammals, therefore closer to the human species than birds, and they are much easier to breed in the laboratory. However, despite numerous attempts and experiments, it seems that mice vocalize innately and show no particular plasticity in their calls.[419] Mouse song is therefore not a substitute for birdsong in the study of language learning. But that's not the point. The ingenuity of scientists knows no bounds. If mouse vocalizations are genetically determined, then they can be used as a phenotypic marker of genetics, just like any visible morphological or anatomical characteristic, such as eye color, for example. Vocalizations are in fact a marker of embryonic development, motor skills, nervous control of muscles, and much more. The study of ultrasound vocalizations in mice thus makes it possible to study cerebral processes and their regulation, emotional states, motivation, and many pathological processes.[420] Did you know that some mice showing signs of autistic or schizophrenic syndromes sing differently from the others? With all the genetic tools available for modifying laboratory mice, it is now possible to unravel the processes leading to these syndromes. I won't take you any further on this subject, but you should know that the bioacoustics of laboratory mice probably has a bright future. As for our striped mice, I bet they will reveal some exciting secrets. Maybe we'll be able to understand the meaning of their song. Maybe they will become a model for studying complex acoustic communication networks. It's quite exciting to think that no one has yet been able to understand what they're saying to each other.

In the meantime, what else can I tell you about ultrasonic vocalizations? First of all, rodents aren't the only animals that produce them. Many insects, some frogs, some bats, as well as cetaceans (dolphins and other toothed whales) are also known to produce and detect ultrasound. While the use of ultrasound by some animals no longer surprises us—after all, ultrasound is simply too high-pitched for our ears—it was only in the twentieth century that it was identified. In the early 1930s, George Pierce, a physics professor at Harvard University who had built an "ultrasound detector," was the first to show that some grasshoppers in the Tettigoniidae family produce ultrasounds.[421] Then, in 1938, Donald Griffin, a Harvard student who was interested in the spatial orientation of birds and bats, brought a few bats into Pierce's laboratory. Using the famous detector, he found that bats emitted ultrasounds when they flew. He then showed that these bats orient themselves in the environment and hunt their prey by relying on the echo of their ultrasonic emissions.[422] Since then, research on bat echolocation has proliferated.[423] I want to be clear that not all bats practice echolocation. But the Microchiroptera, which represents 90% of bat species, uses echolocation, along with fruit bats of the genus *Rousettus*. The other fruit bats (Megachiroptera) are not known for this. The frequency of the echolocation calls varies between 8 and 215 kHz according to the bat species. We are in the ultra of ultrasound at the high end here.

How does echolocation work?[424] The principle is not very complicated. The bat emits a series of calls, each of which is of short duration: "Click-click-click-click-click. . . ." The rate at which clicks are emitted varies greatly from one species to another—from 3 to 200 clicks per second.[425] The rhythm also varies according to the moment: when the bat flies quietly in search of prey, it emits between 3 and 15 clicks per second. Echolocation is a mechanism that primarily allows the bat to navigate *by hearing* in the dark. Very useful for avoiding trees and other obstacles.[426]

When the prey is spotted, the bat approaches the insect. In its terminal phase, when it is about to capture its prey, it considerably accelerates the rate of clicks. Each of these clicks—the term used to designate them is *pulse*—spreads in front of it. If a pulse encounters an obstacle (for example, a tree branch or an insect in flight), some of the sound energy is reflected and an echo is sent back in the opposite direction, reaching the bat's ears. Since the sound propagates at about 340 meters per second, the time between the moment the bat calls and the moment it hears the echo is very short. Imagine that the obstacle reflecting the pulse is 1 meter away from the bat; the sound will travel that distance in 1/340 of a second, or about 3 milliseconds. If we add another 3 milliseconds, corresponding to the time it takes for the echo to return to the bat, we obtain a time delay of 6 milliseconds between the transmission of the call and the reception of the echo by the animal's ears. If you have followed this, you will have realized that by listening to the echo of its call, the bat accesses important information: the distance that separates it from the reflecting obstacle. Numerous experiments have shown that bats are able to gauge extremely short delays between the moment of emission of the pulse and the reception of its echo, and therefore to estimate very close distances. Part of the bat brain is even dedicated to analyzing these *pulse-echo* delays.[427] This auditory brain area is spatially organized, from one end where neurons are found reacting to pulse-echo delays of around 18 milliseconds, to the other end where neurons are specialized in analyzing delays of around 1 millisecond. Those cerebral neurons that perceive pulse-echo delays calculate the distance between the bat and the obstacles around it. The activation of the brain area containing these neurons represents a kind of map of the environment. But the distance to the reflecting obstacle is not the only information the bat extracts from the echo it receives. Experiments conducted in the laboratory on bats trained to distinguish between two reflective targets have shown that their resolving power is extremely high—they are able to differentiate between two objects placed side by side with a time spacing of 2 microseconds (0.000002 seconds).[428] That calculates as $0.000002 \times 340/2 = 0.00034$ meters, i.e., they can differentiate if the objects are separated by a distance of less than half a millimeter. This extreme ability allows the bat to analyze the texture of the reflecting object: the scales on the wings of a moth will not reflect sound waves in the same way as the hairs on a fly—useful information if moths are its favorite dessert. The bat extracts much more information from the echoes it receives, including the size of the reflecting object: the bigger the object, the more energy it will reflect back. This allows it to choose larger prey. Another piece of information is the position in space of the reflecting object, which the bat can determine by comparing the echoes coming to each of its ears. It does this by using the two processes described above: the interaural difference in intensity and the interaural difference in time. We are not going to go much further in the study of bats: it is a complex world that goes beyond our present objective. But I can't resist the temptation to explain to you a little marvel of behavioral adaptation, which I regularly present to my students—always to their amazement. It's a phenomenon known as *Doppler effect compensation*.

You're familiar with the Doppler effect, perhaps without knowing the name. In any case, you've already experienced it. When you are walking around and a car is passing by and honking its horn continuously, or when an ambulance passes you with a screaming siren, you may have noticed that the sound of the horn or siren changes as the vehicle approaches or moves away from you. As the vehicle approaches, the sound seems higher-pitched, and it becomes deeper once it passes you. Yet it is the same horn. When you honk the horn continuously in your car, the sound doesn't get higher or lower over time. It's the fact that the car is approaching and moving away that changes the game. This is the Doppler effect. How do we explain it? Imagine the sound waves of the horn as circles gradually moving away from the car in all directions. If the car is stationary, these circles remain evenly spaced: the wavelength and therefore the pitch (low or high) of the sound remains the same in all directions. On the other hand, if the car moves forward, the sound waves at the front of the car are compressed: the spacing between two waves (the wavelength) is reduced, so the frequency increases and the sound becomes higher. At the same time, the spacing at the rear of the car increases, so the frequency decreases and the sound becomes lower. In short, an approaching sound source will sound higher than a sound source moving away. Let's get back to our bats and look at the next lab experiment. A bat is in the dark, resting on a perch. In front of it is a stationary pendulum—a ball hanging from a string. The animal's calls are recorded ("click-click-click . . .") at a rate of about 15 per second. They are emitted at a constant frequency of 61 kHz, characteristic of this bat species. Then the pendulum is made to oscillate, moving closer, then further away, then closer to the animal again. Surprise! The frequency of the calls then starts to vary with the movements of the pendulum: when the pendulum accelerates as it approaches, the bat calls become more and more low-pitched—down to about 59.5 kHz; then its calls gradually return to their basic frequency (61 kHz) until the pendulum is at its closest. None of this happens when the pendulum moves away: the frequency remains unchanged. It only starts to drop again when the pendulum approaches again. Why this unusual behavior? Because of the Doppler effect. The pendulum is an obstacle that reflects the echo of the calls made by the bat. An echo is always of the same frequency as the original sound. So when the pendulum is stationary, the echoes are at 61 kHz. But when the pendulum approaches, it becomes a moving sound source—like the car coming toward us—and the echoes become higher-pitched. The bat hears it immediately and lowers the pitch of its calls. Its goal: to keep the echo at 61 kHz. Neurophysiological studies have shown that these bats have heightened auditory sensitivity at this frequency. By forcing the echo to remain fixed at this value, the bat's hearing is optimized.[429] As the pendulum accelerates and then decelerates, the bat lowers and then raises the pitch of its calls following the movement to neutralize the Doppler effect. Doppler effect compensation therefore requires hearing the echo from obstacles and correcting the calls produced almost in real time. Let's put ourselves in a real bat's life: When it chases a moth, it must constantly adjust the pitch of its calls according to its relative speed—depending on how fast it is approaching the prey it is aiming at to find out where it is. In the experiment, note that it does not do this when the pendulum moves away, and this is quite normal. If the bat does not approach the reflecting obstacle, it is not interested in it! Other studies have shown the existence of another area in the bat's brain dedicated to measuring the Doppler effect. Concretely, this is where neurons analyze the bat's relative speed in relation to its prey and trigger behavioral responses, such as changing the pitch of the calls emitted, thus enabling the bat to fly and hunt.[430] One last thing: When a bat hears the echo produced by a moth, the Doppler effect also exists on a smaller scale. In fact, during flight, the moth's wings alternately rise and fall back. Hold on tight: if the bat comes from the side, a rising moth wing moves away while a wing that folds down comes closer. So there will be a small extra Doppler effect due to the flapping of the wings. It has been shown that bats are sensitive to this and thus gain information about the type of moth they are chasing: large wings that flap slowly will not produce the same Doppler signature as small wings that flap quickly.[431] Unbelievable, but true.

In the 1950s, a fascinating discovery was made about moths, the favorite victims of many bats: these animals are able to detect the ultrasonic calls of bats and flee from them.[432] These insects have developed "ears" during their evolution that can hear them. This regularly saves their lives. The ears of moths are called *tympanic ears*.[433] What is remarkable is that ultrasound-sensitive tympanic ears have appeared several times, independently, in the history of moths: sometimes on the thorax in Noctuidae, sometimes on the abdomen in Pyrales, or even around the mouth in some Sphinx. The sensitivity of the tympanic organ of a moth species often corresponds to the acoustic characteristics of the vocalizations of bats living in the same place. And the moths with the most sensitive ears are also those that live in environments where bat density and diversity is greatest. These ears are very simple structures: a small membrane to which one to four sensory cells are connected. They are sensitive enough to detect ultrasound before the bat has received its echo.[434] In reaction to the ultrasound, the moth changes direction and sometimes performs acrobatic maneuvers to thwart the attack. In addition, some moths are also capable of producing their own ultrasounds, probably to disrupt bat echolocation by adding noise.[435] These sound signals can also startle the bat or even warn it that the moth is inedible—many species of moths contain toxins. This warning value has been shown experimentally: bats avoid moths that emit ultrasound and prefer to hunt silent species. However, if the sound-producing organ of these moth species is suppressed, bats will start to hunt them just like other moths. Let us note in passing that bats also have their own predators ... and that they can deal with them acoustically! A recent study shows that the greater mouse-eared bat (*Myotis myotis*), when captured, imitates the sound produced by bees or wasps to discourage birds it could be the victim of.[436]

Faced with these adaptations developed by moths in the course of their evolutionary history, bats have not remained without an answer. For example, some bats have lowered or, on the contrary, considerably raised the pitch of their ultrasonic calls so as to move outside the range of frequencies perceived by moths. A remarkable adaptation is that of the western barbastelle *Barbastella barbastellus*: this bat feeds only on moths that are capable of hearing ultrasound and has become capable of transmitting its signals and receiving the echo at an intensity 10 to 100 times lower than other bats. This ability gives it a definite advantage: it will only be detected by the moth when it is very close to it, ready to bite it. Realize that this battle between bats and moths has been going on for more than 60 million years. And it is not unique: other groups of insects—crickets and grasshoppers, mantises and beetles—have also developed ultrasonic ears under bat pressure.[437] While these adaptations allow prey to escape from predators, there are also systems that have evolved to facilitate the work of bats. Some tropical plants have developed umbrella-shaped flower shafts in the form of ultrasonic reflectors, making it easier for nectar-drinking bats responsible for their pollination to spot them. However, if a small ball of cotton wool is placed in the hollow of the flower's pavilion, the ultrasonic mirror is masked, and the plant will not be visited by bats at night and will not be pollinated.[438, 439]

Among the other animals that produce ultrasound, frogs have long kept their ability secret. The first ultrasonic frog was only discovered in the early 2000s by Peter Narins of the University of California, Los Angeles, and his colleagues at the Chinese Academy of Sciences. *Amolops tormotus* is an endemic species of the Huangshan Hot Springs region of China—it is found nowhere else. While most amphibians—including frogs—have limited hearing ability and generally do not hear beyond 12 kHz, *Amolops tormotus* perceives sounds up to 30 kHz, and possibly beyond. This animal produces melodic songs made up of an incredible variety of notes—vocalizations that are unusual for a frog. Its sound productions are closer to the songs of birds than to the usual repetitive croaking of other frogs and toads. Moreover, its tessitura extends far into the ultrasonic range, i.e., beyond 20 kHz (remember that 20 kHz is the audible limit for the human species).[440] Peter and his colleagues wanted to know if these ultrasounds were simply a side effect of a particular production mechanism or if they were a genuine adaptation to avoid the rather low-frequency noise of the torrents near which this frog lives. The scientists conducted playback experiments in the wild with computer-modified songs, and showed that the ultrasonic part of the songs was sufficient to provoke a vocal response from the frog. Electrophysiological recordings made in their brains showed that they are actually able to hear the ultrasound. Ultrasonic communication of *Amolops tormotus* is therefore probably a means of avoiding the noise of torrents.[441] Along with bats, this frog provides a good example of evolution leading to the production of ultrasonic vocalizations.

While most frogs do not produce ultrasound, almost all make extensive use of acoustic communications. Our knowledge of their sonic world is constantly increasing. Recently, Peter, with his vivid imagination, told me about his research on a tiny yellow frog, endemic to Guyana and belonging to the Dendrobatidae family. While we were together at the International Bioacoustics Conference, he took out the digital tablet that never leaves him and showed me pictures of his latest expedition. "Do you see that tiny bug? It's the golden rocket frog. *Anomaloglossus beebei*, as it is called. It croaks in the air, resting on these broad leaves of bromeliads. But its croaking also makes the leaf on which it's resting vibrate." And then he explained that *Anomaloglossus beebei* males are used to emitting powerful vocalizations—a short series of three repeated calls—from the surface of the large leaves. These calls attract females and repel competing males. The surrounding males, resting on other leaves, respond aggressively. Peter and his colleagues hypothesized that, in addition to audible airborne sound waves, this frog's call was translated into leaf vibrations and that these vibrations were signals to other frogs. They brought a portable laser vibrometer into the field (Peter will stop at nothing) and measured the frequency of leaf vibrations when a frog vocalizes. Better still, after real-time processing on a laptop computer, the recorded vibrations were reproduced at the end of a small stick, a sort of minishaker, placed on a leaf about 20 centimeters away from a frog. Peter and his colleagues hoped to make the animal believe that another male was vibrating its leaf. And indeed, all the frogs subjected to this device were fooled: they all moved toward the vibrating end of the shaker and began to produce longer series of calls. They adapted their response to the vibrating stimulus, waiting until it was over before responding. So the little yellow frog uses both airborne sound waves—which we humans perceive—and leaf vibrations that are inaudible to our ears.[442] Probably, airborne sounds allow long-distance communication, such as "Hey, that's my neighbor across the street," while vibrations signal a real intruder ("What the hell is this one doing on my leaf?"). Many other animals, including cicada-like insects, use the vibrations of a substrate—a leaf, a branch, the ground—to transmit sound waves that are imperceptible to us.[443]

For social insects—ants and termites in particular—drumming can be a way to set off alarms in the colony. Let's take a look at termites, those small blind insects that live in colonies that can easily number tens or even hundreds of thousands of individuals. Termites are vegetarian; their life depends on the exploitation of wood resources and plant debris that workers sometimes look for far away from the center of the colony. The surface area used to find food is usually 2000 square meters. Many species of termites grow and feed on a fungus, and the plant debris they search for is then used to fuel the growth of the fungus. A termite colony consists of breeding individuals (queen, king), workers, and soldiers. It is a kind of superorganism based on cooperation between all the individuals that are part of it. All members of this little world communicate, of course, through chemical signals—called pheromones—and mechanical vibrations. Of the 2600 species of termites, it seems that all of them use vibration as a means of communication.[444] Although our knowledge of vibratory communications in termites is still patchy, it is certain that vibratory signals play a major role in alerting the colony to a danger—whether that danger is a predator that eats termites (and there are many) or a simple fungus that threatens to rot the one being grown by the small animals.

It is usually with the head that individual soldiers (and workers in some species) drum repeatedly when disturbed. By banging their heads against the walls of their gallery, they make them vibrate, and this alarm signal attracts other soldiers as it pushes the workers deeper into the nest. Some termites cause their abdomen to vibrate, and the waves are transmitted by leg or body contact with the gallery. The vibrations that propagate along the gallery walls are a kind of micro drum, "Trtrtrtrtrtr-trtrtrtrtr-trtrtrtrtr-trtrtrtrtrtrtrtrtr, . . ." emitting sounds at a rate of 10 to 30 drumbeats per second, depending on the species. In the termite *Macrotermes natalensis,* the soldier raises his head to a height of 1 centimeter and then lowers it extremely quickly, hitting it on the ground at a speed of 1.5 meters per second. It is not known whether these drummings carry accurate information about the level or nature of the hazard. Ultimately, they appear to be fairly constant, except when the danger is major. When the termite mound is attacked by a predator, for example, pheromones are added to the vibrations, and the combination of the chemical molecules and the vibrations increases the aggressiveness of the soldiers by tenfold.

I said that the vibrations produced by termites are transmitted along the walls of the termite mound galleries. Alas, the vibratory waves grow weaker very quickly during propagation: after 40 centimeters, they can no longer be perceived. How can the information manage to travel through the tens or hundreds of meters of galleries in which the colony lives? Well, you see, termites have a relay system. Are you familiar with the Great Wall of China's warning system? When an enemy was in sight, a fire was lit on the top of the watchtower from where it had been sighted. In response to that fire, the guards on the neighboring tower about a kilometer away would light a fire as well, and so on, spreading information along the entire wall and allowing for quick mobilization of troops. Termites follow the same strategy. Soldiers are distributed along the galleries. When one of them drums the alarm, the closest ones perceive the signal and hurry to drum themselves. The information is transmitted rapidly—between 1 and 2 meters per second—in all directions throughout the colony, which responds to the alarm in a few seconds. The attenuation of the signal during its propagation is thus reduced to zero.[445] Isn't it brilliant? This *social amplification of information* has not often been detected in animals.[446] The day I first heard about this strategy, I regretted not having devoted my career to termites. Their worlds are fascinating. One last thing: The termite that perceives an alarm vibration must make an important decision. In which direction should I go? Up or down the gallery? The soldiers will choose to go to the source of the problem, while the workers will run away from it. But how do they all know where the vibrations are coming from? The solution is found in their legs: the termites perceive the time difference between the vibrations arriving at their left and right legs—a process similar to the perception of the interaural time difference we were talking about earlier, except that here the difference is not perceived by the ears but by the legs.[447]

Besides insects such as termites, spiders (which are not insects but arachnids) are also known to make the ground or the leaves they walk on vibrate, especially during courtship. One day, we had the pleasure of welcoming to the laboratory Noori Choi, a student who was preparing his thesis with Eileen Hebets, a professor at the University of Nebraska in the US and an expert in spider communication. To record the vibrations that the spider produces, Noori places the animal on a special type of paper on which the sound waves will propagate and records them with a laser vibrometer, like the one Peter Narins uses with his frogs. The video recordings that Noori showed me were amazing: depending on their species, the spiders had different dances, waving their legs in a regular and rhythmic way and tapping the paper in a seemingly controlled manner.[448]

And what about bigger animals? Can we find this kind of communication in them?[449] In mammals, an example is found in mole rats, whose vocalizations we talked about in the preceding chapter. There are several species of these rodents living in underground galleries. In some, the individuals make the earth vibrate with their heads, while for others it is with their feet. Their ears show adaptations to hearing low frequencies, and these animals respond to the underground vibrations produced by their fellow mole rats by hitting the ground in reply.[450] But rather than detailing what happens in the mole rats, for which our knowledge remains rudimentary, let's see what happens in another fascinating animal—the elephant. And let's move on to the field of very low-pitched sounds, so low that we cannot hear them. These are infrasounds, which are to low-pitched sounds what ultrasounds are to high-pitched sounds.

Elephants, which as everyone knows are big animals, can produce very powerful sounds. Emitting more than 100 decibels (almost the sound level of a jackhammer) is not a problem for them. They can also make very low-pitched sounds: around 20 Hz.[451] Their vocalizations are so powerful that the sound propagates on the surface of the ground independently of air waves and makes the ground vibrate.[452] These seismic waves can be considered—proportionally speaking—as small earthquakes, which can be perceived from a long distance by other elephants.[453] How do they spot them? Two main mechanisms may be involved. On the one hand, the skin on the soles of the elephants' feet has sensory cells—mechanoreceptors—that are sensitive to pressure variations and therefore to vibrations. On the other hand, the vibrations travel along the leg and shoulder bones up to the jaw and inner ear bones, where the sound waves are analyzed by sensory cells connected to the brain. The elephant's foot has a special anatomical feature: a mass of fat and cartilage between the arch of the foot and the toe bones. This mass may act as an *impedance adapter*, facilitating the transmission of vibrations from the ground to the bones.[454] This structure is very reminiscent of the fatty masses found in the heads of whales, which also facilitate the reception of sounds. When an elephant listens to vibrations from the ground, it adopts a particular posture: motionless, it puts more weight on its forelegs. Ground vibrations propagate in principle a little more slowly than airborne sound waves—around 230 meters per second as opposed to 340 meters per second. The further away the sound source is, the longer the time interval between the arrival of these two components will be. This may inform the receiving elephant of the distance separating it from the sender. In addition, seismic waves propagate over a greater distance than air waves, and with less attenuation. Vibrational communication may allow elephants to communicate at distances of more than 2 kilometers without any problem, and maybe up to 10 kilometers. Is there any advantage in being able to communicate over such distances? Elephants are animals with a very complex social organization. Females form family groups in which individuals know each other well. They also know members of other family groups. These recognitions are based on both acoustics and olfaction. Karen McComb, a professor at the University of Sussex in the UK, has shown that an adult female elephant can know the vocal identity of 14 different families, for a total of about 100 individuals.[455] Females of different families stay in contact through their vocalizations, both aerial and ground. Males, on the other hand, may join a herd of females, group together, or remain solitary. They use acoustic signals in the infrasound as well as those that are audible—we all know the trumpeting elephants are capable of—when searching for sexual partners or during competitions between males.

Ovulating females vocalize more, which attracts males. The vocalizations of males ready to breed drive others away. All this can be done at a distance, without the animals seeing each other. Infrasounds, whether propagated by air or by ground vibrations, are excellent vehicles for long-distance information.[456]

Ultrasounds, infrasounds, vibrations of leaves, branches, or the ground —all are sound worlds inaccessible to our senses. We can only imagine the bat perceiving its environment and its prey through the echoes it receives and the elephant communicating with another elephant family a few kilometers away.

14

The laughing hyena

COMMUNICATIONS AND COMPLEX
SOCIAL SYSTEMS

South Gate Campground, Moremi Game Reserve, Botswana. Night. Half asleep in the tent on the roof of the car, I can hear the fuss being made nearby. The trash can in the campground has been knocked over and the garbage is being searched noisily. There's giggling, squeaking, blowing, and whistling. I cautiously open the tent zipper and turn on my torch. In the beam, I see two of them—honey badgers, savagely plundering everything they can. Suddenly, a growl, both muffled and powerful, threatening, long. Freeze frame. The feast has been cut short. The two angry badgers call out in rage and jump back at the sight of a huge spotted hyena, determined to take advantage of the situation. The badgers are formidable beasts that, it seems, do not hesitate to ward off even an elephant; but the jaws of the queen of the night are the strongest, and they know it. The hyena dashes over to a piece of rubbish and then trots away. It'll probably join its family—I can hear them giggling and whooping in the distance. The badgers, relieved, start their ruckus all over again. The sounds of the African night! On another night in a camp further north, I'll be woken up too, but by the mighty trumpeting of an elephant. In the south, at the Polentswa campsite in the Kgalagadi, the lions had roared all night. My camera trap attached to the bumper of the car showed the next day the passage of a lioness followed by a big male—a nightly stroll of the big beasts close by. During these nights, as

Hyena

during the days, far from the hustle and bustle of the university, endless meetings, administrative paperwork, and my computer screen, I always feel in my element. Excited but serene. My second life will be here, I decided one day.

The spotted hyena *Crocuta crocuta* is carnivorous like its two hyena cousins, the brown and the striped.[457] Contrary to what you might think, it is a great hunter and does not deserve its reputation as a scavenger. Spotted hyenas hunt and kill their prey, just like lions, wild dogs (those beautiful striped carnivores), and other predators. I am not saying that a hyena retreats in front of carrion—my nighttime experience shows that it does not hesitate to look for easy food, which is quite understandable. But most of the prey that hyenas devour they've caught themselves. Hans Kruuk, a famous ethologist from the University of Aberdeen in Scotland, studied hyenas in the wild and showed that 80%

of the prey eaten by the Serengeti lions in Tanzania was actually killed by hyenas.[458] So they get robbed more often than they steal. The respective reputations of these two species ought perhaps to be more nuanced . . . the king of the savanna has a few unwitting helpers. Hyenas regularly hunt in groups, skillfully cooperating to isolate and kill their victims. The hyenas can then kill large animals, such as zebras or wildebeests—those antelopes that migrate in vast herds that you've probably seen in animal films crossing rivers where crocodiles are waiting for them.

The spotted hyena is not a solitary animal; it loves large families—very large indeed since a clan of hyenas numbers between 6 and 90 individuals.[459] Family meals are obviously very hectic. Not only do you have to eat before a lion comes and ends the feast, but you have to contend with the members of your family in intense competition over the carcass of the unfortunate wildebeest. Each family member devours as much food as possible. The giggles from frustrated individuals who are struggling to find a place at the table are everywhere.

In many ways, the social organization of spotted hyenas resembles that observed in many species of monkeys of the Old World (Cercopithecinae—baboons, macaques, mandrills, etc.). The females born in a clan in principle remain there all their lives, while the young males leave to join another clan. This system is coupled with a very strict matrilineal hierarchy; in other words, in clans of spotted hyenas, the females lead. The males, if you have followed me, are all outsiders. This social organization has been particularly studied by Kay Holekamp, a professor at the University of Michigan and a specialist in the spotted hyena. I was also able to see the hyena's social organization during my stay at the University of California, Berkeley. The university's field station, hidden in the hills overlooking the famous campus, was home to about 30 spotted hyenas. This colony had been established several decades earlier by the late Stephen Glickman, then a professor at Berkeley, to study the hormonal basis of social dominance and aggression. Glickman was particularly interested in the mechanisms of masculinization that affect the genitalia of female hyenas.[460] The latter have a pseudo penis, which makes them look like a male hyena, the rest of the body being very similar between the two sexes. However, females are on average larger and more aggressive than males.

I was introduced to Glickman by my colleague Frédéric Theunissen. Glickman was then a respectable gentleman, of an equally respectable age, who had kept his scientific spirit alive. "Frédéric told me about your joint project to study the vocalizations of hyenas. As far as I know, no one has really looked into the matter. Yet, as you will see, acoustic communication is essential for hyenas." Glickman was obviously right. During the first few hours I spent observing the Berkeley colony, I was constantly surprised by new types of calls. In addition to the famous giggle, of course, the "uh-uh-uh-ooh-ooh-ooh-ooh-hi-hi," there's also the whoop, "wooooooooo . . . woo-hoop!," which individuals use when they don't see each other, not to mention multiple variations of growls, rumbles, and whispers that allow hyenas to finely tune their social interactions. Experts agree that there are at least a dozen different basic calls. In addition, each of these calls can be modulated (louder—lower- or higher-pitched) depending on the circumstances. These are called *graded vocalizations*. This feature allows the sender to refine the information it encodes in its call. "I'm just a little frustrated, I'm going to giggle softly." Or "my frustration is mounting because I am very hungry and this very aggressive female is preventing me from getting to eat; I'm going to giggle very loudly." All this at the risk of bringing in lions from elsewhere, which is another story.

Holekamp showed that the hierarchical position of a hyena in its clan is roughly established around 18 months after its birth.[461] For females, daughters of mothers high up in the hierarchy acquire the same rank— they're born with a silver spoon in their mouths. This is probably due to a combination of their own level of aggression (they have been bathed before birth in a particularly testosterone-rich uterine juice) and the attention their mothers give them by protecting them from other members of the clan. For lower-ranked females, it's not the same picture, and the spoon is made of plastic. The situation is quite different for males. Here it is the order of the male's arrival in the clan that roughly determines his rank. Of course, things can be more subtle and reversals can happen sometimes, especially for those who are smart or clever enough to foment coups by forming coalitions with fellow hyenas. The social system of spotted hyenas is more than a simple linear hierarchy. As in ape groups, affinities are created. So are enmities. This one will ally itself with that one, these three will form a coalition against that one, and so on. In short, the organization of a hyena clan is more like a mafia than a military regiment. But the most general rule is that these hierarchies, once well established, last over time.

Imagine that you are a female or male hyena in a clan of 80 individuals. Complicated to know who is who, who is allied with whom, who has just arrived, who is in a good mood today, who you'd better not rub the wrong way, how to behave with this individual or with that one, etc. To answer these questions, you fortunately have an important tool at your disposal: the communication signals that others will send you and those that you are capable of producing. And among hyenas, they are legion: I repeat, more than a dozen basic calls, with variations. But you also have to add in the influence of pheromones, those chemical molecules produced by various glands, as well as the use of many postures that serve as visual signals—for example, the attacking posture (head held high, ears and mane erect, mouth closed, and tail upright) and the fleeing posture (ears flattened against the skull, mane smoothed against the body, and tail hidden between the hind legs). No doubt about it. Communication in hyenas is multimodal, that is to say, it uses several channels of communication, from chemical to visual to acoustic.[462]

But Frédéric and I are bioacousticians, as you must be well aware by now. Our only ambition was to decipher the information contained in the hyenas' vocalizations. One question in particular was bothering us: Do hyenas have a voice? An identifiable voice, like the ones we are able to recognize on the telephone with barely three spoken words? In other words, was it possible to identify an individual signature in a hyena's giggle and to find that same signature in another vocalization? This question may seem trivial. Yet, in the animal kingdom, there is no known example of a well-established voice. Even in the human species, it is not as obvious as we think. Would you be able to recognize someone by their laughter, shout, or whisper, if you have previously only heard their spoken voice? Of course, many animals recognize each other individually on the basis of vocalization. In previous chapters, I have detailed several examples: the white-browed warbler that recognizes its neighbors by listening to their song, the elephant seal that is able to identify each of its opponents by their "clork . . . clork," etc. But these recognitions, however precise and effective they may be, most of the time relate to a single type of vocalization. To my knowledge, only rare studies have explored individual recognition through the vocal repertoire of a species.[463] One of the most thorough was carried out on the zebra finch by Julie Elie. Do you remember her? She is the expert on the vocal repertoire of this little Australian bird. Julie showed that the zebra finch is only able to identify a fellow finch, whatever the type of call it emits, if it has first learned the individual vocal signature carried by each of these calls.[464] In other words, the bird is not able to learn to recognize an individual from one type of call and then transpose it to another call. Why not? Because zebra finches don't have a true individualized voice.[465] If it is assumed that the same is true for hyenas, then just because Kadogo recognizes Ursa or Winnie (three of the Berkeley residents) by hearing their giggles, that doesn't mean it will recognize them when they whoop or growl, unless of course it has learned to recognize them for each of their different calls.

So off I went to record our hyenas. Helped by the technical staff at the field station and by Aaron Koralek, a friendly Berkeley student recruited by Frédéric, it was not very complicated. Mary Weldele, the head of the station, had spent so much time with her animals that she knew each one by heart. So she knew the context in which to put this one or that one so as to have the best chance of hearing it giggling in frustration or whooping to call a mate. A few years before my arrival, Mary had suffered a serious injury: while trying to retrieve a piece of meat that had fallen on the ground, she had her hand bitten by a hyena. Hyenas have tremendous jaws, able to crush big bones like you crush an eggshell. Experiments on hyenas at Berkeley have shown that they can easily develop a force of 1422 newtons, which is enormous.[466] The jaws of hyenas are only surpassed by those of lions (3400 newtons), tigers (3000 newtons), and bears like the polar bear (2400 newtons), all much more imposing animals. By the way, the power of our human jaws is 650 newtons, which allows us to easily crush roast chicken! Mary had been wise not to try to force her hand out of the animal's mouth, which would have certainly made the situation worse, and the hyena finally released her. Mary's hand, however, bore impressive stigmata. So caution was called for—above all, being extra careful when transferring animals from one enclosure to another. Entering one of the enclosures was out of the question.

The easiest vocalization to provoke is the giggle, so we decided to start our investigations there. All you have to do is present a hyena with a piece of meat without letting it have it immediately, and it will giggle with frustration. A few seconds of recording, and voilà! The released piece of meat is gulped down. Extensive acoustic analysis, followed by solid statistical processing—something Frédéric is an expert at—revealed to us that the hyenas' giggle contains previously unsuspected information.[467] First of all, each hyena giggles in a very personal way. It's a bit like in humans—there are frank or hesitant giggles, raucous or discreet, singing or monotonous. As a result, it's not so difficult to distinguish between Kadogo's giggle and Ursa's or Winnie's. But that's not all: the giggle also gives information about the age and hierarchical rank of the sender. Older individuals have lower-pitched giggles. A dominant individual produces a slightly variable giggle ("ho-ho-ho-ho") while a subordinate—who will giggle much more, because it is more often frustrated—has a giggle akin to singing ("oh-ooh-ooh-ooh-ooh-hi-hi-hi-hi"). This information about hierarchical rank could be very useful for a hyena trying to access a carcass. Should I stay or should I go? We then saw that the major information conveyed by the whoops ("wooooooo . . . woo-hoop!") represents the sender's individual identity, age, and gender. Not very surprisingly, this vocalization is emitted to establish contact with distant kin or clan members and can be heard from several kilometers away. It's important to know who you're dealing with.

To test whether hyenas could recognize an individual's voice, we decided to adopt an operant conditioning protocol. Remember? That's the approach Colleen Reichmuth from the Pinniped Laboratory in Santa Cruz uses to test the hearing abilities of seals. The principle is to teach the animal to respond to a sound emitted from a loudspeaker (the target stimulus) with a particular behavior. Finding out how, in practice, to adapt this kind of protocol to hyenas was quite a challenge. Luckily, this challenge fascinated Julie, the zebra finch expert, who had also joined us in Berkeley. There were many attempts, some less successful than others as the hyenas brutally destroyed any device within their reach. Julie devised a system in which the test subject first had to trigger a sound by pulling a rope and then had to move to a food dispenser that only worked if the sound was the target stimulus. Yes, okay, it's not that simple, but a system had to be found. Imagine the scene: one of the hyenas was isolated in a paddock. In one corner was the rope; at the other end of the paddock was the food dispenser. We wanted the hyenas to learn how to pull and pull again on the sound-emitting rope until they identified the target stimulus and went to the dispenser in response—triggering a conditioned reflex, so to speak. What were the stimuli used? We started out by emitting whoops from different individuals, only one of which triggered a food reward. So the hyena had to learn that a whoop from Kadogo, for example, was a reward, while it would get nothing from all the others. We hoped that it would pull the rope to trigger successive whoops without going to the food dispenser until it heard Kadogo's whoop. But it was hard to get anything from the hyenas. While four of the five hyenas included in the experiment obviously understood what we expected of them, they soon became bored and lost motivation after a few tries. Maybe they weren't hungry enough to spend energy playing with us. The meager results of these experiments are interesting, however. First of all, since most of the animals passed the test, we can conclude that hyenas are able to distinguish their fellow hyenas by their whoops. Second, by using recordings of giggles instead of whoops, we found that hyenas can also distinguish between two individuals by this vocalization. The individual signatures carried by both whoops and giggles are therefore perceived by hyenas. On the other hand, no hyena succeeded in the last phase of the experiment, i.e., recognizing an individual's giggle when it had simply learned to identify its whoop. A hyena that learned to identify the vocal signature of a fellow hyena contained in one type of vocalization does not seem to be able to recognize it in another. It would seem that our hyenas are in the same lot as the zebra finches. Unable to transpose their knowledge from one vocalization to another, they must learn to recognize the vocal signature for each of an individual's calls. As I have already said, it is not certain that we ourselves have this ability: Does the same person who speaks, laughs, or calls have the same identifiable vocal signature in each case?

Although they don't seem to have more individualized voices than zebra finches, hyenas nonetheless have a complex and effective system of sound communication. We now know that these animals use acoustics to recognize clan members, to identify a stranger, and to obtain all manner of information about the sender, such as age, sex, hierarchical rank, and even its mood at the time. The diversity of their vocal repertoire is probably due to the complexity of their social organization. This situation is found in all animal species living in groups where strong and lasting social interactions (alliances, cooperation, well-structured hierarchies, etc.) develop. Let's explore some of these worlds together.

Let's start with the example of meerkats (*Suricata suricatta*).[468] You probably know these beautiful and dynamic mongooses of the African savanna, standing on their hind legs with their arms at their sides, their little nose pointing toward the camera in wildlife documentaries. They live in groups of up to 50 individuals, hunting scorpions and beetles, raising their young cooperatively. Their vocal repertoire is particularly extensive. Marta Manser, from the University of Zurich, who studies meerkats in the Kalahari Desert, has been able to distinguish more than 30 different calls. Meerkats are therefore great users of acoustic communication, whether it is during episodes of competition for food, during mutual grooming sessions, or when traveling in groups. A meerkat has a very special life; it spends between five and eight hours a day digging in the ground for food. With its head half buried in the sand, it can't see what's going on around it for most of the time. Sound signals are the main means of keeping in contact with the other members of the group. Sentinel individuals, standing on their hind legs, are on deck keeping watch for the arrival of danger so they can alert burrowers with calls. These small animals are very vulnerable and are a food source for a whole cohort of predators—eagles, jackals, and other carnivores. In addition, our small earthworkers are often several dozen meters from the first burrow. Meerkats take risks! And they pay a heavy price; on average, 60% of adults perish in a single year. This is not only because of predators; it must be said that conflicts between groups of meerkats also cause damage. Under this multiform pressure, the meerkats have developed a sophisticated alarm call system allowing all members of the colony to be quickly informed of danger and above all to know what behavior to adopt. Woe to the distracted individual, who's a goner for sure.

In the late 1990s, Marta spent several months observing 18 groups of meerkats along the Nossob River in what is now the Kgalagadi Transfrontier Park, the vast semidesert region straddling South Africa and Botswana, and further south near the Kuruman River. The beds of these two rivers are mostly dry, except during floods—the famous flash floods—which follow the violent thunderstorms that sometimes occur over the Kalahari. In spite of the drought, the relative proximity of the groundwater allows the existence of some vegetation with its associated food chain. Marta reported that the meerkats often use their alarm calls; during observation sessions, she rarely spent more than 45 minutes without hearing one.[469] To test whether the calls contained any information about what caused them, Marta provoked them, walking a dog on a leash, dragging a stuffed jackal across the field, or flying various objects suspended from balloons. Acoustic analysis of the recordings showed that the meerkats' alarm calls vary depending on two factors: the nature of the predator and the level of threat.[470] With regard to the first factor, predators appear to be identified as either *ground* or *air* based on the corresponding calls made by the guards. For example, those calls emitted at the sight of an aerial predator extend over a wider band of frequencies. As to the second factor, the level of threat, the calls may vary in several ways. For instance, the greater and more immediate the danger, the louder the corresponding calls will be: they take on a chaotic character, a bit like when a human baby goes from a soft cry to an angry scream. The vocalization may escalate from "or ... or ... or ... or ... or" when the sentinel spots a jackal on the horizon to "ourrh-ourrh-ourrh" when the jackal approaches and becomes really threatening. Another vocal clue to the threat level is that the more immediate the danger, the shorter the time between calls. In aerial alarm calls, the more urgency the situation requires, the fewer calls the individual sentinel makes. If a martial eagle is spotted, the sentinel can't stay out in the open for long. Another vocal variation is the *recruiting call*, which, as its name suggests, calls the other meerkats to the rescue. It is emitted when the sentinel sees a snake or identifies something that it finds odd (such as traces of urine, droppings, or hair from foreign meerkats or predators). The final variation, the panic call, is used as a last resort. This death call is emitted once or twice very quickly by the sentinel, either in reaction to alarm calls from birds (Horror! The birds see something, and I don't see anything!) or when a predator, ground or air, is spotted at the last moment, already very close to the group. This call does not give precise information on the identity of the predator; it is an "everyone for themselves" call, provoking the immediate and desperate flight of the whole meerkat family, diving into the burrows to take shelter.

With the alarm calls of the meerkats, we therefore find a real code, combining a lot of useful information for decision making. The playback experiments conducted by Marta and her team have shown that the meerkats know this code like the back of their little furry hands. They rarely make mistakes and react to different calls in a perfectly appropriate way. They either gather around the sentinel who gave the alert and examine the source of the danger in chorus (if it is not immediate) to evaluate the best strategy to adopt (Is it better to form a group to go scold the trespasser? To wait a little? To hide on the spot?) or they run like mad to take cover if that is obviously the right decision.

Returning to the idea that these calls carry information about the identity of a predator, even if they are not the equivalent of words in human language since they are not constructed on the basis of a combination of syllables, they are nevertheless strongly reminiscent of them. This kind of sophisticated system is called *referential communication*.[471] In other words, a call means or represents something in the environment. These are far from simple calls of fear emitted by reflex. In fact, Marta has shown that young meerkats improve their response to alarm calls over time, suggesting some learning in their behavior.[472]

In recent decades, several referential communication systems, most of them related to alarm calls or calls indicating the presence of food, have been identified in mammals—in particular, in several species of monkeys. For the record, Charles Darwin anticipated this. The father of the theory of evolution was very interested in animal communication. He related that one day he presented a stuffed snake to some monkeys at the London Zoological Gardens, and observed the monkeys emitting "calls of danger that were understood by other monkeys."[473] It took a marriage between bioacoustics and primatology to demonstrate the existence of referential communication in monkeys. The first and most famous example is certainly the one observed in the vervet monkeys (*Chlorocebus pygerythrus*). Do come with me; I'll take you to discover this communication system. It has become mythical.

The vervet is an African monkey that produces three alarm calls: a first ("kof") for leopard-like land predators; a second ("uaho") for aerial predators like the martial eagle, and a third ("cla-clack . . . cla-clack") for snakes—very useful because large pythons can attack this monkey. Just as meerkats show behavioral responses consistent with the alarm call heard, so do vervets. They climb a tree when they hear the leopard call. When it is an eagle call, they look to the sky and hide under the nearest bush. And in response to the snake call, the adults stand up on their hind legs and scan the ground, grouping together to harry the trespasser. Thomas Struhsaker, a professor at Duke University in the US, published these observations in 1967.[474] But the experimental evidence was to come in 1980, and it made a pair of primatologists famous: Dorothy Cheney and Robert Seyfarth, two students of Peter Marler, the discoverer of the dialects of the white-crowned sparrow. Cheney, Seyfarth, and Marler confirmed through playback experiments conducted in the field in Kenya that the vervets fully understood the meaning of their various calls. Their study was published in the prestigious journal *Science* and received worldwide media coverage.[475] Monkeys were proving that semantic communication was not restricted to human language! Semantics, a big word, perhaps? Let me explain. Whereas until now it was considered that animal vocalizations were merely a record of the emotional state of their sender, the experiments of Cheney and colleagues showed that animal calls could represent elements of the environment, just as the word *table* means to us the object on which the cutlery is placed. Seyfarth and Cheney reported the details of their investigations in a book that I recommend to all ethology enthusiasts: *How Monkeys See the World: Inside the Mind of Another Species*.[476] Since then, a lot of water has flowed under the bridge and many people have struggled to find out if this remarkable characteristic of human language can really be attributed to vervets. We don't know the cognitive mechanisms that drive the production of calls in meerkats or vervets, and biologists are afraid of being anthropocentric! In my opinion, we are bound to see striking similarities between the way in which animals' calls indicate a type of danger and the way in which human words represent an object in the environment. David Premack, a former primatologist at the University of Pennsylvania, has suggested that a communication system based solely on the reflex expression of emotions may well evolve into a referential system.[477] His argument is this: if you know that I squeal with delight when I see strawberries and scream with fear when I see a snake, you will soon learn to associate each of my calls with the right trigger. In other words, my vocalizations give reliable information about the objects around me. Does perceptual semantics go hand in hand with conceptual semantics in meerkats or monkeys? That is, does a vervet monkey climb a tree because it has really understood that a leopard is approaching? Let's dare to go even further. Could the vervet imagine, in its little monkey head, an image of a leopard? Or is the explanation for its behavior more basic? It climbs a tree because it reacts to a particular acoustic signal ("kof!") and it knows it must climb when it hears it? It's not an easy task to answer these questions. However, a study with Diana monkeys *Cercopithecus diana* by Klaus Zuberbühler, a primatologist at the University of Neuchâtel in Switzerland and an associate of Cheney and Seyfarth, provides some interesting insights.

The Diana monkey inhabits the rainforest of West Africa, between the Gambia and Ghana. A bit like the vervet, the adults produce two different alarm calls, depending on whether the predator approaching is an eagle or a leopard. But strangely enough, here both males and females have their own calls. Moreover, the females react to the male alarm calls with their own alarm calls, which suggests that females might understand the meaning of the male calls. Does an eagle image spring to mind when they hear males call "eagle approaching"? To answer this question, Zuberbühler tested whether the response of Diana females was caused solely by the acoustic characteristics of the male calls or by the existence of a predator concept. Let's look at the protocol of the experiment; hang on tight, because it's a bit complicated. Zuberbühler first recorded or obtained from a sound archive four different calls: an eagle call from a real eagle, a roar from a real leopard, an eagle alarm call from a male Diana monkey, and a leopard alarm call from a male Diana monkey. In each experiment, a group of Diana females heard two of these calls, separated by five minutes of silence. In the so-called baseline condition, the two calls were exactly the same, acoustically and therefore semantically identical (e.g., an eagle call followed by the same eagle call). In the test condition, only the semantics were the same (e.g., an eagle alarm call by a Diana male followed by an eagle call). Finally, in the control condition, the stimuli differed both in their acoustics and meaning (e.g., a leopard alarm call by a Diana male followed by an eagle call). Zuberbühler hypothesized that if the Diana females formed a mental representation of the predator based on the alarm call, they would only show surprise if the call following the alarm call did not match. However, if there was a match between the two successive stimuli, or if the two stimuli were identical, the Diana females would show less interest in the second stimulus. He was right. In the baseline experiment, when the Diana females heard the first eagle call, they responded with their own eagle alarm call. But at the second eagle call, their response was less intense. Since no birds were visible nearby, they seemed to lose interest. Same story with two successive leopard calls. In the test condition, the females initially reacted strongly to the eagle alarm calls from the Diana male by emitting their own alarm calls. However, they hardly called out at all in response to the second stimulus, which was a real eagle call. Presumably, by the time the second stimulus was given, they had already integrated that an eagle was there and had the image of it in their heads. Finally, during the control condition, when the first call was a Diana male's leopard alarm call, the females initially responded with their own leopard alarm call. In addition, when the male's call was subsequently followed by an eagle call, the females responded very strongly to the eagle call with their eagle alarm call. Are you still following me? We can interpret these results by listening to Zuberbühler: Diana females get an idea of the predator by listening to the male's call. If the predator does not match what the males announced, they sound the alarm again, but with the "right" call.[478] Nothing says whether they blame the males for making a mistake, or whether they then take retaliatory action.

Referential communication systems have so far been identified in several other primate species, such as the Campbell's monkey *Cercopithecus campbelli*,[479] the chacma baboon *Papio cynocephalus ursinus*,[480] the black-fronted titi monkey *Callicebus nigrifrons*,[481] the chimpanzee,[482] and some lemurs (for example, the particularly remarkable forest-living ringtailed lemur *Lemur catta* with its orange eyes and long black-and-white striped tail[483]).

Sometimes it is not the acoustic structure of the call that carries information about the type of predator but a combination of calls that encodes referential information. The leopard alarm call of the *Colobus guereza*, a beautiful monkey with a long black-and-white coat, is a succession of many repeated "rrooarr-rrooarr-rrooarr" sequences. Their eagle alarm call is a succession of a few sequences, each containing many roars.[484] And there are more complex systems. In the Campbell's monkey, each of the three identified alarm calls are graded (like hyenas' calls, they vary according to the motivation of the sender), and, surprisingly, they can be given suffixes (i.e., a short call that is added after the alarm call and changes its meaning). Human language is teeming with examples: a motor/a motorist; the suffix *ist* changes the meaning of the word. In response to different contexts or events, the Campbell's monkey emits the calls "boom," "krak," "hok," or "wak." A team of primatologists led by Zuberbühler and Alban Lemasson, a professor at the University of Rennes in France, has shown that each of these calls can take the suffix "oo" and become "wak-oo," "krak-oo," etc. Moreover and amazingly, the monkeys combine these calls in sequences. It is now known that calls with suffixes and call sequences do not appear to be random but could correspond to particular contexts and events, such as a tree falling, a neighbor arriving, an increase in urgency, etc. The basic principle is that each individual call carries specific information ("leopard," "eagle"). If a call is repeated alone, the meaning does not change—it is still "leopard" or "eagle." If a suffix is added, the meaning becomes more general ("predator"). If two calls with suffixes are combined, some of the observations suggest that the overall meaning may change drastically: "leopard" with suffix + "eagle" with suffix is emitted when a tree falls, for example.[485] In the absence of playback experience with Campbell's monkeys, it is currently impossible to say whether this syntactic complexity actually carries any information for the receiver individuals. The future may tell us. The only small piece of evidence that we have today of the role of suffixes comes from playback experiments with male Diana monkeys. They respond more strongly to calls without suffixes (indicating a leopard or an eagle) than to calls with suffixes, whose meaning is less precise.[486]

I've talked a lot about the alarm calls. Don't think I'm done with them yet. It is indeed in the context of predator signaling that referential communication has been most often highlighted. But be aware that there are a small number of animals in which referential communication has been found to signal food. Marmosets, chimpanzees, and bonobos emit calls or particular sequences of calls in the presence of food. For example, a chimpanzee produces specific vocalizations when it discovers "high-quality" food, and its colleagues are particularly sensitive to this.[487] In bonobos, it is instead a certain sequence of calls that indicate the quality of the food.[488] Aside from primates, there are few other examples. The calls made in the presence of food, although they often provoke the arrival of other individuals, do not really disclose the nature of the food.[489]

Referential communication systems do not only exist in meerkats and monkeys.[490] Ground squirrels and marmots also have a repertoire of several calls in response to the arrival of different predators. The Gunnison's prairie dog *Cynomys gunnisoni* has four calls, referring respectively to a bird of prey, a human, a coyote, and a dog.[491] In addition, more than a dozen species of birds are known to produce referential signals.[492] The first example identified was in chickens, which emit a particular call to announce the presence of food and two distinct calls to warn of the arrival of either a land predator or an air predator.[493] (We mentioned this example in chapter 11 when we looked at the audience effect.) Funny thing: The aerial predator has to move sufficiently fast to trigger the chicken's call.[494] If the suspicious thing doesn't exceed a certain speed, the cock won't crow!

Let's stay with the birds for a while. In recent years, research on referential communications has flourished in birds, with astonishing results.[495] Three main systems were highlighted. Like the chicken, some birds use alarm calls that point out this or that type of danger. Others vary the number of notes, the speed of repetition of the calls, or even finer acoustic characteristics, such as the duration of the call, its pitch, or its intensity. Finally, there are those who use combinations of notes. Let's take a few examples to illustrate. The first case involves a study with my usual adventure companion, Thierry Aubin, in the field in Australia. We went there to meet a very small bird, the superb fairy wren *Malurus cyaneus*, known to use an acoustic alarm system with two calls.[496] Off we go to the other side of the world!

The superb fairy wren is one of the favorite study models of Robert Magrath, a professor at the University of Canberra in Australia. Robert has been interested in the vocalizations of the fairy wren for years, and when Thierry suggested that we study the coding of information in its alarm calls, he willingly accepted. Fairy wrens are very common in the southeast of the continent. Imagine two little balls of feathers—one simply dressed, with a brown back, lighter on the chest; the other sporting a more complex outfit, consisting of an iridescent light blue metallic head, a line of jet black encompassing its eyes and widening along its neck, deep blue on the throat, back, and tail, and finally a chalk-colored belly. Guess who the female is. On each ball of feathers is a never-ending tail, often pointing upward. The female and male fairy wren live as a pair on a territory that both defend bitterly with songs and by always being ready to ruffle a few feathers of any overly daring intruders. The breeding adults often team up with helpers—youngsters from the previous year who participate in the rearing of the chicks. Cooperative rearing is not uncommon in birds.

The fairy wren loves eucalyptus forests. We studied it in the Canberra Botanical Garden, which is full of them—a huge and very beautiful park but very different from the wild places where we usually traveled. I was a little disappointed. I hadn't imagined Australia like this, as I dreamed

Fairy wren

of vast stretches of wilderness rather than urban parks. Besides, I had hay fever. It was November, and it hadn't occurred to me that the Australian spring would not escape the grassy outbreaks. I hadn't brought any appropriate medication with me, and I had a strong tendency to be gloomy. That said, the place was pleasant, not very crowded, and I appreciated being able to get out of the university routine once again. Here a kangaroo, there the nest of a bowerbird decorated with colorful bottle caps. Much improved by the exotic animals, the vegetation and the climate, and the pleasure of working together again, our stay was nevertheless excellent. Not to mention that our experiments yielded good results. "Here is the map of the botanical garden; you will find many birds banded by us," he said. Robert Magrath is a highly meticulous person. He let us work alone, but we felt that letting us approach his protégés worried him vaguely.

Thierry was eager to tackle the two calls of the fairy wren. Issued in the presence of a danger—a predator most of the time—they provoke diametrically opposed reactions from the others present on the territory. The mobbing call, as its name suggests, attracts other fairy wrens to harass the predator as a group. The flee call incites everyone to take immediate refuge in the bushes. Why choose one of the calls over the other? Because sometimes it is better to hide from a predator that is too skilled than to try to scare it away by harassing it. The two calls have important acoustic similarities: they are both extremely high-pitched trills, lasting about 90 milliseconds. The flee call ("trreee") is around 9000 Hz. The mobbing call differs from it in two ways. First of all, a hyperfast and rapidly rising high-pitched whistle ("ps") is added in front of the trill. On the spectrogram, it draws a kind of hook on the front of the call, a primer if you like: the sound starts around 7500 Hz, goes down immediately to 6500 Hz, then goes up no less immediately to 9000 Hz. Then the trill starts at around 9000 Hz, just like the flee call, but it goes down regularly to 7000 Hz. The mobbing call could be translated as "pstrreeoo," where "ps" represents the hook and "trreeoo" the declining trill. Because they are short and very high-pitched, the two calls are difficult to distinguish for an untrained human ear. Our problem was to understand how the birds choose between going together to face the danger or hiding quickly. Given the structure of the two signals, we had three hypotheses: either the presence of the initial "ps" was the critical cue that allowed the birds to know immediately which call it was; or the cue was given instead by the frequency of the descending slope of the trill; or the combination of the "ps" and the trill slope were necessary for the recognition of the call.

You are now familiar with our experiments, so you already know our methodology: playback experiments, of course. And fortunately for us, fairy wrens are fairly easy to test. With a loudspeaker placed in the middle of their territory and a bit of waiting hidden behind bushes, we observed the reaction of the fairy wrens to four different signals: the flee call ("trreeee") and the mobbing call ("pstrreeoo") as the control measures of their behavioral reaction, a mobbing call without a hook ("treeoo"), and a flee call to which we had added the hook ("pstrreeee"). As always, we expected these artificially modified signals to give us the key to the information coding. And so they did. As expected, the birds' responses to the flee call and the mobbing call were exact opposites: fast flight to the nearest bush and absolute silence for the flee call; approaching the loudspeaker, erratic flights above, and many calls for the mobbing call. When the mobbing call was missing the hook, it did not lose its meaning: the birds continued to approach and noisily express their displeasure. On the other hand, when the hook was present but the trill remained around 9000 Hz (as in the flee call), the birds would dive into the bushes. The hook does not encode the nature of the predator and therefore does not carry information about the decision to be made: to harry the intruder, or to dive into a bush and lie low. It is the trill that fulfills this role. One hypothesis that we had not yet tested was that the hook provides an excellent location cue for the bird emitting the call. This would not be surprising since we know that a sound that varies extremely rapidly in frequency is easily locatable, and this would be particularly relevant for fairy wrens, which need to know quickly where to go to scold an uninvited guest. One can even venture to imagine that the flee call, the "trreeee," first appeared in the ancestor of the fairy wren, saving many lives. Then it was transformed over time to a "treeoo" that proved to be a rallying cry, allowing an effective collective struggle in the face of danger. When the "ps" was added to give "pstrreeoo," it further facilitated the operation by helping the birds locate each other. This pretty description quickly sums up what probably happened. The evolution of communication systems, like that of any biological system, is rarely so simple and linear.

The system used by the fairy wren is therefore based on the use of two calls, each designating a different type of danger. Many other systems are based on the use of a single call with varying characteristics. One of the best-known examples is that of the black-capped chickadee *Poecile atricapilla*, the most common chickadee in North America. As we discussed in chapter 11, this bird changes the number of repetitions of the note "dee" in its call "chick-a-dee" to signal to its fellow chickadees the size of a bird of prey perched motionless on a branch. Playback experiments have confirmed that other chickadees understand the message: When the recording of the sender chickadee produces numerous repetitions ("chick-a-dee-dee-dee-dee-dee-dee-dee"), signaling a small owl or other small predator, the chickadees come close to the speaker, scolding bitterly and continuously. If the number of notes is limited ("chick-a-dee-dee"), the chickadees know that it is a large owl, and they are less motivated to attack it because they are more agile and quicker than the owl, so it represents much less of a danger.[497] I admit that, with this chickadee as with the fairy wren, we are in somewhat incomplete systems of referential communication. The chickadee doesn't tell us exactly who the predator is. But the information is precise enough to induce appropriate behavior on the part of the chickadees listening.

Finally, the last system encountered in birds is similar to the combinations of calls I was telling you about in some monkeys. Many species of birds combine calls or notes in sequences of greater complexity to indicate the type of predator or how far away it is, and the receiving individuals respond appropriately. Take the Japanese great tit *Parus major minor*, which looks a lot like the European great tit. It produces alarm calls (called ABC) when it detects the threat of a predator, a recruitment call (D) when it wants to attract other tits in a seemingly harmless environment, and a combination of these calls (an ABC-D sequence) when it wants to motivate other tits to harass a stationary predator, such as a bird of prey perched on a branch. Toshitaka Suzuki, an assistant professor at Kyoto University, recently showed through playback experiments that individuals receiving these signals respond to ABC by scanning the surroundings, to D by approaching the speaker directly, and to ABC-D by mixing these two responses—i.e., by approaching while scanning. If the order is reversed, by presenting D-ABC to the birds, they seem to be annoyed because they do not express either response.[498] These results suggest that tits also use the order of the calls like a kind of syntax, to extract information. I'm not implying that the process observed in tits is strictly identical to the syntax of words in human language. A tit's call is not a word—not in its structure and probably not in what it provokes in the most intimate part of the bird's brain. However, a recent study has shown that Japanese great tits who have listened to the alarm call they usually make at the sight of a snake are more likely to spot a snakelike object later on.[499] These results show that animals can have complex composition rules for their acoustic communication units. The language of the human species is a particularly remarkable example of this, but it is not the only one.[500] Indeed, there are still many more surprises in store for you. I'm going to tell you a few more stories that will make you even more aware of the complexity of acoustic communication systems.

A sophisticated example of information encoded in a sequence of calls is given by the southern pied-babbler *Turdoides bicolor*. Big as a blackbird, black and white, with a slightly curved beak, it lives in the savanna of southern Africa in social groups of 3 to 15 individuals. In one group, only the dominant pair reproduces. The other individuals help defend the territory and the nest, incubate the eggs, and feed the young. The birds also feed in groups, searching the ground for various invertebrates. Like the meerkats, they spend most of their time with their heads more or less buried in the ground, so acoustics are their preferred means of communication. Sabrina Engesser and her colleagues from Simon Townsend's team at the University of Zurich have observed that babblers produce a first type of call, a slightly hoarse one, in response to a sudden but not very intense threat (another approaching animal, for example), and another call, a repeated hissing sound, to attract other animals on the move.[501] Remarkably, southern pied-babblers combine the alert call and the recruitment call in sequence to lead their companions to harass a land predator. Encouraged by this initial result, Engesser and Townsend went much further.

Human language is made up of combinations of meaningless units that we can easily discern by ear. These units can be combined to create a potentially unlimited set of information-carrying signals. For example, consider the two words *arc* (a/r/c) and *car* (c/a/r), which have dissimilar meanings. What differs between the two words from an acoustic point of view is the combination of the sound units /a/, /r/, and /c/, each of which has no meaning in isolation, but which in combination comprise the respective meanings of *arc* and *car*. Do we find similar phenomena in nonhuman vocalizations—meaningful signals consisting of combinations of meaningless sounds if they are isolated? This is what Engesser and Townsend sought to find out with the chestnut-crowned babbler *Pomatostomus ruficeps*. Their work was published in the prestigious magazine *PNAS* in September 2019.[502] Let's take a closer look. The chestnut-crowned babbler has two very similar calls in its vocal repertoire: the flight call, which serves to coordinate group movement, and the prompt call, which is used when feeding nestlings. The flight call is composed of two successive sound units, F1 and F2. First, to find out whether the birds were able to distinguish between F1 and F2, Engesser conducted an experiment known as habituation-dishabituation. If a bird is repeatedly made to listen to the F1 unit, it eventually loses interest in the loudspeaker. But if F2 is then emitted, it looks again in the direction of the speaker, showing its ability to distinguish between the two units. So far, it's clear, but pay close attention to what happens next because things get a bit complicated. The next step was to test the bird's ability to distinguish between the two F units and three other units, P1, P2, and P3, which in combination characterize the prompt call. The results showed that the birds perceived F1 to be identical to P2, and F2 to be indistinguishable from P1 and P3. In other words, this babbler builds its calls with a vocabulary of two sound units (on the one hand, F1 and P2 are equivalent; on the other hand, F2, P1, and P3 are also equivalent). If we denote the first sound unit as A and the second as B, the flight call will be written AB, while the prompt call will be written BAB. So these two calls are different combinations of the same two units. In a playback experiment, Engesser checked that each of the calls induced a different behavioral response. However, when the sound units were played alone, there was no noticeable change in the listeners' behavior. Conclusion? The calls make sense while the units do not—just like in *arc* and *car*. Some nonhuman animals can therefore combine sound units to construct vocalizations that make sense in a way that is reminiscent of how we construct our words.[503]

Besides syntax (combining units into meaningful sets), human language is famous for what is called pragmatics. You may never have heard this term, and yet you know what it is because you practice it every day— just as the famous character Monsieur Jourdain from Molière's play was practicing prose.[504] *Pragmatics* encompasses the processes by which the context, or more generally our knowledge, influences our understanding of words or phrases. Some examples of pragmatics? Here are two. If a parent says to a child, "Your room is a mess," the child will understand "You have to clean your room," because he/she knows that the parent expects no less. If a friend had said the same exact thing, the child might not have come to the same conclusion. Here's another example: "Did

you know that Anthony has quit smoking?" asks your companion. If you
don't smoke, and you know Anthony, you can deduce a whole range of
more or less probable information: Anthony must have made a lot
of effort, he's probably in a difficult mood these days, be careful not to
upset him, or anything else that seems relevant to you. On the other
hand, if you smoke, you will probably understand what is not being said:
"You should do the same!" And, sheepishly, you'll put out your cigarette,
or you'll shrug off the information like water off a duck's back. It will
depend on a whole set of parameters. In short, the same sentence, the
same acoustic signal, can give you different information depending on
the context and can play differently on your feelings and behavior. So in
human language information is not just the acoustic signal. In other
words, in accordance with Shannon's mathematical theory of communication that we discussed in chapter 4, the sender is not the only one involved; the receiver is an integral part of the information transmission
chain. The passage of information from the sender to the receiver requires two steps: coding, which is the job of the sender who produces
the sound signal; and decoding, carried out by the receiver by integrating
the information given by the signal and the context. What about nonhuman communication systems? Obviously, at first glance, it seems complicated to explore the world of pragmatics in animals. However, we now
have evidence to safely assert that it does play a role. And by no means
just any role—a role that is all the more important in animal species living in groups with complex social organizations, since information exchange between kith and kin is more developed and indispensable. Let
us again follow Dorothy Cheney and Robert Seyfarth, whose research
on vervets we spoke of earlier in the chapter.[505] Cheney and Seyfarth
studied chacma baboons from the Okavango Delta in Botswana for
years, and they came home with an incredibly detailed knowledge of
their social life and communications.[506]

Chacma baboons live in groups of 50 to 150 individuals. If a baboon
becomes isolated from the group, even the bravest male feels lost. And
not without reason, because it is in great danger of falling prey to a leopard. I remember seeing one, near the Khwai River in Botswana, wandering alone on the edge of the woods after its group had been dispersed

by rangers because the monkeys had savagely attacked a camp. The poor male barked once or twice a minute in a loud voice. Its calls lasted most of the night, which moved us a lot. Among chacmas, recognition happens at a distance by voice. The call repertoire is relatively limited, even though each call can vary (the repertoire is graded as in hyenas): grunts, threat grunts (emitted by individuals of high rank toward those of low rank), barking of fear, and screams (emitted by individuals of low rank toward those of high rank). Our lone male, regardless of his initial rank, must have been barking out of fear and looking for his group.

Field experiments conducted by Cheney and Seyfarth showed that the communication system of baboons has three characteristics. First, when a baboon hears a call, it evaluates whether the sender's call is addressed to it before responding. Cheney and Seyfarth give the following example. If two animals have just finished fighting, and then one hears a threatening growl from the other, it will respond as if the threat was intended for it. But if both had previously been enjoying some mutual delousing, the listener will behave as if the growl was directed at another monkey. Same call, two different contexts, two different interpretations. The pragmatics are already there. Second, calls facilitate social interaction. If a female approaches a mother with her cub by emitting growls, the mother will eventually be able to entrust her cub to her. But if she approaches silently, the mother will run away with her baby. It is likely that, in this case, the growls signal peaceful intentions. Again, the receiver's response depends on its assessment of the sender's intentions, like the child being told, "Your room is a mess." Third, the receiver monkey combines the information encoded in the acoustic signal (the type of call, the identity of the sender, etc.) with information from its memory of past events ("we just had a nice grooming session, it's not me that Paul is aggressively talking to"; "I remember Paul, what he says is trustworthy"). Also involved is the knowledge acquired about the sender's relationship with the receiver and with the other members of the group ("I have already confronted Paul—he is stronger than me"; "Paul is Peter's ally, who is the ally of Mark, who is my ally. So it is probably not to me that Paul addresses his threats"). We can see that baboons have a very detailed knowledge of the social network in which they are integrated.

If an individual knows that the female Anna dominates the female Julia, and yet it hears Julia make a threatening growl followed by Anna barking with fear (a situation that Cheney and Seyfarth, and others after them, have obviously mimed by playback), the receiving individual marks its surprise. Wow! It's a bark that violates what it knows about the hierarchical relationships of the group. Many similar experiments have been conducted, always with the same conclusion. If we consider the isolated vocalizations of baboons, they are rather generic signals, associated in a rather vague way with affectionate or, on the contrary, aggressive behaviors, or with fear or alarm. But, as Cheney and Seyfarth point out, these vocalizations are not produced in a social vacuum. Quite the contrary. Each one occurs in a context where sender and receiver know each other personally and share a rich, common history. When one vocalizes, the other completes the information given by the vocalization with information taken from the context. A call carrying very general information will then become very informative. Pragmatics, I told you!

Whether in baboons and other monkeys, meerkats, birds, or of course our beloved hyenas, the communication systems described in this chapter are based on sound units, which the sender combines in a more or less complex way to encode information, according to rules specific to the species but with a large degree of freedom. On the receiver side, the decoding of information depends on the acoustic structure of the signal and practical inferences based on social knowledge. Although each animal species, including humans, has its own communication system, there are common rules. Can we talk about languages? In chapter 18, the last chapter of the book, I answer this question. And I adopt a position. But for now, let's continue our sound journey by going to see how these animal vocalizations—like ours—can express emotions.

15

Ancestral fears

THE ACOUSTIC EXPRESSION OF EMOTIONS

Crocoparc, Agadir, Morocco. Late afternoon. The Nile crocodiles are lying on the banks, enjoying the last rays of sunshine. The weather is still good. Jasmine exhales its sweet fragrance. All is calm. Suddenly a human baby cry breaks the silence: "WAAAAaaa-WAAAAaaaa!" A curious wailing. And our crocodiles, especially the females, throw themselves into the water to swim vigorously toward the guilty speaker. There are 5, 10, and soon 20, irresistibly attracted by the distress signal of a kiddo who is not of their world. But the modulations of the little human's cry—its rapid, high-pitched crescendos, its rough-hearted side—evoke for them the calls of their own children, and from the depths of their hearts comes their protective instinct. Quickly! We must go and rescue this baby fast, it is calling for help! . . . I am in Luc Fougeirol's zoological park, with my colleagues Gérard Coureaud and Nicolas Grimault from the CNRS, our students Julie Thévenet and Leo Papet, and Niko Boyer, our technician. A few hours earlier we were in the middle of a discussion: Can we understand the emotions encoded in the vocalizations of species other than our own? Of course, it is not very difficult to guess from your dog's voice if it is afraid or in pain. But what about crocodiles? They are much more different from us. Do you think they perceive emotions encoded in mammalian vocalizations? I had recordings of human babies' cries on my computer, and the rest was obvious:

"Why don't we ask them?" When we saw these Nile crocodiles grouped in a star shape around our loudspeaker, we had our answer.

You may remember that crocodile mothers (and sometimes fathers in some species) stay with their young after hatching, protecting them from predators. When a small crocodile is isolated or under threat from a predator, it gives a distress call, and the mother drops everything to come to its rescue. I have observed this behavior many times during field playback experiments with different species, from the jacaré caiman to the Nile crocodile, the spectacled caiman, the black caiman, and the Orinoco crocodile. During a trip to Venezuela with Thierry Aubin, we observed that the Orinoco crocodile and the spectacled caiman respond indiscriminately to the distress calls of the young of either species.[507] The acoustic differences, both real and audible to our ears, are certainly not huge. That an Orinoco crocodile understands the distress encoded in the calls of a black caiman is not surprising, that's for sure. After all, they are very similar species—their common ancestor does not go back very far. But the experience at the *Crocoparc* in Agadir with these Nile crocodiles reacting to the cries of human babies suggests that this inter-species communication can extend much further, between very distant animal species. Is this an exceptional case, a curiosity? Or do general rules of coding exist for certain information, especially emotional information, that would cross the animal kingdom? This is what we are going to talk about here.

In *The Expression of Emotions in Man and Animals*, Darwin describes many species that use sounds to express their emotions.[508] He points out that animals that are usually not very talkative, such as rabbits, will produce shrill calls when they are in pain. He says that many also vocalize in comfort situations, such as when they find a lost companion. He was also aware of the isolation calls from mammalian mothers and their young as well as the threatening calls that make an opponent back down in the face of obvious rage. Finally, he also emphasizes that vocalizations are not the only acoustic manifestations of emotion. The porcupine threatened by a predator makes its quills vibrate with fear or rage, we don't know for sure. In short, in sketching a picture of animal sound signals, Darwin suggested that the sound codes used to express emotions may be widely shared, at least within mammals and perhaps beyond (he reports that the buzzing of bees changes when they are angry). Over the past two decades, research on this topic has expanded considerably. It has a practical objective—that of understanding the mental health of animals kept in captivity in order to improve animal welfare. It is possible to estimate how an individual feels by recording and analyzing its vocalizations, without manipulating it and in the absence of invasive procedures (such as blood tests that would otherwise have been necessary to measure stress hormone levels, for example). So what are these acoustic markers and how do they code for emotions?

Elodie Briefer, a professor at the University of Copenhagen in Denmark and former doctoral student of Thierry Aubin, is a world-renowned specialist on this question. Her aim is to identify acoustic markers and assess the extent to which they may be valid between species.[509] According to Elodie, "Emotions are defined as intense, short-lived affective reactions to specific events or stimuli of importance for the organism. Their crucial function is to guide behavioral decisions in response to these triggering events or stimuli, through approach or avoidance. . . . An emotion can be described by two main dimensions: its valence (positive or negative) and arousal (its level of alertness)." To describe emotions, Elodie supports the bipolar model, with an axis of negative and unpleasant emotions ranging from depression to fear, sadness, and anxiety, and an axis of positive and pleasant emotions ranging from a feeling of calm and relaxation to joy and excitement. Elodie and other scientists have shown that this classification of emotions is expressed through measurable neurophysiological and behavioral indicators. Some indicators, most of them neurophysiological, are more indicative of the level of arousal or alertness: heart rate and its variability, breathing rate, skin temperature, skin conductivity (an emotion can lead to an increase in sweating, which, even if weak, results in the skin's ability to conduct electrical current), the secretion of certain hormones, etc. In addition, and as you have probably already seen with dogs, behavioral postures can also provide information about the animal's emotional state and are more linked to the emotional valence. Ear and tail position, bristling fur, eye movements—all of these behaviors leave little doubt as to the mental state of the animal, relaxed or aggressive, as it approaches you. If the dog whines gently or if on the contrary it growls, there will be little room for doubt.

Let's be precise. How can vocalizations encode information about the valence and arousal of an emotional state? An experiment conducted by Elodie with horses is very informative here.[510] Elodie placed horses in four situations, inducing different levels of arousal with negative or positive valences. As everyone knows, horses are very social animals that, when left in the wild, live in harems (a stallion, mares, and their foals) or in packs of young single males. Elodie and her colleagues have worked with groups of horses. Within each group, the individuals knew each other well and had formed social bonds over a long period of time. The horses emitted a varied repertoire of calls, including neighing, snorting, and squealing. Scientists have focused on neighing, which is known to be produced by horses both when they are separated from their companions (negative valence) and when they are reunited with them (positive valence). We have all heard it before. A typical neigh begins with a rather high-pitched introduction ("huuuh . . ."), then continues with a long, strongly frequency-modulated "climax" ("huh huh huh huh huh . . ."), and ends with a low-frequency, low-intensity "finale" ("BRRrrrrr . . ."). Sounds familiar, doesn't it? In a first experimental situation, neighing was recorded when an animal was totally isolated from its companions—a particularly negative situation for an animal that loves company. Other times, the horse was recorded while it was separated from its best companion, while most of the group remained with it. Still a negative situation, but a little less stressful. Finally, horses were recorded during reunions, either with the entire group for those who had been completely isolated, or with their best companion. These last two situations were therefore of positive valence with, Elodie hoped, a different degree of arousal. To check this, she measured the level of arousal objectively, recording the heart rate of the horses in each case. This measurement indicated three levels of arousal: a low level when a horse was reunited with its favorite companion (heart rate around 44 beats per minute), a medium level when its favorite companion was taken away or when all horses from the group came back after a separation

(50 beats per minute), and a high level when the horse was isolated from all of its fellow horses (56 beats per minute). What about neighing? The analyses carried out showed that the acoustic structure of a neigh is complex. It is a kind of double whistle, known as *biphonation*. It's as if two sound sources are combining their effects. The two frequencies are called F0 and G0.[511] In the horse, complex vibrations of the vocal cords are probably responsible for biphonation. Neighing produced at high levels of arousal has a higher pitch, with a higher F0 than that produced at low levels of arousal. This result is consistent with what has been observed in other animals, such as the pig, the squirrel monkey, and even the zebra finch: when an individual experiences a strong emotion, its voice normally becomes higher-pitched.[512] The novelty of this horse study is that Elodie and her colleagues observed neighing of shorter duration, with a lower G0 frequency in emotionally positive situations in comparison to emotionally negative situations. In addition to expressing the level of arousal, the horse's neigh thus contains information about the valence of the emotion experienced by the sender. This dual coding of information about the sender's emotional state, including *arousal* and *valence*, is likely to exist in vocalizations of other animal species (it is found in goats, pigs, and wild boars, for example).[513] It is probably enough to look for it to find it! More work to be done.

While expressing emotions through vocalizations is probably a fairly common feature of animal sound signals, are there common rules for encoding this information throughout the animal kingdom? Deciphering the universal encoding of emotions is the challenge that scientists like Elodie are tackling. This challenge is not new, since Darwin himself proposed, in another of his books, the idea that the vocal expression of emotions could have its roots in the ancestor of terrestrial vertebrates, and that today's animals share its key characteristics.[514] Based on the premise that all vertebrates living in the aerial environment have lungs and a trachea for inhaling and exhaling the outside air, Darwin suggested that, when these early vertebrates were particularly excited and contracting their respiratory muscles tightly, they would inevitably produce sounds. Darwin hypothesized that these primitive sounds may have been useful in signifying to the others the excited state of the individual sender. The process of natural selection to increasingly sophisticated speech devices would thus have been set in motion. A century later, Eugene Morton, a researcher at the Smithsonian Institution in Washington, DC, stated the principle that the acoustic structure of a sound signal should reflect the motivation of the sender (motivation-structural rules).[515] Morton's principle is simple: birds and mammals use rather low-pitched and "rough" sounds when they have hostile intentions, and rather high-pitched and pure tone sounds (whistles, if you like) when they are frightened or animated by friendly intentions. Morton was not really talking about emotions, but he emphasized the relationship between the physical structures of sounds and the motivation behind their use. Recent studies by Elodie and other scientists have clarified these relationships. The vast majority of these studies show that vocalizations are louder and produced at a faster rate, higher-pitched, and more frequency-modulated when arousal increases. These variations in sound correspond closely to the effects of anatomical and physiological changes linked precisely to variations in intensity. Moreover, the valence of an emotion is also encoded by acoustics, with positive vocalizations shorter and less modulated than negative vocalizations. As we know, Darwin was an excellent observer and a brilliant visionary.

Coding emotions into vocalizations is one thing. Being able to *decode* them when you hear acoustic signals is another. If it is easy to distinguish between the threatening bark of an angry dog and the disarming whimpers of your favorite Rex, can we finely decode the information about emotions in the vocalizations we hear? And on what acoustic basis? Research has mainly focused on the distress calls of babies, human and nonhuman, when they feel in danger, bringing parents or any other individual in charge of their protection back to them. For convenience, let's call them cries. This term should not surprise you if you have ever heard the plaintive wails of a lamb or puppy taken away from its mother. They're definitely crying. One of the research projects I'm doing with my colleague David Reby (Remember?), a specialist in mammalian voice production, is looking at the cries of human babies. We want to identify the information carried by these cries, analyze what acoustic traits they are encoded by, and study how the individual receivers (parents among others) decode them—in other words, how information is transmitted from the baby sender to the adult receivers. Alexis Koutseff, then a postdoctoral student at our laboratory, and my colleague Florence Levréro recorded babies aged two to three months while they were bathing (not all babies cry during their bath, but we managed to record some of them anyway) and the same babies when they were vaccinated at the pediatrician's.[516] In addition, the babies were recorded under two different conditions. Some were given a first injection of a vaccine that was supposed to be painful and then a second injection of another vaccine that would be less painful. Other babies received the same two injections, but in reverse order: the mildly painful first, followed by the more painful one. Acoustic analysis showed that the recorded cries can be placed in a two-dimensional acoustic space, and that they move around in that space according to the pain the baby is experiencing. Imagine two lines (two axes) that intersect at right angles and delineate four quadrants: upper right, lower right, lower left, upper left. The horizontal line represents the *roughness* of the cry, ranging from the most harmonic ("ouuinn") on the left to the most guttural ("iiirraahh!") on the right. The vertical line represents the pitch of the cry, ranging from the lowest-pitched cries at the bottom to the highest-pitched at the top. The cries recorded during the bath are in the quadrant at the bottom left ("ouuuiinn"). Let's follow the trajectories of the cries caused by the vaccines. Those following a first injection of the painful vaccine are resolutely on the right ("iiiiRRRRhh!"). Once the injection is finished, the cries move to the point where the axes cross, becoming less rough ("iiiiAaaarr!"). The pain subsides a little. The second vaccine, which is supposed to be less painful, is then injected. The cries start again on the right, without, however, reaching their first position ("iiiiRhhh!"), then return to the center of the acoustic space ("iiiiAaaarr!"). The acoustic characteristics of the cry therefore reflect the level of pain experienced by the baby. Roughness is obviously the best marker: the more intense the pain, the more the cry grates on the ears. The second, less reliable marker is the pitch of the cry, the fact that it is more or less high-pitched, with a tendency to be higher-pitched when the pain is stronger.

The group of babies who received the vaccines in the reverse order (least painful first) would teach us that the order of injection influences the baby's pain. The first cries caused by the least painful vaccine are slightly shifted to the right from the vertical axis. It hurts, but not that much. Once the injection is finished, they cross the vertical axis to the upper left. That's better! The second injection, of the painful vaccine this time, makes them come back to the center, so much less to the right than when this vaccine is injected first. You get used to everything.... The last cries are in the lower left quadrant, similar to the bath cries, harmonious and lower-pitched. The baby is still not happy, but it is no longer in pain. The moral of the story is that the least painful vaccine should be injected first. That's a useful bioacoustic experiment, isn't it?

To see if adult listeners can extract pain information as efficiently as our acoustic analysis, we conducted psychoacoustic experiments. The principle is simple; it is a matter of making people listen to cries and asking them to rate them on a pain scale between 1 (no pain) and 7 (very strong pain). Our results showed that adults are able to evaluate the pain perceived by a baby based on the acoustics of its cries. They could easily distinguish between cries caused by an injection and those emitted during bathing. However, the tested adults were not very good at distinguishing variations in the level of pain; they did not detect the difference between the cries provoked by the two vaccines. Perhaps this is because they heard cries from unfamiliar babies. I'm fairly confident that people who know the baby they are caring for can measure the pain level very accurately by hearing their own baby's cries.

Rules for encoding emotions across the animal kingdom suggest the possibility of communication between species. A recent study has confirmed that we humans are able to distinguish between vocalizations marking different levels of arousal, even if these signals come from animals as varied as a monkey, a pig, a panda, an elephant, a bird, an alligator, or . . . a frog![517] Susan Lingle's work at the University of Winnipeg, Canada, points in the same direction: Mule deer and white-tailed deer mothers are attracted to the baby cries of a variety of mammalian species.[518] Unsurprisingly, all these vocalizations follow Morton's principle. But then again, are we able to accurately judge these vocalizations? That's the

Chimpanzee

question Taylor Kelly, one of my students when I was a visiting professor at Hunter College in New York, and I asked ourselves. Taylor made adult women and men listen to the cries of human babies, as well as the cries of baby chimpanzees and bonobos.[519] The results showed that the people tested assessed the degree of distress expressed by the vocalizations based on their pitch (high/low). They applied the same rule to all cries, those of both baby humans and baby apes. Because bonobo calls are extremely high-pitched, all were consistently rated as expressing severe pain. On the other hand, the cries of baby chimpanzees, which were quite low-pitched, were recorded as indicating low levels of distress. These results show that, in the absence of familiarity (our listeners had probably never heard baby chimp or bonobo cries before), our ability to assess the emotional content of our closest cousins' vocalizations is biased. The interspecific value of the signals coding for emotions, if it is real, should not be overestimated.

What about our crocodiles? They have no tail posture, no eye rolling, no bristling fur. Encouraged by our first observations with the Niles of Agadir responding to the cries of human babies, we repeated the experience with the cries of baby chimpanzees, then baby bonobos. Each time the crocodiles reacted, running toward the loudspeaker. On closer inspection, however, we found that the number of crocodiles approaching the loudspeaker and their readiness to move varied according to the sound stimuli. The cries of human babies recorded during a vaccination session were the most attractive; those recorded during a bath were much less effective. In conclusion, crocodiles also perceive the level of stress encoded in the cries of human babies. By comparing their reactions to human and ape cries, it appears that the *rougher* the cry, or more unpleasant to hear, the more irresistible it is.

These reactions to signals that do not belong to one's own species raise the question of a possible contagion of emotions through vocalizations. We humans know this very well. A call or a cry can give us goose bumps or bring tears to our eyes. A happy laugh will put us in a good mood. We are accustomed to these phenomena of empathy—when we feel what others feel. Now, do our female crocodiles, who rush to the cries of a human baby as well as to the calls of their own babies, really feel the distress expressed in these vocalizations? I think it's likely. There is no objective reason why evolution has not allowed or even promoted empathy in crocodiles (although it is possible, of course, that crocodiles are attracted to crying human babies for less friendly reasons—to eat them, for example). It seems to me that a mother Nile will be more willing to spend energy to come to the rescue of her young if she actually feels the distress. To support this point, there is growing experimental evidence in mammals for emotional contagion through the voice.[520] This contagion even appears indispensable in animals living in groups and having to coordinate or cooperate. It is reasonable to assume that the fright experienced by the vervet monkey or meerkat seeing an eagle and triggering the alarm spreads to other members of the group. Primatologist Julia Fischer, director of the German Primate Center, has observed that chacma baboons pay more attention to alarm barks produced in response to dangerous predators than to other alarm vocalizations.[521] You may remember the chacma who had lost his group and whom I heard barking plaintively for hours. The emotion I felt when I heard him cannot be far removed from what one of his group mates would have felt if they had been able to hear him.[522] The contagion of emotions through the voice is probably a very common phenomenon in mammals. It is certainly highly developed in those with complex social relationships, such as monkeys, apes, meerkats, or hyenas, when understanding how the other individual feels is particularly important.

And in other animals? In birds, for example? When we were studying zebra finches in our laboratory, my colleague Clementine Vignal had the idea of looking into this question. Do you remember from chapter 11 that this bird cannot do without its fellow birds? I have already told you that, when we tested the ability of males to identify their partners by their voice, we had to introduce companions into the experimental room. Alone in the room, the male responded poorly to the playback of its female's calls, and did not differentiate between its partner's calls and those of other females.[523] We then hypothesized that the stress caused by loneliness must be significant and must explain this difference in behavior. To test this hypothesis, we repeated the same playback experiment, but this time we paid attention to the blood level of a stress marker hormone: the corticosterone. In this new experiment, some males that were with companions (control group) heard calls from females while in the presence of mates. Other males (isolation group) heard them when they were alone. Finally, other males ate a few seeds soaked in the stress hormone just before hearing the playback signals in the presence of mates (CORT group). Clementine's doctoral student recorded the test males throughout the playback. She then analyzed the acoustic structure of their calls. As expected, the males in the isolation group called less than those in the control group. More importantly, their calls were different—higher-pitched and longer. The CORT males called just as much as the controls, but their calls resembled those of the birds in the isolation group. Blood tests showed that the isolated birds had more of the stress hormone than the control birds. Let's summarize the situation. In response to calls from females, a male that feels stress (either because it is stressed from being alone or because it has just ingested the stress hormone) produces calls that are higher in pitch and longer than a bird accompanied by other birds and therefore not stressed.[524] So the calls of the zebra finch provide information about its emotional state. But the story doesn't end there. According to other experiments, if the female listens to the calls of her stressed male mate, the amount of stress hormone circulating in her blood increases.[525] You read correctly: the partner's voice is enough to stress the female! She doesn't need to see him to feel the same way he does. When she hears her partner call out to her in an anxious voice, the female zebra finch is also frightened. The contagion of emotions through the voice exists in birds. Quite an interesting discovery. Mammalian vocalizations are not the only ones that express emotions.

The fact that emotions are coded according to rules that transcend the animal kingdom has interesting applications. To end this chapter, let's look at two of them: communication between humans and domestic dogs and the use of distress signals to frighten birds.

At the risk of disappointing you, I have no particular fondness for pets. (Nobody's perfect!) However, when I was a professor in New York, I had to find subjects to study. New York is an extraordinary city, of course, but in terms of wilderness it's not ideal. One day, Tobey Ben-Aderet, one of the students whose master's thesis I was supposed to supervise, suggested that I take an interest in the communication between humans and dogs. "To pay for my studies, I work in a kennel. It would be easy for me to record them. It is a subject that fascinates me." So off to the dogs we went! I sought advice from my colleague David Reby, who is a true dog lover. Then we thought about the scientific question that Tobey could tackle.

When we talk to our babies, we use a particular voice register, characterized by a higher and variable pitch, a slower rhythm, and clearer articulation. Why this habit? Perhaps because of the emotion caused by the toddler's tender face. Maybe because it makes it easier to learn language through hearing. Or maybe both? In any case, what we find is that we talk to our dogs in a slightly similar tone. Is this a consequence of the juvenile appearance of some of our four-legged companions, or an attempt to interact with a living being who is not gifted with language and whom we judge to be of limited intelligence? In the first case, we should restrict "baby talk" to puppies. In the second case, we should continue to talk to adult dogs in this way. Interesting hypotheses. Tobey's scientific objective was well established.

First, we had to record "dog-directed speech." Tobey presented photographs of dogs to people who had to say the following sentences,

Dog

while imagining themselves talking to the dog in the photo: "'Hi! Hello cutie! Who's a good boy? Come here! Good boy! Yes! Come here sweetie pie! What a good boy!" The same people then had to say those sentences again, but in a neutral tone. Acoustic analysis of the recordings confirmed that our way of talking to dogs is very similar to "baby-directed speech." And above all . . . that we use this register regardless of the age of the dog.

You're probably thinking, What do the dogs think? Well, Tobey asked them straight out. With the help of another student, Mario Gallego-Abenza, she played her recordings to puppies and adult dogs. Can you imagine the scene? The dog is in a comfortable room, and a loudspeaker on the floor is playing the phrase pronounced in "dog-directed speech" and then the same sentence pronounced in a neutral tone. The results of the experiments have the merit of clarity. When the tone was neutral, both adult dogs and puppies showed very little interest in the loudspeaker. On the other hand, the puppies became very excited when listening to the "dog-directed speech," barking and approaching the speaker, while the adult dogs ostensibly ignored the recording.

What conclusions can be drawn from this? First, dog-directed speech is effective for engaging an interaction with a puppy, but loses its effectiveness for an older dog. The older dog is probably waiting for other signals from us, such as gestures or facial expressions. Or it may not respond to an unfamiliar voice coming out of a loudspeaker. Yet we talk to adult dogs like babies. The reason may be that, consciously or unconsciously, we may feel that the dog may not understand us. As if it were one of our babies, we hope it can make progress. If we talk to our dog in this way, it is because we would like the dog to one day respond to us . . . by talking.[526]

With this example of dog-directed speech, we can see that there is still progress to be made in terms of communication between species.[527] However, sometimes precise knowledge of the coding of information makes it possible to finely, and I would even say artificially, control an acoustic communication system. This is what is done by using distress calls and their synthetic avatars to scare birds at airports.

Birds have always been a challenge for airplanes—unless it's the other way around, you might say, but that's beside the point. On September 7, 1905, Orville Wright's plane inaugurated what was to be a long series of collisions between birds and airplanes in flight.[528] These accidents occur mostly at low altitude or when the aircraft is taking off. The cost is considerable, several hundred million dollars a year, not to mention the loss of human (and bird) life. How can we limit these collisions? The method considered the most effective and least expensive is to use distress calls to disperse the birds. Distress calls are vocalizations emitted by most birds when they are captured or injured by predators. They differ from the alarm calls that we have discussed extensively, both in the circumstances in which they are emitted (an individual producing an alarm call is not in the claws of a predator) and in their acoustic characteristics. Distress calls appear as complex, highly frequency-modulated sounds (i.e., rapidly changing from high-pitched to low-pitched and vice versa), lasting about half a second ("Zeeoooop!"). When birds hear a distress call, they react very strongly, fleeing on the wing. Sometimes, as with seagulls, they fly over the area where the call comes from before flying off into the distance. But the result is there, very clearly; the distress calls are amazing bird scarecrows.

You may remember that I prepared my doctoral thesis in Jean-Claude Brémond's laboratory, where Thierry Aubin also worked. These two researchers devoted part of their careers to the study of these distress calls. The acoustic details differ according to the species of birds—the distress call of the black-headed gull is not exactly the same as that of the starling, for example. However, Aubin and Brémond observed that the similarities are strong and that there is a true acoustic convergence between the distress calls of different bird species. This convergence is accompanied by an interspecific information value: gulls react to starlings' distress calls as they would to those of other gulls. Distant animal species, such as deer or wild boar, do the same. In fact, different species use the same law for decoding distress information.[529] Based on this common code, Aubin and Brémond developed computer-made signals, mimicking distress calls but exacerbating the acoustic parameters carrying the distress information, thus amplifying their effectiveness and limiting habituation. What a sight it was to see how they were able to test the effectiveness of their signals: stationed in a car some 150 meters from a group of about 40 herring gulls, black-headed gulls, lapwings, ravens, and starlings, with the loudspeaker on the roof suddenly emitting about 60 distress calls per minute. All the birds flew away, then dispersed into the distance—exactly the desired effect on an airport runway before a plane takes off.[530]

From the calls of baby crocodiles to the distress signals of birds, from the cries of baby humans to the neigh of joy of the horse reunited with its companions, you can see that acoustic signals powerfully encode the emotions of the sender. Besides the static information related to the stable characteristics of the individual (its size, age, sex, genetic inheritance, etc.) that define its vocal signature, sound signals also carry dynamic information that reflects the individual's current state. The challenge for the receiver is to decode all this information in order to correctly interpret the received signal.

16

The booby's foot

ACOUSTIC COMMUNICATIONS AND SEX ROLES

Village of San Blas, Nayarit Province, Mexico. The rendezvous on the beach where we are to board was scheduled for six o'clock in the morning. But the beach is empty, and the village next door is still asleep after yesterday's party. We are waiting patiently. The sun rises quickly. On the horizon a flight of pelicans materializes—at least they are awake and going fishing! A few hours go by before two men arrive. Not very talkative. One carries jerricans, approaches a boat, and starts to fill the tank, with a cigarette in his mouth. "*Aqui esta tu barco. El es el que te conduce*," he says, pointing to his massive companion. Moments later we are on our way, quickly lost in the blue vastness. From the village to the island is 70 kilometers. Four hours in the boat! It all started with a phone call from Fabrice Dentressangle, a former student of whom I had very fond memories. "Nicolas, I'm preparing my doctoral thesis with Roxana Torres in Mexico. On the reproductive behavior of the blue-footed booby. I'm working on an island paradise!" And he continued, "In the blue-footed booby, the female and the male have totally different calls. They take turns at the nest, and I'd like to know if they recognize each other vocally. Wouldn't you like to come and pay me a little visit?" You bet! I immediately took the opportunity. A phone call to my friend Thierry, plane tickets, and a few months later, we were there. "Did you see that?" asks Thierry suddenly. "A shark's fin! These waters must be infested with them." We're happy to be away again, disconnected from the world. As in a dream, the mysterious *Isla Isabel* approaches.

The blue-footed booby *Sula nebouxii* is a surprisingly good-looking seabird. You may already know them because they are often seen in documentaries about the Galápagos Islands, where they are present in large numbers. It is a large bird, with a long dagger-shaped beak, perfect for harpooning fish; a big head and neck speckled with brown; a light eye with a distinct black pupil; a brown back; and a white breast. As its name suggests, its feet are blue. You must see the female and male displaying together. They walk side by side, each raising their feet high, their toes as far apart as possible to spread their azure webbing. And each one calls out. The female makes her "quack-quack," the male his "phiew-phiew." This difference in their calls is almost the only clue to telling them apart. Along with pupil size: a pinhead in the male, rounder in the female—not easy to spot from a distance. Sexual dimorphism comes down to very little in the booby.

Isabel Island is a volcanic islet less than a kilometer wide and a kilometer and a half long, where thousands of seabirds breed. Thousands of frigate birds, brown boobies, red-footed boobies, blue-footed boobies, brown noddies, seagulls, and pelicans flock here among the iguanas, omnipresent on the island. The blue-footed boobies settle on the top of the beach, a nest every hundred feet. They are a monogamous species, where female and male share the incubation of the eggs and the rearing of the young. The parents take turns on the nest. While one fishes at sea, the other stays ashore. The reunion is sonorous—both partners start calling before the returning fisher has even landed.

As soon as we disembark, we find Fabrice. Present on the island for many months, he has become the manager of the camp where about 15 students are each working in their own area of study. This small colony is a real-life *Swiss Family Robinson*: a long table for meals taken together; a stove, a few pots and pans regularly visited by the iguanas, an old fridge serving as a pantry; water reserves; solar panels to recharge the batteries of the equipment; tents to sleep in, between which blue-footed boobies wander happily. The atmosphere is both relaxed and hard-working. Thierry and I have to be efficient because we have only planned a fortnight

Blue-footed booby

♀ ♂

on the spot. Fortunately, Fabrice and his team have prepared the work well. The nests are already carefully labeled and the youngsters are all banded so that we can easily identify the animals. From the first morning, the recordings begin. Equipped with a long pole, we put the microphone right next to the nest, wait for the return of the female or the male that has left for the sea, and record the vocalizations of the pair during their re-union. About 20 pairs are "in the can" within a few days. All that remains is to conduct the playback experiments. That is another kettle of fish.

Placing the loudspeaker near the nest is not complicated, especially since the birds seem to be completely unaware of it. Unroll the cable (we didn't have wireless speakers), put ourselves at a distance, and wait for the right moment—that is, the moment when the bird left in the nest, whose partner's calls (or those of an unknown individual of the same sex) we are going to play, is calm. When the young are not beg-ging. When there isn't a big iguana coming to get who-knows-what that's scaring all these little guys off. . . . Here we are. First try. The fe-male call comes out of the speaker. Surprised, the male glances briefly up at the sky. Then . . . nothing. "Here's a response that is going to make life difficult," says Thierry, used to the exuberant vocal demonstrations of penguins. Invoking one of his favorite phrases, he declares, "We have to make a decision." On the way back to camp walking on the beach, the decision is made: we will conduct these experiments rigorously but blindly. I will play the signals from the tape recorder, without passing comment. Thierry will evaluate the behavior of the bird being tested, without knowing whether it is the call of the partner or that of a stranger that we are playing—seriously frustrating for Thierry who, as the days go by, will never know if the results of the experiments validate or not the hypothesis of a partner's voice recognition. But it's the only way to be sure not to bias our data.

In the blue-footed booby, female and male have very different calls. The female's "quack-quack" is weakly modulated but extends over a wide frequency range. It sounds like a cartoon duck. The male's "phiew-phiew" is wheezy, more muffled, like a failed attempt by someone practicing a whistle. Our acoustic analysis showed that these two calls can identify individual boobies with a success rate of about 45%. According to our calculations, chance would result in only a 6% success rate, so both the female and male calls have a fairly accurate individual signature. Our playback experiments have confirmed that birds in the same pair use these signatures to recognize each other. Of those individuals we tested, 11 of the 14 females and 11 of the 15 males showed a more pronounced behavioral response to their partner's voice. Parity could not be expressed more clearly.[531]

I left *Isla Isabel* with my head full of thousands of birds, whales jumping out of the water just a stone's throw from the beach, friendly conversations with people who know the value of calm, and images of flamboyant sunsets (which earned me a few laughs from Thierry: "You could have taken the same photos in Brittany!"). A feeling of paradise. One day Fabrice suggested we go down into the crater of the volcano. There, forget the sound of waves. Only the calls of the superb frigates— modern-day pterosaurs, hijackers snatching their fish from other birds—broke the silence. *Isla Isabel* takes you back in time to a world where the human species was absent.

Throughout these pages, we have seen that sexual dimorphism—the fact that female and male have morphological and behavioral differences—is quite widespread in the animal kingdom.[532] Sexual selection has a lot to do with that, with its two components, *intrasexual selection* (competition between females or between males) and *intersexual selection* (specific preference for the characteristics of the other sex). Both encourage the expression of extravagance, especially in males: imposing body size (think of elephant seals) or complex sound signals (some birdsongs), to name but a few. They contribute strongly to the exaggeration of differences between females and males. In birds and insects, it is often the males that sing to attract females and repel intruders. But let's not forget that one of the major characteristics of the living world is its diversity. This differentiation in the appearance of females and males and their behavior is not an absolute rule. Far from it. And we are not always aware of it. Among scientists, as everywhere else, the way we perceive things is tainted by ideological biases. This is even more true when it comes to studying and understanding the behavior of females and males.[533] A 2011 study published in the journal *Animal Behaviour* shows that the vocabulary used by scientists to describe female and male behavior is incredibly stereotypical, generally referring to females as passive while males are said to be the real actors.[534] In the blue-footed booby, however, the obvious vocal dimorphism does not translate into asymmetrical recognition. Female and male are equal actors in acoustic communication. There are many examples of this equality. Let's observe the shearwaters, seabirds in which the balance between the roles of females and males during communication is astonishing.

Do you remember Charlotte Curé? I told you about her research in chapter 6 when we were with the whales and dolphins. Well, Charlotte started her career with a thesis on the acoustic communications of Mediterranean shearwaters. Shearwaters are agile seabirds that only come ashore to nest. They choose cliffs where it will not be necessary to move around on the ground because they cannot walk. The nest is usually a burrow, at the bottom of which female and male take turns brooding and caring for the young. When Charlotte asked us to supervise her thesis, Thierry and I were very interested. How does a shearwater find its burrow? Previous work had shown that olfaction could play a role, as it is highly developed in these birds. But the fact that shearwater colonies are noisy suggests that acoustic communication plays a

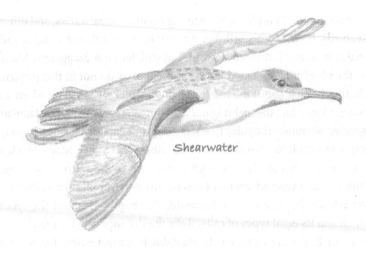

Shearwater

role. Sight, perhaps?, you might think. Wrong! Shearwaters return to their nests . . . at night.

Charlotte had not chosen an easy subject. Crawling over the cliffs in the dark, risking breaking her bones at any moment, putting her microphone or loudspeaker here and there at the entrance to the burrows to record the occupant or test its ability to identify this or that sound signal: that was her daily life! The shearwaters' call is a kind of mewing, high-pitched, strangled "eyoh-eee-eyoh-haaa." Charlotte patiently recorded and analyzed the vocalizations of many individuals and several species of shearwaters. Each time, females and males had different voices.[535] Following the example of what we had observed in boobies, Charlotte showed vocal recognition between the partners of the pair (female and male respond to each other) and detailed the acoustic parameters that supported it.[536] But in her experiments, she observed something unexpected. When she had a bird listen to calls from a stranger, the bird would answer if the stranger was of the same sex, but it would not respond if the stranger was of the opposite sex. In other words, I respond only to my partner and to calls from strangers of my sex. No extramarital relationships among shearwaters. Both female and male only speak to same-sex intruders. Intrasexual competition is mediated by acoustics.[537]

All right, that's understood. Among boobies, shearwaters, and other seabirds, there is a beautiful parity between the roles of females and males in acoustic communication. But still, let's not exaggerate. Many of the songbirds discussed in previous chapters do not fit this pattern, do they? What about the robin and the warbler we see everywhere in our gardens? And the nightingale? And the canary? For all these familiar species, we must recognize that it is indeed the male that sings, bravely exposing itself to predators, and the female that waits, wisely hidden away in the bushes. It is enough to confuse anybody! But the song, however demonstrative it may be, is not the only vocalization in the bird repertoire. Remember our discussion in chapter 11 about the zebra finch and its eight types of calls? Take the distance call, for example— the one that allows partners to reestablish contact when they've lost sight of each other. Here again, female and male produce different versions, and both versions allow equally effective vocal identification of the partner. It was Solveig Mouterde, whose thesis I was supervising with Frédéric Theunissen, who explored this question.[538]

We already knew that distance calls carry an individual signature, but we didn't know how far that signature could be transmitted. To find out, Solveig and I conducted propagation experiments. The principle is simple: a loudspeaker and a microphone are placed at different distances from each other (2, 16, 64, 128, and 256 meters). The biggest difficulty of the experiment was to find the right field—flat and free from road noise. The results showed that, even after more than 200 meters of propagation, there is still enough information in the calls for us to recognize the individual, whether it's a female or a male.[539] Our acoustic and statistical analyses revealed that a call, quite short, and whose intensity is made very weak by a journey of several hundred meters, can still be identified as Jane, Lucy, Matthew, or Mark. But what about the birds? Can they identify it?

Solveig tested female zebra finches. She asked them to distinguish between the calls of two males, according to two different experimental protocols. In the first, the females learned to distinguish between calls from two unknown males recorded at a short distance (2 meters). Once they had successfully learned this, Solveig had them listen to calls from the same males recorded at greater distances, up to 256 meters. In the second protocol, the females were tested several days in a row with different pairs of males and the calls propagated at different distances. This meant that they did not have the opportunity to practice distinguishing between the two males. The results showed that, even in the untrained situation, females were able to distinguish between calls from males up to 128 meters apart. In the first protocol, training increased the performance of the females and they were able to distinguish between the two calls from the males even if they had previously been transmitted over 200 meters. In short, female zebra finches are experts at identifying males by means of a call, even from a great distance.[540] For time reasons (a thesis must be done in three years in France), Solveig could not do the symmetrical experiments, with males. But as the individual signatures coded in the calls remain well pronounced at long distance for both male and female calls, it is quite likely that males would pass the test as well.[541] Moreover, Solveig and Frédéric showed in other experiments that the neurons allowing individual recognition of the partner are equally effective in both female and male brains in zebra finches.[542]

Let's get back to the songs—and ideological bias with regard to male singing. Even Darwin was guilty of it, suggesting that the first role of female birds was to listen and choose "the most melodious and beautiful males."[543] However, the situation is actually highly contested. A study published in Nature Communications reports that the species of birds where the female sings are numerous.[544] You want figures? Female song is found in 229 out of 323 species of songbirds, or 70% of the species. That's hardly negligible. By reconstructing the history of songbirds, the authors of the article show a high probability that both females and males could sing in the ancestral species, about 40 million years ago.[545] During the diversification of bird species, most of the lineages have retained singing females and males. Most of the species where females no longer sing live outside the tropics, establish territories for a short period of time, and show pronounced sexual dimorphism both morphologically (e.g., feather color) and behaviorally (females and males have different roles). These characteristics are often found in temperate environments, characterized by a reduced duration of spring and also reduced availability of food resources—constraints that explain the task differentiation between females and males. However, even this does not prevent some birds in these regions from having singing females, such as in the European robin *Erithacus rubecula*.

In many tropical birds, females and males produce equally complex songs.[546] Males and females sing to defend food resources or to attract a mate.[547] Sometimes females and males sing together, forming duets to defend a territory or simply to maintain contact, especially during breeding.[548] Let's go listen to one of these duets in the home of the plain-tailed wren *Pheugopedius euophrys*, which lives in certain regions of the Andes in South America, between two and three thousand meters above sea level. In the dense bamboo groves where these wrens are found, the duets are emitted at a high amplitude (over 80 decibels) and can be heard at over 200 meters. You can't lose your partner! But how do they coordinate when singing together?

During my visit to Eric Fortune's laboratory at the New Jersey Institute of Technology, I was very impressed. Eric is a neuroethologist, and among other topics he details precisely how animal brains can interact with each other during acoustic communication. With the plain-tailed wren, he has an excellent study model. The spectrogram of the song scrolls on his computer screen: "Woopweewoopoopwee . . . woopweewoopoopwee"; the notes flow continuously. "One bird? . . . No, two!" says Eric. "The female and her male! Incredible, right?" Incredible indeed. A remarkable duo of speed and precision, exactly as if a single individual were running through its song. Females and males alternate notes, each for less than 300 milliseconds with intervals of about 20 milliseconds. "Woopwee," begins the male; "woop," answers the female; "oop," takes over the male; "wee," ends the female—"woopweewoopoopwee." And the pair repeats this sound pattern a hundred times for two minutes. Often birds improvise with variations, a change of syllables or their order. You have to know your partner well to succeed in such a tour de force. Newly formed pairs get a bit muddled up, miss notes, and make shorter duets. Like two pianists in front of a four-handed piece, it takes practice. The first attempt is rarely a success—one goes too fast, the other too slow. It is only by

listening to the four hands together, many times, that the duet is successful, as if there were only one musician. When wrens sing in a pair,
they can hear themselves and their partner. This acoustic feedback
allows them to constantly adjust their vocal production.[549, 550] In fact,
Eric has tracked the activity of wren brains during these moments,
and has shown that the singing area (the HVC) is activated chiefly
when the bird hears a full duet, including all the notes. One wren
knows not only its own score but also that of the other. Sound recordings that Eric and his team have made in the wild also show that the
male makes more mistakes than the female. Every once in a while he
forgets a note, especially when the duet lasts a long time. It's as if he
can't keep up. The female continues singing anyway, but with longer
rests between notes—like the piano teacher slowing down the tempo
to give the student time to get back into the melody. (Believe me, I've
been through this before.) However, Eric and his team do not say if
the female wren gets tired of the males who make mistakes over and
over again . . . ! Nevertheless, as you can see, the respective roles of
the female and male in acoustic communication in birds are much
more varied than is usually assumed.

Now let's jump to the human species. I've been studying the cries
of babies for a few years now, and there, too, surprises await us. It was
my colleague Philippe Gain, with his thousand ideas per minute, who
had strongly encouraged me to take an interest in babies' crying.
I quickly noticed in the myriad publications on this subject the same
two questions: how to use babies' cries to identify diseases or dis-
abilities and why babies often cry unexplainedly. Very few scientists
had looked at the cry as a communication signal—the ethologist's eye
had not been focused there. That was enough to arouse my curiosity.
Inspired by the research carried out with Isabelle Charrier on baby fur
seals, I decided to start by working on the recognition of the baby by
its parents. Is it possible to identify one's own baby just by its cries?
I talked to David Reby, who was enthusiastic about the idea, and we
decided to start a study. We were in luck: Hugues Patural, a former
high school classmate, had become the head of the neonatal ward at
Saint-Etienne Hospital and was fascinated by the project. Florence

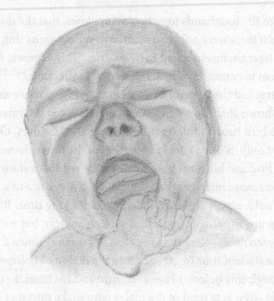

Levréro, who was always interested in the smallest little primate, was also on board. Then we welcomed a postdoc researcher, Erik Gustafsson, who agreed to join us in the laboratory. The new "baby cries" team could get down to work.

Of course, you've heard of maternal instinct—the idea that mothers everywhere have a special gift for caring for babies, a quality that is innate, written in the genes; whereas this gift is absent, or at least very underdeveloped, in fathers. This opinion is widespread, both among the general public and among biologists. Yet nothing particularly obvious supports it. In mammals, many different systems of social organization exist, from species where females and males meet only very briefly, with the female remaining alone to raise her young, to cooperatively bred species, where all members of the group are likely to participate. Take monkeys, for example, where there is great diversity.[551] In yellow baboons, males have very little interaction with their young. But in siamangs, tamarins, marmosets, titi monkeys, and owl monkeys, fathers are very involved, and sometimes other individuals (such as the previous year's young) as well. Among the titis, it is the males that carry the young from the age of three weeks. The baby titis prefer to go with their father rather than with their mother.

What about the human species? Of course, we know that pregnancy, with its considerable hormonal modifications, has important effects on the maternal organism, including the brain, but the extent of these effects and their true nature remain poorly understood.[552] In addition, the mother is usually not the only person to take care of the baby. Among the few remaining hunter-gatherer societies in the world, cooperative breeding is the norm. In the pygmies of central Africa, an average of 14 people take care of the same baby. By the age of four months, the baby will have been carried less by its mother than by other people.[553] But in our "modern" societies, things are different. The basic family is nuclear, and the perception tends to be that it's primarily the mother who takes care of the baby—at least in general, right? Well, it's not as obvious as that. Following the birth, the mother doesn't hesitate to entrust the baby to the good care of the hospital staff; in other words, to leave it in the hands of people she doesn't know. And when she returns home, it is not uncommon for the father, siblings, grandparents, and even neighbors to be able to give the baby bottles, change diapers, or bathe him or her. I'm not even talking about what happens next: day care, school, leisure activities . . . multiple people will take care of the baby and then the child. In short, let's stop beating around the bush—we are a kind of primate with cooperative breeding.

Does this characteristic of the human species have consequences for our ability to recognize a baby by its cries alone? Our first hypothesis was that both mothers and fathers should be able to do this, as equals. However, two previous studies conducted some 40 years ago suggested that mothers were much more gifted than fathers.[554] Not really convinced by these results, we decided to start from scratch, doing what we would have done for any animal species: recording cries (at the time of the baby's bath), looking for the presence of vocal signatures, and testing mothers and fathers through playback experiments. Acoustic analyses of the cries of about 30 babies first showed that the individual signature was indeed there. You probably expected that. The acoustic characteristics of each human baby's cry, as in a fur seal baby, thus make it possible to identify reliably who is crying. To test whether parents can detect and use this vocal signature, we had them listen to the cries of different babies. Of the 30 recordings heard, only 6 were from their offspring. The test results were conclusive: the parents correctly identified more than 5 of the 6 cries on average.[555] The only parent who had some difficulty identifying their baby's cries was spending less than four hours a week with the baby. In other words, they never heard it. (I'll let you guess whether they were a mother or a father.)

In a cooperatively breeding primate species, individuals other than the mothers and fathers are therefore involved with the babies, and this is the case in the human species. But then—if we follow this reasoning—everyone should be able to learn to recognize any baby by its cries. We had to test that hypothesis. Aurélie Cantais, an intern working with Hugues at Saint-Etienne Hospital, and Hélène Bouchet, a postdoc who had just joined our ENES team, were about to do just that. In the first part of the study, Aurélie recorded newborns in the days following their birth and did playback experiments with the mothers. The results showed that the mothers increased from 40% to 80% recognition of their babies between one and three days after birth. I was very amused by this result. Remember: it takes between two and five days for mother-young recognition to be established in the fur seal on the island of Amsterdam. Hélène invited adults—women and men without children—to the laboratory and played them a baby crying. "Pay attention! This is your baby's cry!" she told them. Each person had to listen to a different baby, of course, and a number of cries varying between one and six sequences of about 10 seconds each. People willingly played the game, listening carefully to the cries of a baby they had never seen before. A few hours later, they returned to the lab for a playback test, where they were asked to identify "their" baby every time they heard it. Two main results were obtained. First, those who had cared for a baby in the past performed better, recognizing "their" baby more than 60% of the time compared to 42% for the others. All they had to do was hear two or three sequences of a few seconds each of "their" baby's cry to be able to distinguish it from any other baby. Such rapid learning is remarkable. And there was no difference between women and men.[556]

With these studies, we showed that recognizing a baby's identity through his or her cries can be learned. What about our ability to assess the level of pain encoded in cries? With Siloé Corvin, a PhD student at the lab, we recently hypothesized that, again, experience matters. To check this hypothesis, we tested different categories of adults: people with absolutely no experience with babies; people with a little experience (having babysat or having had younger siblings); parents of children over five years of age; parents of very young children, less than two years of age; and professionals in the field of pediatrics. Each person was first asked to listen to eight cries, recorded during bath time, of a baby who would be considered that person's "familiar baby." The familiar babies were different from one person to another. The next day, each person again heard bath cries of "their" baby (but not the same recording sequences) and also cries of the same baby recorded during a vaccination session at the pediatrician's office. They were also played the bath and vaccination cries of a baby they did not know. For each of these cries, the question asked was simple: Would you say that this cry expresses simple discomfort (bath cry) or real pain (vaccine cry)? Well, what did the results say? That the listeners' ability to categorize cries as "discomfort" or "pain" depends on their past and current experience with babies. And the results couldn't be clearer. People who had no experience with babies decided at random whether the cry they heard was one of discomfort or pain. Those with moderate experience showed some ability to identify the cry, recognizing only the pain cries of the familiar baby. In contrast, adults with strong experience with babies, i.e., those who were parents or pediatric care professionals, were able to identify the familiar baby's discomfort and pain cries. Remarkably, parents of very young children were also able to do the same for the unfamiliar baby! Gaining experience is therefore the key to successfully decoding babies' cries. This is true for both men and women.[557]

We then moved on to explore adults' perceptions of cries by looking at the influence of gender stereotypes. We all know that men have a lower-pitched voice on average than women. Does this knowledge influence how we perceive the cries of baby girls or boys?[558] The first step in our experiments was to see if there are differences between the cries of baby girls and baby boys. In particular, are baby girls' cries higher-pitched than those of baby boys? Acoustic analysis showed that this was not the case. In a second step, we asked adult listeners to give the sex of the babies whose cries they were listening to. Surprise! Instead of answering that they didn't know, people classified the cries as girl or boy based on the pitch of the cry. The higher-pitched the cry, the more likely it was classified as a girl's cry; the lower-pitched it was, the more likely it was classified as a boy's cry. So people did indeed base their sexing of cries on what they knew about adult human voices. It is nevertheless rare that one is led to define the sex of a baby by using its cries; there are other more direct ways! But two experiments followed.

This time, people listened to babies' cries whose sex they knew. More exactly, we told them what the sex of the baby was. They had to assess the baby's degree of masculinity or femininity just by hearing the cry. Again, the pitch of the cry was critical. Whether they were listening to a girl's or a boy's cry, adults rated babies with higher-pitched cries as more feminine or less masculine. What they didn't know was that we played them the same cry, telling them that it was either a girl's or a boy's. You can see that here things are more subtle. Imagine taking care of a baby girl or a baby boy. Depending on whether the cries are higher- or lower-pitched, you may attribute more or less feminine or masculine characteristics to her or him without being aware of it (these are called *gendered traits*). We also asked people to rate the degree of distress coded in these cries. The higher-pitched the cry, the more people thought the baby was in pain. Most importantly, the same cry was rated differently by male listeners depending on whether we told them it was a girl's or a boy's cry. Female listeners did not make a distinction. It is interesting and important to know, therefore, that the perception and interpretation of a sound signal of major importance to our species may be subject to cognitive bias.

Our studies on baby cries are ongoing. We should know more in a few years. With our colleagues Roland Peyron, Isabelle Faillenot, and Camille Fauchon from the University of Saint-Etienne, and François Jouen from the École pratique des hautes études (EPHE) we are exploring, for example, how the brain perceives the cries. We are using the techniques of functional magnetic resonance imaging (fMRI) and thermal imaging. Initial results show that specific areas of the brain are activated by the sound of a crying baby—especially the areas of empathy, the ability to make us feel what the other person is feeling. The differences observed at the brain level between men and women are minimal. However, parents' brains appear to differ from those of nonparents, likely reflecting learning from their baby's cries. We are a cooperatively breeding species, you know that. On the other hand, I am willing to bet that the brain of one of those huge male elephant seals doesn't blink much when it hears a baby elephant seal crying.

17

Listening to the living

ECOACOUSTICS AND BIODIVERSITY

Désert de Platé, French Alps. It's freezing. Hindered by our snowshoes, we slowly climb the slope on the hardened snow. Yesterday we arrived in a storm, as is common here in May. On the track, the Land Rover quickly gave up, and we climbed on foot to the hut that would serve as shelter for the night. We left it at three o'clock in the morning. Frédéric walks in front of me, with his long ice axe in his hand: a confident mountain man. He extends his arm in the direction of Mont Blanc. The giant is decorated with a long light trail that scars its flank—city-dwelling mountain climbers lining up to conquer its summit, their headlamps winking as their heads move. We stop on a rocky promontory, drenched in sweat. With his usual foresight, Frédéric pulls on a fresh T-shirt. The wait begins, cold and long. "Rrrrrr-rr-rr"—a first song breaks the silence. A hoarse, rattling sound, right in front of me, soon followed by another one further to the left. I note on the counting sheet the estimated position of the two rock ptarmigans. Back at the hut, all the counters will pool their observations: five or six birds for the whole area. The ptarmigan is disappearing from our mountains. Here, as elsewhere on our planet, mornings are falling silent.

You already know Frédéric Sèbe from chapter 11. He was the one who spent a year observing the communication networks of the screaming pihas in the Amazonian forest. Frédéric now uses bioacoustics as a tool for monitoring biodiversity. One of his research projects concerns

Rock ptarmigan

the rock ptarmigan *Lagopus mutus*, a kind of snow partridge, white in winter and brown in summer. Typical of the Scandinavian tundra, it is still found in France at high altitudes in the Alps and the Pyrenees. Frédéric's aim is to be able to estimate the size of ptarmigan populations, i.e., the number of individuals, by using automatic sound recorders. These are very weather-resistant tape recorders that Frédéric leaves at the test site to record ambient sounds at regular intervals. To validate the method, human observers check the presence and number of ptarmigans visually and especially by ear. Then, in the laboratory, sophisticated techniques for analyzing the recorded acoustic signals must be developed. Since tape recorders remain in place for several weeks in a row and record for several hours a day, I'll let you imagine how many hours of recording have to be processed. Developing analysis techniques was the core of our PhD student Thibaut Marin-Cudraz's work for three long years. He spent many hours in front of his computer, but his tenacity was rewarded. He managed to show that this automatic counting by the bioacoustic method is reliable. He proved that it avoids double counting, as when two observers count two different birds when they both hear the same individual. This technique also compensates for those possible moments of inattention in some observers.[559]

Identifying and counting birds or other animals automatically by recording them rather than listening to them is one of the practical applications of bioacoustics.[560] This technique is also used in the ocean to spot whales and even fish.[561] But we can also work on a different scale. Why not take a look at all the species that produce sounds? In other words, why not consider soundscapes as indicators of the animal species living in the same environment?[562] This approach, which is currently in full expansion, forms the discipline called *ecoacoustics*. The idea is simple: to use bioacoustics to measure biodiversity and assess the condition of an ecosystem.[563] All we had to do was think of it.

Almost at the end of our sound journey, you now have a good idea of the extraordinary profusion of sound signals used in the animal world to exchange information. In aerial environments, birds and insects are at the top of the list, followed by frogs and mammals. Below the water, shrimp, fish, dolphins, and whales compete in their ingenuity to produce sounds. The tropical forest, with its incredible diversity of species, rustles with a thousand sounds. Coral reefs, the tropical forests of the sea, are equally rich.[564] It's true that a soundscape only partially represents the diversity of life, but everything comes together in nature. Where the species of insects and birds producing acoustic signals are diverse and numerous, so are other animals, plants, and all other living things. The widespread and ongoing declines in bird populations on Earth cause changes in soundscape quality.[565] The soundscape reflects the diversity of life.

Do you remember Frédéric Bertucci? This is the enthusiastic PhD student I told you about in chapter 10, who developed playback experiments in aquariums when we were trying to test how the fish *Metriaclima zebra* reacted to the sound productions of its fellow fish. Once he had finished his PhD, Frédéric left the semimountainous climate of Saint-Etienne (where our ENES Bioacoustics Research Lab is located) to work in the tropics with Eric Parmentier (the piranha man) and David Lecchini, a specialist in coral reef fish. It is on the island of Moorea, in the Pacific Ocean, that the *Centre de recherches insulaires et observatoire de l'environnement* (CRIOBE) is located, a superb field station installed on the edge of a lagoon bordered by superb coral reefs.[566]

If coral reefs represent only 2% of the total surface of the oceans, they represent 30% of their biodiversity—and a very interesting sound ambience. In Moorea, the coral reefs are dominated by only a few sounds (between 2 and 6) during the day, while at night it rustles with more diverse signals (up to 19 different sounds). Dusk and dawn are the busiest times. Attributing these sounds to specific fish species is still a difficult task, as our knowledge of the repertoire of each species remains superficial. However, Frédéric and his colleagues were able to identify that the most sonorous fish belong to the families Balistidae, Pomacentridae, Holocentridae, and Serranidae. Some sounds, such as the "whoots," a long and strongly modulated signal, still keep their mystery.[567] Coral reefs are fragile environments because, to survive and develop, they need particular conditions of light, temperature, and acidity level. Any imbalance results in the death of corals, which begins with their bleaching.[568] A bleached coral is a coral that has lost the microscopic algae with which it is in symbiosis and which are essential for its survival. No coral means no shrimp or fish—nor any other marine organism, for that matter. If conditions become favorable again, a bleached coral reef takes about 10 years to recover its colors. To save coral reefs, we must first be able to measure their health. This is what scientists are trying to do using ecoacoustics.[569]

Ecoacoustics is a recent discipline, defined as the ecological study and interpretation of environmental sound. The main idea is to measure and monitor the biodiversity and ecology of a living environment through sound.[570] While the objective of classical bioacoustics, as we have seen, is to study the transfer of information between individuals, the objective of ecoacoustics is to "consider sounds as a component and indicator of ecological processes."[571] Bernie Krause, who has spent more than 50 years recording soundscapes around the world,[572] coined the terms *geophony, biophony,* and *anthropophony* to describe soundscapes.[573] *Geophony* represents all the noise caused by nonliving natural phenomena—the thunderclap, the rustle of the wind in the trees, the murmur of a stream, the roar of a waterfall. Although they may be of short duration, they often provide a remarkably continuous background to the landscape. Sometimes they mark changes in the time of day. In spring in the mountains, the torrent is silent in the morning when every-thing is frozen. Later, when the heat of the sun melts on the peaks, the water starts to cascade again, and it becomes thunderous. *Biophony* encompasses all the sound productions of living beings—with the exception of those of the human species, which constitute *anthropophony*.[574] Krause separates anthropophony into two constituents: first, what he calls controlled productions—the speech, the songs, the music; and second, the incoherent noises emerging from human activity—air and sea transport, construction activities, mining and industrial activities, wind farms, etc. There is no shortage of sources of anthropogenic noise. Since the Industrial Revolution and the world population boom, anthropophony has become massive. It continues to grow. Both on land and underwater, the places on Earth where it is absent are becoming rare.[575] In Europe, for example, only particularly remote areas are preserved. A new phenomenon on the geological time scale, anthropophony is to be considered as pollution, in the same way as chemical pollution or ocean acidification.[576]

As I write these lines in my house located in a rather quiet neighborhood, a common Eurasian blackbird *Turdus merula* sings loudly a few meters away from me ("tooodeee-too-tooo-deeee"). A few sparrows chirp ("tchip-tchip-tchip"). Every two or three minutes, a car drives along the street. Its noise does not completely cover the blackbird's vocalizations, but it considerably diminishes its apparent power. One of my neighbors in the distance is mowing his lawn. An insect in a hurry is humming as it whizzes by. Ah! Another car. The blackbird is silent, perhaps discouraged? Anthropophony affects biophony. Knowledge accumulated over more than 20 years confirms this with certainty.[577] Arthropods, amphibians, fish, birds, mammals—all are affected,[578, 579] at all ages.[580]

Noise due to human activities affects animals' communication, their distribution in the environment, their feeding behavior, and of course their physiological condition and survival,[581] both in the air and underwater. They even affect animal species that do not use acoustics to communicate, such as mollusks.[582] The disturbances caused by noise depend of course on their acoustic characteristics. The continuous noise of highway traffic will reduce the range of birdsong and perhaps cause chronic stress to birds, as it does to many humans.[583] A sudden mining explosion may destroy hearing systems or cause irreversible damage to various organs of the body, including the brain. Underwater, the problem of anthropogenic noise is particularly critical.[584] As you know, sounds travel much faster and farther in the aquatic environment. Where a car will bother singing blackbirds nearby, the sound of a cruise ship can disturb whales or dolphins for miles around. Remember, cetacean sound signals are as low-pitched (up to 20 Hz) as they are high-pitched (over 300 kHz). Any anthropogenic sound, whether the lowest or the highest, will fall within this frequency range and be a serious parasite.[585] Did you know that every year there are numerous underwater explosions along tens of thousands of kilometers to find oil reservoirs in the ocean floor?[586] Whales and dolphins are victims of these sound explosions.[587] But they are not the only ones. Recent studies have shown that anthropogenic noise pollution drastically alters the soundscape of coral reefs. This pollutant, whose consequences are probably underestimated, can mask communication among reef organisms, such as certain fish, and represents a stress factor whose effects have not yet been fully discovered.[588] The increase in anthropogenic noise in the oceans is such that it is estimated that North Atlantic right whales (*Eubalaena glacialis*) may have lost two-thirds of their communication space.[589] Studies show that these underwater sound waves can even kill zooplankton—the small shrimp, larvae, and other animalcules that float in the water and are the main source of food for countless animal species.[590, 591]

Some animals are more or less comfortable with anthropogenic noise.[592] As when we find ourselves in an environment where the noise level is high, one of their strategies is to increase the intensity of their own signals (the so-called Lombard effect, named after the French doctor who described this phenomenon more than a century ago). In the city of Berlin, for example, the nightingale sings more quietly on weekends, and much more loudly during the week when road traffic from the working world increases.[593] Underwater, the humpback whale also increases the power of its song when the background noise due to human activities increases.[594] A recent study shows that the bearded seal is also able to sing louder in the presence of noise. However, scientists who have documented this behavior found that the seal lacks the ability to increase its voice level when the ambient noise becomes too loud.[595] Since anthropophony is characterized by rather low-pitched sounds, frogs, insects, and birds produce higher-pitched sound signals to make their signals stand out from the anthropophonic cacophony.[596] This phenomenon was first observed by comparing the songs of urban great tits with those of individuals living in the countryside. City birds sing higher-pitched songs.[597] In addition, for biomechanical reasons, high-pitched vocalizations are often emitted with greater intensity (to sing high-pitched, one must sing loudly).[598] This strengthens their signal even more against ambient noise. These adaptations can come from individual behavioral plasticity, where the singer modifies the way it vocalizes depending on the background noise, or they can be true evolutionary adaptations. It seems that, in the great tit, it is exposure to noise over several generations that leads the birds to modify the pitch of their song.[599]

A study recently showed that birds start to sing less loudly if anthropogenic noise is reduced. The population containment measures imposed during the COVID-19 crisis in the spring of 2020 resulted in a drastic decrease in human activity, in particular road traffic. Elizabeth Derryberry and her team at the University of Tennessee have studied the consequences of this on the vocal activity of the white-crowned sparrow—Peter Marler's bird, which, we learned in chapter 12, has different dialects depending on its location in the San Francisco Bay Area. Derryberry and her colleagues have shown that, during the COVID-19 crisis, this bird sang less loudly.[600] There was no longer any need to sing loud in an environment free of noise pollution: the range of their song more than doubled. And if the range of each individual's song doubles, four times as many individuals can be heard at the same time, which also explains why people reported hearing more birds than before the crisis. In addition, the birds sang some lower-pitched notes, as did their recorded ancestors in the 1970s, when the San Francisco Bay Area was still relatively quiet. Marler's finches somehow filled the new acoustic space that was available to them. Nature abhors a vacuum, as we all know.

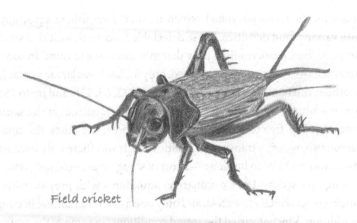

Field cricket

Insects, too, are capable to some extent of adapting to the presence of noise caused by human activity. By combining observations and field playback experiments, Mario Gallego-Abenza, in collaboration with David Wheatcroft and me, has shown that the response of field crickets (*Gryllus bimaculatus*) to anthropogenic highway noise depends on their previous experience with cars. Male crickets living on the edge of a highway reduce the rhythm of their chirping when a car passes by. Mario compared the response of these crickets to car sounds emitted from a loudspeaker with that of other individuals living further away from the highway (up to a mile and a half). The farther the crickets lived from the noise, the more they decreased the rate at which they chirped.[601] So the highway crickets have become a bit used to the noise. That's good for them. However, the study does not say if they are more stressed, if they find partners just as easily, or if they are more or less resistant to predators than their countryside cousins. In female crickets, it was recently discovered that ambient noise disturbs their male preferences. While they are usually attracted by a rather fast song—a sign that the male is vigorous—they no longer show any preference when the sound environment becomes annoying. The noise makes them lose their judgment in a way.[602]

At our laboratory in Saint-Etienne, my colleague Vincent Médoc is studying the effects of boat noise on the behavior of freshwater fish. In one of his experiments, Vincent and his colleagues put European minnows (*Phoxinus phoxinus*), which are small river fish, in aquariums with varying food densities. These delectable food items included small larvae of the *Chaoborus* fly, a prey that minnows love to hunt. In some aquariums, very few prey were served, only 8 *Chaoborus* larvae per tank. In others, the number of prey was higher: 16, 32, 64, 128, and up to 256 larvae—simply huge. Two fish were put in the aquarium at the same time. One was free to move around to hunt prey; the other, the companion fish, was in a transparent plastic tube. It was there only because minnows don't like to be alone—a sort of stooge that could only watch as lunch passed by. Let's summarize: aquariums with prey densities ranging from scarcity to excess, and only one fish with a hunting license. In addition, Vincent varied the sound conditions: in some of the aquariums, it was dead calm; in others, a loudspeaker regularly reproduced the sound of a motorboat passing by. Vincent let the fish feed for an hour; at the end of this time, he counted the number of prey that had been eaten. As one might expect, the results showed that the noise of the boat slowed down the hunting of the minnows. What's remarkable is that the lower the larvae density, the greater the effect. To give you an idea, with 32 larvae in the aquarium, the minnow swallowed about 15 in the absence of boat noise and only about 10 in the presence of noise. The noise seemed to spoil their appetite. On the other hand, if there were 256 larvae in the aquarium, the fish ate as many with or without the noise. However, it must be said that larvae were everywhere and easy to find.[603]

What Vincent's study shows is that the effects of anthropogenic noise on fish behavior depend on other environmental characteristics, in this case, prey density. But how do things work under natural conditions, in a river with minnows, *Chaoborus*, and a whole bunch of other, different prey likely to be eaten by fish? We can hypothesize that the passage of boats may change the hunting strategies of minnows depending on the relative density of the different prey. In calm conditions, minnows may hunt this or that prey instead, because they prefer it, for example. In noisy conditions, stressed minnows may hunt their preferred prey less efficiently, especially if this prey is at low densities. They will then fall back on other, more abundant prey, even if it is less tasty. These variations in hunting strategies will lead to variations in the relative densities of the prey, which in turn will lead to variations in the relative densities of the species the prey feeds on, and so on.

Are you familiar with food webs—the idea that each living being present in an ecosystem is the food of another, and that an ecosystem can be seen as a set of interactions between all the living beings that it comprises? The presence of anthropogenic noise could possibly alter these webs across hitherto unsuspected levels, even to living beings that hear nothing but that undergo modifications from higher levels of the network (e.g., plants or bacteria)—because at the end of the line, the first foods consumed in a food web are not animals. The result, which is a bit frightening, is that it is the whole ecological balance that could be affected by anthropophony. Today, we don't know much about these domino effects.[604]

A change in the sound environment caused by anthropogenic noise is not the only one to impact living organisms. An irruption of silence can have equally deleterious effects. Do you remember that the larvae of coral reef fish lead a planktonic life away from the reef of their birth? When they reach the appropriate age, these larvae are attracted to the reef background and choose to settle here or there, according to their musical taste, so to speak. I explained this in chapter 6, where we had our ears underwater. What happens when the reef is degraded and its soundscape is changed? Well, young fish have a harder time coming to the reef, either because they don't find the soundscape pleasant any-more or because they just don't hear it. Moreover, fish have a delicate ear: it does not take much change for larvae to decide to desert a reef with a depleted sound environment. A study conducted on the Great Barrier Reef in Australia showed that a sound environment that has lost 8% of its attractiveness will result in 40% fewer fish larvae on the reef.[605] Another study conducted by Bertucci and Lecchini, in collaboration with other scientists, has shown experimentally that other larvae—those of corals this time (the small immobile animals of the sea anemone group that build the reefs)—that are usually attracted by the odor emitted by some coralline algae are reluctant to approach these algae in the presence of boat noise.[606]

A key idea of ecoacoustics is that natural soundscapes are *structured*. In other words, the organization of the ecosystem, with its undisturbed trophic chains, is perceptible through the soundscape, which can be recorded and analyzed. The interactions between the organisms that make up an ecosystem would result in the structuring of soundscapes. This is the "great animal orchestra" cherished by Bernie Krause, where each species is like a musical instrument in dialogue with the others. On the other hand, when balances are disturbed by human activities, the soundscape would become chaotic. The structuring of the biophony would be lost. The biophonic orchestra would be stifled, become less complex, or even disappear completely. Anthropophony would take center stage. The soundscape is evidence of these changes. As Krause says: "The soundscape is the most reliable way to assess the state of a natural habitat."[607]

The idea that soundscapes are structured is based on two assumptions, that of the *acoustic niche* and that of *acoustic adaptation*. The *acoustic niche hypothesis* states that each species of animal occupies an acoustic space, different from that of other species living in the same location (known as *syntopic* species). In other words, each of the syntopic species would produce acoustic signals occupying different frequencies from those of the other species, or emitting them at different times. This would avoid acoustic competition. They would not interfere with each other, if you like. This would make the communication mechanisms between individuals within each animal species optimal.[608] This hypothesis is directly inherited from the concept of the *ecological niche*: in an ecosystem, there cannot be two animal species using the resources of the environment (food, space, etc.) in exactly the same way.[609] If this happens, the two species are in competition, and eventually one will exclude the other. Most often, the ecological niche of one or both species will change—the individuals of one species change their diet, for example—and the competition diminishes.

In its original version, as stated by Krause in 1993, the *acoustic niche hypothesis* considers sound frequencies as a limited environmental resource.[610] Animals should therefore "share" the frequencies. For example, some species will have very high-pitched vocalizations, others a little less, others will be quite low-pitched. When you look at the spectrogram of a recording taken in the rainforest, for example, it is easy to see that the different animals actually use different frequency bands. Jérôme Sueur of the *Muséum national d'Histoire naturelle* in Paris, and his collaborators, recorded the soundscape at the biological station of the Nouragues Nature Reserve in French Guiana on December 12, 2010, at 6:30 a.m.[611] On the spectrogram showing the recording, three levels are clearly visible. In the low frequencies, between about 0 and 1 kHz, an almost continuous color band is visible. These are the growls of howler monkeys. Above that, between 1 and 5 kHz, colored spots indicate birdsongs—at least two species, one making trills, the other monotonous whistles. When the trills are there, the whistles are difficult to perceive. Finally, in the upper part of the spectrogram, the one that corresponds to high-pitched sounds, between 5 and 15 kHz, we have insect chirps. Again, there are several species. Some produce chirps between 5 and 7 kHz, which sweep the spectrogram from beginning to end. Other chirps, centered on very high frequencies around 11 kHz, are more rhythmic and mark pause times.

This example from the Nouragues is quite demonstrative. There is indeed a kind of frequency sharing between animal species, but this sharing is imperfect. Many birds, for example, are found in more or less the same frequencies of the spectrogram. Does that mean that they get in the way? That they are competing for acoustic space? Perhaps, to some extent. But not necessarily. The low or high pitch of a vocalization isn't everything—its temporal structure is also very important for encoding information. And when the songs of different species of birds use the same frequencies (i.e., equally high-pitched, equally low-pitched, or equally in the midrange), their different temporal structures—their different rhythms, if you like—allow them to be distinguished from one another. Jean-Claude Brémond showed this phenomenon very well with the European wren *Troglodytes troglodytes*. This tiny little bird with its raised tail is common in European woods and forests and produces a song lasting a few seconds, composed of very fast trills. Brémond tried to artificially mask the song of the wren by mixing it with the song of other species, such as that of the Eurasian blackcap *Sylvia atricapilla*, the willow warbler *Phylloscopus trochilus*, or the dunnock *Prunella modularis*, which has a song close to that of the wren. In playback experiments, the wrens reacted vigorously to these noisy signals, showing that identifying the song of their species mixed with that of another bird was not a major challenge for them. Brémond carried on with his noise experiments by mixing the song of the wren with artificial noise, always in the same frequency range. Once again, the wrens identified the song from their species without difficulty.[612] So a wren is well able to locate the song of another wren even if the frequency band it uses is occupied by other sounds.

The hypothesis of the acoustic niche is therefore nuanced, and it is not surprising that studies about it give conflicting results.[613] As I was describing in the recording of the Nouragues in French Guiana, it is true that there is a sort of layering of the animals' vocalizations from low to high notes. But this layering is rather crude because many animals occupy the same frequency range, and its source is only partially explained by acoustic competition.

Let's remember the drumming of the woodpeckers, which I talked about in chapter 10. This illustrates a rather common situation, in my opinion. Each species of woodpecker drums in its own way. However, the closer two species are to each other, the more similar their drumming is, because this signal is largely determined by genetics. In a forest, the woodpecker species that cohabitate are mostly distant cousins. Two species that are too similar would compete not so much for acoustic space but mainly for food resources: they would look for the same insects on the same trees, for example. Does this mean that acoustic competition does not exist at all in woodpeckers? No, of course not! Because, as I told you, if two sister species with similar drumming patterns live in the same place, then there is a character shift, as evolutionary biologists say. Their drumming evolves and becomes a little bit more different than it was in the beginning. But remember: these changes are rarely rapid. They usually occur on a geological time scale.

So that's the acoustic niche hypothesis. Now let's move on to the acoustic adaptation hypothesis. The *acoustic adaptation hypothesis* states that the sound signals of an animal species are optimized to be transmitted as well as possible in its habitat. Again, you've heard this hypothesis before. I first considered it at the very beginning of our sound journey, when we were with the white-browed warbler and all the other birds of the Brazilian Atlantic Forest. As you may recall, the results of our propagation experiments were not very convincing: while some species of birds did produce songs adapted to high propagation by the height from which they were vocalizing, others did not. But perhaps our study focused on particular cases, giving a distorted view of the general situation. For example, I have told you how great tits singing at a higher pitch are more successful when confronted with the noise of the city. As a lot of work has been done on this issue, it should be possible to get a clear picture.[614] Let's take a look.

First of all, it is not really surprising that animal sound signals are somewhat adapted to the physical properties of the environment. Imagine a situation where an individual has to transmit information over several hundred meters in a forest environment. A low-intensity ultrasonic signal will not do the trick because the ultrasound only propagates a few meters.[615] This type of signal, not adapted, will therefore have no chance of being preserved by natural selection. On the other hand, such a signal will be perfect in a situation of very short-distance communication, transmitting information well and, as another advantage, not likely to attract a predator.

In a more straightforward way, one can make predictions as to which acoustic features should be favored in a given environment. In forests, the ideal long-distance signal would require a relatively low frequency, long duration, and little modulation—a kind of long hissing sound, impervious to the reverberations caused by tree trunks and adept at passing through obstacles. When vegetation is absent, short, highly modulated signals (going from high-pitched to low-pitched or vice versa very quickly), such as trills, would be more appropriate.[616] Since sound waves of very low frequency are strongly attenuated when propagating close to the ground, frequencies above 0.5–1 kHz should be favored. Finally, in a noisy environment, sound signals should be tuned to frequencies as far away as possible from those of the noise, e.g., very high-pitched if the animal lives close to a torrent whose tone is in the lower frequencies.[617]

Slower time rates and lower modulations would also be more conducive to successful signal transmission and reception.

Things can be subtle, of course. For example, in a forest, the acoustic conditions are different depending on whether you are vocalizing from the ground, from midheight, or from the top of the canopy. The microhabitat of the animal must therefore be taken into account if we really want to test the acoustic adaptation hypothesis. That's what we did in the Atlantic Forest, with mixed success, as I mentioned a few moments ago. That's also what Sandra Goutte and her colleagues from the *Muséum national d'Histoire naturelle* did by working with frogs.[618] And their results are in line with ours. They combined data from 79 species of frogs to test the hypothesis that frogs living near streams produce vocalizations adapted to a noisy environment. They looked for correlations between the acoustic structure of the signals and the fine characteristics of the habitat (size of the rivers, slope of the bank, temperature, density of vegetation, ambient noise measured at the exact location where the frog was recorded, etc.). To sum up, although frog vocalizations are mainly determined by genetics, they found that the croaks of frogs living near streams have a higher frequency than the others. On the other hand, their temporal rhythms are not different, and the signals are more frequency modulated, thus varying rapidly between high and low frequencies, which is contrary to predictions.

The study by Sandra Goutte and colleagues, I feel like repeating, only partially validates the acoustic adaptation hypothesis. Similarly, an analysis of the song of 5085 bird species found no relationship between song frequency (whether it was low- or high-pitched) and the density of vegetation in the environments inhabited by the different species.[619] Rather than being an adaptation to travel far, depending on whether one is in a forest or an open area, song pitch is essentially explained by the body size of the bird: the largest species are those that sing the lowest, the smallest the highest. In fact, the adaptation of the characteristics of a sound signal to the acoustic constraints of the propagation environment should not be seen as the only element that counts in the evolution of a communication system. First of all, every system has its own constraints: an animal species is the inheritor of its ancestors—of their genes and of the anatomical, morphological, and physiological constraints that accompany them.[620] A frog cannot modify its croaking at will. Second, the important thing is that the transmission of information between senders and receivers must be sufficiently stable, and not necessarily optimally so. Life is the result of chance and necessity. I bet no engineer has ever thought of that! Nothing is perfect; everything is a precarious balance. Finally, as we have seen with the wren, the capacities of the individual receiver to decode the information of a signal, even if strongly degraded by propagation, can be astonishing, and can compensate for an imperfect acoustic adaptation.

Does the fact that ecoacoustics is based on simple and imperfectly confirmed hypotheses invalidate the overall approach? Certainly not. This does not prevent ecoacoustics from being a powerful tool that is proving more and more indispensable day by day to assess biodiversity and ecosystems—perhaps one of the best tools available. Because two things remain true: each environment has its own sound signature, and it is possible to characterize it. A soundscape integrates a considerable amount of information—much more than a photograph, for example. As Bernie Krause says beautifully, "If a picture is worth a thousand words, a sound is worth a thousand pictures." Ecoacoustics also has a considerable advantage: we can follow the evolution of things over the long term, without the time, effort, expense, and disruption of having observers go out in the field every day to measure biodiversity or assess the presence of this or that animal species. A few discreet recorders, left on site for long periods of time, are a much more advantageous solution.[621]

It was in Piran, Slovenia, at a conference of the International Bioacoustics Society, that Jérôme Sueur and I were outside at a café, chatting.[622] Usually cheerful, Jérôme was pessimistic that day. "I think our work is useless. Studying the songs of cicadas and birds doesn't bring much benefit. Couldn't we do some useful research?" Although I didn't agree with him, I didn't have many counterarguments. Selfishly pursuing my passion was enough for me. But, deep down, I was quite disturbed. "Jérôme is right," I thought to myself, "but what can I do?"

Jérôme found his solution. He is now one of the leading figures in global ecoacoustics. One of his main activities is to participate in the development of *acoustic indices of biodiversity*.[623] Imagine that you want to report on the biodiversity of an environment, such as a forest, through bioacoustics. You're going to place automatic sound recorders in different places and then let them record every day, for weeks, months . . . years! How, from this raw data, can you obtain measurements, figures, notes, representing biodiversity? It would be impossible to identify and count all the animals participating in the soundscapes that succeed one another over time by observing kilometers of spectrograms. You would need to define more global measurements, and this is what we are doing with these famous indices. There are two main families of indices: alpha-type indices, which measure the biodiversity of a given soundscape, and beta-type indices, which compare two soundscapes. I won't go too much further here. You'll find them elsewhere if you want more detail.[624] Just be aware that they allow us to describe the richness, complexity, heterogeneity, regularity, and composition (in terms of geophony, biophony, and anthropophony) of soundscapes, both on land and underwater.[625] No single current index can capture all aspects. Every year, new indices are developed by scientists.[626]

To have a clear idea, let's go back to the coral reefs.[627] I think everyone is up for such a trip! Rather than trying to identify which animal species produces which sound on the reef, scientists can use acoustic indices calculated directly from the soundscape, and this approach allows a global assessment of the state of the reef ecosystem. In their study of the Bora Bora lagoon in French Polynesia, Frédéric Bertucci and his collaborators used two indicators: the signal intensity (sound pressure level, in decibels) and the acoustic complexity index.[628] They found, not surprisingly, that the signal intensity is higher where there are many fish and that the complexity index is higher when the reef is more developed.[629] Acoustic indices thus give varied information about the recorded reefs. Let's be more precise: high sound intensity values or a high complexity index in the low frequencies of the sound environment (< 1 kHz) indicates that the reef is healthy, with diverse and abundant populations of fish and other organisms. However, if these indices are high when considering the high-frequency portion of the soundscape, the ecosystem is degraded, with many dead corals and shrimp feeding on detritus.[630] In addition to information on the composition of the biodiversity of an environment, acoustic indices can provide what biologists call *functional information*. To illustrate this, let's take the example of a study carried out by the researcher Simon Elise and his collaborators in the coral reefs of Europa Island, a small piece of land in the southwestern Indian Ocean. These scientists tested whether it was possible to get an idea of ecosystem *functions* using acoustic indices calculated from the reef soundscape. In other words, instead of simply focusing on the diversity and abundance of fish and coral cover, Elise wanted to use acoustics to report on the activity of fish and other animals on the reef. Europa Island being particularly remote, this was quite an expedition and was the subject of a film that I would recommend you watch.[631] To record the underwater sounds, the scientists deployed a large metal tripod onto which they fixed a hydrophone. The tripod was gently placed on the bottom, with the hydrophone suspended at about 1.5 meters. All sounds between 0 and 50 kHz were then recorded for a minimum of two hours. This operation took place at several locations on the reef. At the same time, the scientists carried out meticulous photographic surveys to identify the organisms present, from corals to fish, including shrimps and other shellfish. This was followed by an in-depth acoustic analysis of the recorded sound signals, which required long hours of work in front of the computer. At the end, no less than six different acoustic indices were calculated.[632] The results are remarkable, since Elise and his colleagues have shown that the six acoustic indices can be considered as indicators of six ecosystem functions. They reflect coral cover, with encrusting corals on one side and nonencrusting corals on the other, habitat complexity, grazing fish (which graze or scrape the coral), planktivorous fish (which eat plankton floating in the water), and tertiary consumers (defined in the study as fish-eating species smaller than 50 cm). It is assumed that acoustic cues miss algae, plankton, and invertebrates such as sea slugs, all of which are very quiet. Some acoustic indices are closely correlated with the functions listed above. For example, the acoustic complexity index discussed earlier increases with the amount of grazing fish. The number of tertiary consumers affects the index of temporal variability: the more tertiary consumers that are active, the higher the index. In fact, the values of the six ecosystem functions given by the photographic surveys and the visual counts are closely correlated to at least one of the acoustic indices calculated from the sound recordings. Acoustics can therefore be used to illustrate the functioning of a reef ecosystem.

Of course, these acoustic approaches are not yet fully mastered, and we still need many years of research to refine the way to extract information from acoustic recordings.[633] But things are progressing fast, and it is likely that one day coral reefs will be acoustically monitored everywhere, providing firsthand data for the implementation of effective conservation policies. This monitoring will allow us to evaluate the risks and effects due to various stresses, both environmental (e.g., storms and hurricanes, or arrival of invasive animal species) and human (anthropic noise, chemical pollution, etc.).[634] The recordings made on the coral reefs of Puerto Rico during Hurricane Maria have shown that the shrimps remained silent—and thus inactive—during the storm and only gradually resumed their normal life in the days that followed. On the other hand, the vocal activity of the fish increased, perhaps because the increased turbidity of the water pushed them to use acoustics to communicate or because they became difficult to spot visually and thought they were safe from predators.[635] When listening is done in real time, as is beginning to be the case in some places, the sound ambience of the reef can give an early warning of any change affecting the ecosystem.[636]

The greatest challenge in ecoacoustics is to make sense of very long recordings—to deal with big data, as they say.[637] Researchers like Jérôme spend a lot of time inventing solutions to automate the analysis of recordings. Ecoacoustics depends crucially on new technological developments and is very demanding in terms of mathematical processing capabilities. We are probably in the prehistory of this discipline. In 20 or 30 years, perhaps much less, we can hope to monitor the health of a tropical forest or a coral reef from a distance. The fact that recorders have become reasonably priced, that data transmission systems are more efficient and reliable, that artificial intelligence and its calculations are becoming more accurate by the day, is rapidly increasing our ability to analyze soundscapes and therefore to analyze ecosystems by listening to them.

Soon, only arrays of microphones connected to high-powered computers by radio waves will be needed. Listening to forests and reefs, lakes and mountains, will tell us in real time, and for a modest cost, how they are changing, whether new animal species have joined them, whether others have disappeared, and whether human disturbances are affecting them.[638, 639] Let's only hope that there will still be forests and coral reefs, lakes and mountains left. Bernie Krause has sounded the alarm.[640] According to him, more than 50% of the habitats he has recorded over the last 50 years have already disappeared or have been profoundly modified.

At the sunny table of the café, Thierry has just joined us. "You never know what fundamental research is going to be used for. It's totally unpredictable! Did you know that a sound analysis method my student Ruben Mbu Nyamsi and I had developed to study bird calls was later used by a car manufacturer to identify problems with electric windows on an assembly line? And think of Nobel Prize winner Pierre-Gilles de Gennes, whose fundamental research on the physics of soft matter inspired the development of shampoos and helped maximize the grip of car tires!" [641, 642] Point taken. I was reassured. Jérôme and Thierry are both right.

Some time has passed since this conversation at the Piran café. Jérôme continues to work on the development of acoustic ecology to better understand biodiversity and its dynamics. As for me, I realize that my fundamental research on acoustic communications illustrates the value of the conservation and preservation of biodiversity. I now dream of the moment when artificial intelligence tools will allow us to link soundscapes and information exchange processes between individuals. It will then be possible to describe the structure of soundscapes, to quantify the information flows that cross them, and to extract the network of relationships between the levels of organization of an ecosystem. Bioacoustics has already started this transition toward studying the dynamics and interactions between organisms that characterize ecosystems. The exploration of acoustic biodiversity has only just begun.

18

Words . . . words

DO ANIMALS HAVE A LANGUAGE?

Tanjung Puting National Park, Borneo Island, Indonesia. Yesterday we flew over some gigantic palm oil plantations—primary forests sacrificed to the crazy population explosion and human consumption. Today, under the foliage of the still intact trees, I try to forget the disaster that is taking place over there, and I let myself be engulfed by the concert of insects and birds. Dawn is breaking already. From far away, like long echoes, almost human songs come out of the forest: "Woop woop woop woop wooopooooppoooop woooooooop woooooooooop wooooooopp. . . ." The vocalizations follow one another, at first slow, then faster and faster. Simple calls expressing emotion that the sender cannot repress? Or complex vocal construction skillfully mastered? Two individuals seem to answer each other. What mysterious rules of conversation do they follow? What information is exchanged in this way? What a bizarre *language* is that of the Bornean agile gibbon!

When questions come from the audience at the end of my conferences, there's always someone who asks, "Finally, can we say that animals have language?" During the course of the book, you were able to develop your own opinion on this subject. But perhaps you feel that everything is not so clear-cut and that answering this question would leave you uncomfortable. The problem is that there are two sides to this question. The first is, Can animals express anything other than their current emotions through calls that they have no control over? The second is more specific: Do animals have a language of the same type as our spoken language, with rules that allow us to exchange information as we do, through sequences of words and sentences?

Since we cannot enter a nonhuman brain, it is complicated to decide on that first question. However, whether it is the isolated male zebra finch that only responds to its female's calls in the presence of another pair of zebra finches, the bonobo that informs its companions of the presence of food through a series of vocalizations, the superb fairy wrens that cooperate by calling out to scare off a predator, the orcas that adjust their hunting behavior by whistling, or the chacma baboon that responds to the vocalizations of its fellows according to its memory of past events, all suggest that the production of sound signals in animals may be more than an uncontrolled reflex mechanism. Without taking too many risks, we can therefore affirm that at least some animal vocalizations represent something other than an emotional state expressed in a purely spontaneous manner. Especially in species with complex social lives, where acoustic communications allow individuals to manage their interactions astutely, animals can certainly control all or part of their sound production. So why, then, would we forbid ourselves from talking about *language*? It is, after all, a convenient term for an acoustic communication system in which the sender produces sounds voluntarily to send information to the receivers.

However, we must take a precaution and not put all animal species in the same basket. Remember the first chapter of the book. I told you that each animal is in its own world, the human species as well as the others. There is not a single animal language. There are as many animal languages as there are animal species that use sounds to communicate. The spoken language of humans is one of them.

Let's come to our second question: Is there a language in animals comparable to the *spoken language* of humans, that is to say, with an organization based on a combination of units such as words or sentences? For the vast majority of people, scientists included, the question of the originality of spoken language in the living world does not really arise: it would be unique and more complex than any animal acoustic communication system. This led the authors of a famous article in the magazine *Science* some 20 years ago to say that a Martian landing on planet Earth would notice that the spoken language of humans is the only one on Earth to be so complex and to allow the communication of an almost infinite amount of information.[643]

However, as Yosef Prat of Tel Aviv University points out in a recent article, before coming to this conclusion, the Martian would have to identify in the continuous flow of human speech the units that make sense (such as phonemes, syllables, or words) and then understand how these units come together in groups that make sense (expressions, sentences).[644] For us, who have been immersed in spoken language since our earliest childhood, these operations are automatic and easy to perform. But what about the Martian? If it could make sense of the spoken language of humans, maybe it could do it just as well for the song of the humpback whale and the song thrush. Who knows? Maybe it would find in these animal languages degrees of complexity that we don't yet suspect.[645] Or rather would it be disappointed by all its earthly observations—for example, by noting that spoken language takes a long time to say simple things like "Could you pass me the bread, please?" when the transmission of a Martian thought at 10 times the speed of light is sufficient. The Martian would then put us in the same acoustic basket as the other animals, considering all this small terrestrial world as a group of inferior beings quite similar to each other and unable to measure up to Martian standards.

Enough joking. We are among serious people. Here again, we will not take many risks by saying that human spoken language is, on our good old Earth, the one that allows us to communicate the most information. As it is trivially said, bonobos have never built a library and do not seem to give long philosophical speeches. The social and cognitive complexity of the human species has probably evolved along with the complexity of its communication systems, of which spoken language is one of the paragons. The story of the Martian simply points out the difficulty—and the bias—of analyzing animal languages as *external observers*, while we understand human spoken language *from the inside*. Our ability to analyze what other animals mean, what they have in their heads in some way, is limited.

Therefore, we must be wary of drawing conclusions about the sophistication of nonhuman languages. As early as the sixteenth century, Michel de Montaigne had a good sense of this. He wrote about animals: "This defect that prevents communication between them and us, why should it be reserved for us? Whose fault is it that they do not understand us? Because we do not understand them any more than they understand us. Therefore, animals may consider us to be unintelligent, just as we ourselves consider them to be unintelligent."[646]

Another point to consider is that spoken language is only a small part of the acoustic space that humans use to communicate. Through our vocalizations, we do much more than just talk. We call out in fear, joy, and pain, we sing songs or opera arias, we sigh with sadness or impatience . . . these unspoken vocalizations could represent living fossils of our ancestors' vocalizations when spoken language did not yet exist.[647] This diversity is found in nonhuman animal species. Thus, when we compare their languages with human language, we must consider all our vocalizations, not just words. Then the perspective changes. For example, as Prat's article notes, an acoustic analysis would perhaps show that our fear calls contain less precise information than those produced by vervet monkeys, which inform their fellow monkeys of the precise nature of the predator—eagle, leopard, or snake. To compare the relative complexity of human and nonhuman vocalizations, it is therefore essential to use the same methods of analysis.

When employed, this comparative approach highlights interesting similarities. Let's call the Cape penguin (*Spheniscus demersus*, also known as the African penguin) for assistance. The first time I saw African penguins in their natural environment was at their breeding colony on Boulders Beach in Simon's Town near Cape Town, South Africa. About 60 centimeters high and weighing in at 3 kilograms, it's a small penguin. On the beach, beak outstretched toward the sky, neck stretched as far as it can go, and wings spread, the males trumpet their parade call. It's all about pleasing their loved ones and impressing their mates. Nothing very original, is there? This call is a succession of several syllables. First a series of short (A) syllables, increasing in intensity ("honk! honk! honk! hoNK! HONK!"), followed by a long (B) syllable, very strong ("HUURRRR!!!!")—all of which are emitted by expelling air from the lungs. Another lungful of air, and the series cycle of A followed by B starts again—sometimes with short (C) syllables produced when the animal catches its breath between the A syllables or between A and B. No monotony—it is important not to put the females to sleep.

A study led by Livio Favaro, a professor at the University of Turin in Italy, shows that the calls of the African penguin follow two laws, known to characterize human language: *Zipf's law* and the *Menzerath-Altmann law*.[648] Zipf's law, or the *law of brevity*, states that the more frequent an element of a signal is, the shorter it will be. In human spoken language, words like "yes" and "no" that are used very often are indeed very short. Menzerath's law, on the other hand, says that the larger the whole, the smaller its constituents. Thus, in a text, the longest sentences have on average shorter words than short sentences. These two laws are predicted by Shannon's mathematical theory of communication, which we discussed in chapter 4: *information compression* (shortening the duration of the signal) increases the efficiency of coding information and its transmission from the sender to the receiver.

By analyzing 590 songs from 28 penguins, Livio and his collaborators found that the longer a song sequence is, the greater the proportion of short syllables A and C, and the shorter the duration of the C syllables even if they are already very short. In the African penguin's call, the duration of the syllables is therefore inversely correlated with the number of times they are repeated. Zipf's law is respected. Moreover, the number of syllables in a sequence is inversely proportional to the average length of the syllables. Menzerath's law is also present. If you're a little lost, just remember that we find in the calls of the African penguins the principle of information compression: the more we want to say, the shorter the signals are. Other teams of scientists have shown that these same laws can be found in the vocalizations of various animal species, particularly in certain primates, such as the indri lemur *Indri indri*, the gelada *Theropithecus gelada*, the Formosan macaque *Macaca cyclopis*, and two species of gibbons (*Nomascus nasutus* and *N. concolor*).[649] Information compression seems to be a universal principle shared by all languages, human and nonhuman.

African
penguin

Let's come to the rules that characterize vocal exchanges. When two humans enter into a conversation, they follow the unwritten rules of practicing alternation (each one speaks in turn) and observing avoidance (one does not speak when the other speaks).[650] These rules of alternation and avoidance are respected, more or less, as you can see in many examples if you watch televised debates. But when they are broken too often, the conversation stops (except on TV!). During our sound journey, we met animals that vocalized almost alone (the deer that bellows occasionally in solitude) or, on the contrary, all at the same time (the frog choruses). In these two cases, it is not necessarily easy to spot a conversation rule. Besides, are they conversations in the strict sense? However, we have also observed in the plain-tailed wren the duet in which female and male alternate in a perfectly synchronized manner.

Let's not forget the territorial disputes of chickadees, where a bird can wait until its neighbor has finished his song before starting its own. But if it does interrupt, it is precisely to let the other bird know that it intends to dominate the situation. Like all phenomena where the concept of rhythm intervenes, this rule of alternation ("turn-taking") is much studied in animals.[651] It is particularly searched for in nonhuman primates (with the risks of bias that this entails . . . we always end up finding what we are looking for). As a conclusion, we can say that the turn-taking rule can be found just about everywhere, from the starling to the bonobo[652] (but not the chimpanzee[653]), via the elephant and the meerkat[654]— which is not very surprising if we go back to the basics. The mathematical theory of communication predicts that the transmission channel will be noisy if signals interfere with one another. In order for information to be transmitted, it is better to send the signals one after the other and each one in turn.

But the question that interests us is more subtle: Do individuals who practice the turn-taking rule consider it to be a social rule that everyone must respect? In other words, are they surprised when they hear others violating the rule? To find out, my colleague Florence Levréro and a team of researchers conducted playback experiments with gorillas. The idea was to play the apes exchanges of calls between familiar individuals, by creating three different situations. In the first one, the sequence broadcast by the loudspeaker was a succession of two calls between two adults, the two calls being separated by a 500-millisecond silence, which reproduced a normal and familiar situation. In the second situation, a voice exchange was broadcast, still between adults, but where the second call started while the first call was not finished. This is not at all normal in gorillas. Finally, in the third case, the vocal exchange emitted by the loudspeaker was a series of two calls separated by half a second of silence, but the first call was that of a young individual while the second was that of an adult—an event that apparently rarely happens in natural situations in gorillas (when gorilla children talk, they almost never get an answer from the adults).[655] In short, one playback mimicked a completely conventional situation, while the other two were out of the ordinary and supposed to surprise the gorillas. The results of this experiment are a bit confusing but still interesting. First, there was evidence that the gorillas paid more attention to the first sequence (an ordinary succession of two adult calls) than to the second (where the calls overlapped). Perhaps this overlap caused some confusion, and the gorillas preferred to ignore this unexpected vocal exchange? Does this situation make them uncomfortable? We don't know, but it is clear that not following the alternation rule makes a difference to them. Further, scientists did not observe any significant difference between the gorillas' response to the third situation and their response to situations one and two. Apparently, a young gorilla talking before an older gorilla does not shock them that much.

To be clear, let's summarize our answers to the original questions: (1) Can animals express anything other than their current emotions through calls they have no control over? and (2) Do animals have a language of the same type as our spoken language? As to the first question, the answer is yes! It is now established that some animals can express something other than their emotions of the moment, as we do; they therefore control some of their vocalizations according to the context and the memory they have of the events and their interlocutors. The answer to the second question is also yes! But it's not a single language. There are in fact as many languages as there are animals using sounds to exchange information. These nonhuman languages are more or less complex and follow general rules that are very similar to ours. However, the structure and organization of the sound signals used for communication remain specific to each animal species, including humans.[656]

I can feel one last question burning on your lips: That's all well and good, but how did human spoken language come about? My answer is going to disappoint you a little; our certainties in this matter are few, and I'm not going to write a long essay on the subject here. But hang in there; there are a few things worth considering.

First of all, it is clear that our ability to master spoken language differentiates us from other apes.[657] Our ability to express any concept goes far beyond what any other animal species can do.[658] Relieved, right? During the second half of the twentieth century, however, there were a number of attempts to teach apes to talk.[659] None of them were very successful. Here is an outline, although I find these experiments ethically questionable. The animals lived in conditions that were poorly adapted to their species, separated from their fellow apes, sometimes dressed as humans. In short, these experiments were far away from the Tinbergen-style ethological approach.

In the 1950s, Cathy and Keith Hayes were among the pioneers. They raised a female chimpanzee to monitor her intellectual development, and in particular to see if she could learn to speak. Her name was Vicki, and the Hayes's raised her as if she had been their child. Although they present things in a rather positive way in their published paper,[660] it was a failure: after six years, Vicki still couldn't pronounce words correctly (she was able to say four words, with the greatest of difficulty).[661] Then other scientists—Allen and Beatrix Gardner, David Premack, and Duane Rumbaugh, to name a few—conducted the same experiments with other chimpanzees, using different approaches. For example, instead of trying to teach their chimpanzee (a female named Washoe) to talk, the Gardners taught her sign language. They had relative success: Washoe actually learned to produce about 30 words in 22 months of training.[662] Did the chimpanzee go further, combining words to construct expressions or phrases, as the Gardners maintain? On that point, things are not very clear.

To get to the heart of the matter, Herbert Terrace, a professor at Columbia University in New York, also raised a chimpanzee, Nim.[663] After long training sessions, Nim could name over a hundred objects—and seemed to be able to combine words. But alas! By carefully watching the videos of the training sessions, Terrace realized that, in the vast majority of situations, the person training Nim was giving him clues. Nim was content to imitate his coach, hoping to be rewarded. This was a great disappointment for Terrace. He decided to watch the films made by the other scientists who had taught other apes to talk. Terrace says that, again, in most situations, the person training the monkey gave them clues.[664]

Does that mean apes are incapable of humanlike language? Not exactly. All these experiments and others, such as those with the bonobo Kanzi or the chimpanzee Ai,[665] have shown that they are able to learn to associate objects, colors, numbers, and even concepts such as "the same" and "different" with words. They are even able to name them in sign language or using graphic symbols (after extensive training) or to associate these learned signs in original combinations to express their current desires.[666] But, in any case, no acoustic production! The apes don't articulate a sentence, or even a word. They are not well equipped for that.[667]

The same kind of experiment, in which an attempt was made to teach an animal to speak "human," was carried out with a parrot named Alex. Irene Pepperberg, a professor at Brandeis University near Boston, had managed to teach it a hundred words. Alex would point to objects and colors and answer questions such as "How many yellow cubes are there?"[668] However, Alex wasn't having big conversations. Only innovations in the history of the human lineage could therefore lead to the emergence of spoken language. The question is, Which ones? Because when it comes to vocal communication, we have a lot in common with other animals.

First of all, we share our auditory abilities, which aren't very original. The hearing machinery is quite similar in all four-legged vertebrates, from the anatomy of the ear to the information-processing mechanisms in the brain, including the ciliated neurons that convert sound waves into nerve signals.[669] So our hearing apparatus probably predates the development of the ability to speak. It was already in place.

Human originality must therefore lie elsewhere. For a long time, people thought it was due to the low position of our larynx.[670] Remember the larynx? The organ with the vocal cords that allows mammals to make sounds? In reality, other mammals also have a descended larynx, so at first glance it is not technically impossible for them to be able to speak.[671] Moreover, the larynx is not absolutely essential for conversing: human beings can talk . . . by whistling. *Whistling languages*, which imitate the acoustic form of the words of the locally spoken language, are indeed used for long-distance communication in forests or mountainous areas.[672] Another quirk of our species is the absence of laryngeal sacs, the diverticula that allow monkeys to make their vocalizations lower-pitched and more powerful. But no one knows if this played a role in language acquisition.

Of course, researchers have also looked at genetics. Whoever finds the "gene for language" will win the Nobel Prize for sure. And some people thought they had reached the grail when the FOXP2 gene was identified as an important contributor.[673] *Forkhead box P2* is its full name. This gene makes it possible to manufacture a protein made up of 715 amino acids that is involved in the regulation of DNA transcription. In other words, it is a gene that regulates the activity of other genes. Some of its mutations, one of which has been identified in a British family, lead to severe language deficits.[674] Present in all vertebrates, and even in the small *Drosophila* fly, FOXP2 exists in a modified form in humans. But not modified by much—only 2 amino acids differ from that of the chimpanzee and 3 from that of the mouse.[675] Apparently, FOXP2 is especially important for language acquisition. It plays a key role in the synaptic plasticity that accompanies learning, when neural networks form and connect.[676] Experiments with songbirds have shown that a mutation of FOXP2 renders them unable to imitate their tutor—a feat of science.[677] It has also been possible to isolate and analyze the FOXP2 gene from our Neanderthal cousins, who died out about 30,000 years ago; they possessed the same version of FOXP2 as we do, except for a very small part of the gene that allows it to regulate its activity—its switch, if you will. Maybe the switch that changed everything . . . and the beginning of The Word?

But FOXP2 is probably just one of many genes that played a role in language acquisition.[678] Recently, researchers have started to do big things on the genetic side, and we may be on the verge of major discoveries. For example, a team of scientists led by Eric Jarvis of the Howard Hughes Medical Institute has compared the activity of all the genes (yes, all the genes) in the brains of humans and different species of birds. They showed that the brains of songbirds (those that learn their song by imitating a tutor) showed gene activities in some of their structures that are very similar to those of humans. In particular, the genes in the neurons of the region involved in the control of the syrinx (the vocal organ of birds) are activated in a manner quite similar to the genes in the region of the human brain involved in the control of the larynx.[679] Although this is all very promising, we are still far from having put all the pieces of the puzzle together.[680]

While waiting for advances in genetics and new fossil discoveries, it is probably the neural control of speech that we should be looking at. In any case, this is the hypothesis defended by Tecumseh Fitch, a professor at the University of Vienna and a specialist in the evolution of language[681]—a hypothesis in the footsteps of Darwin, who said, "The fact that superior apes do not use their vocal organs to speak depends, no doubt, on the fact that their intelligence has not progressed sufficiently."[682] It is true that two interesting innovations for spoken language distinguish the human brain from that of other primates. First, our species has direct connections between the part of the brain called the motor cortex and the neurons that drive the muscles involved in the act of speaking. Second, our brain is characterized by a considerable development of neural connections between the auditory and motor areas. In other words, there are very many neurons that make connections between what we hear and what we say. That seems to be unique in the animal kingdom. It's good to feel unique, isn't it?

Over the course of human history, our brains have gradually organized themselves to give us the capacity to speak. This did not happen in a regulated way—there was no design office with the plan in hand to oversee its realization—but in a way in which every small change to better manage our vocal communications with others gave our ancestors an advantage over those who could not. In our human lineage, natural selection has favored smooth talkers.[683]

This cerebral reorganization has been accompanied by anatomical modifications of the larynx, the organ of vocal production. For example, with the loss of air sacs present in our ancestors, the human larynx has become simpler, allowing it to produce more stable and more easily controllable sounds.[684]

In conclusion, the acquisition of spoken language by the human line probably required relatively few innovations. Most of the structures and mechanisms involved in human spoken language have very ancient roots.[685] Everything was there before. Spoken language simply required some neurophysiological (and a bit of anatomical) reorganization, which probably went hand in hand with the increasing complexity of our social relationships.

Robin Dunbar at Oxford University thus defends the idea that spoken language (as well as laughter and singing) is our way of engaging in social grooming[686]—you know, the habit monkeys have of picking lice out of each other's fur that serves as a way to build and maintain social bonds. One thing is obvious: the more people to be groomed, the longer the grooming takes. Dunbar defends the idea that it would be impossible to groom all members of a human social group one by one in the monkey fashion because there's not enough time and the group is too big. So language would have been a good substitute.

"Communication is the glue that holds animal groups or societies together, and in general sociality goes hand in hand with sophisticated communication systems."[687] Given the complexity of our human societies, that glue must be strong. Between the emergence of the first *Homo* about two million years ago and modern humans *Homo sapiens*, the size of social groups has increased considerably. In traditional societies, the number of individuals that a human is connected to is around 150. However, according to calculations by Dunbar and his colleagues, direct grooming connects to 50 individuals at best.[688] By telling stories around the fire, we can "groom" many people at the same time, better, and for longer.[689] And what about the possibilities offered today by social networks—a super-grooming!

There are many other hypotheses about the origin of human spoken language. Indeed, disputes between scientists have been raging for a long time on this subject, and it's not over yet. To give you a little anecdote, you should know that the prestigious *Société linguistique de Paris* already wrote in its statutes of 1866 that it would not accept any communication dealing with the origin of language. We can guess that in the mid-1800s it was already necessary to calm down opinionated zealots. Since the 1990s, scientists have been working to unify positions, or at least to look at the bigger picture by avoiding overly simplistic points of view.

As Chris Knight and Jerome Lewis of University College London argue, it may not be possible to base a theory on the origins of human spoken language by isolating it from theories on the origins of morality, law, religion, and so on.[690] For Knight, spoken language is out of the ordinary and it is not enough to study it as a "classic" acoustic communication system. Considering that, alongside the real world governed by the laws of physics, humans build a world based on beliefs (in deities, in the value of money, etc.), he proposes the idea that spoken language is there to navigate this virtual world. Words and grammar allow extraordinarily developed levels of interaction within a human group, as well as the establishment of rituals more or less disconnected from the real world. The question is complicated, and if you want to delve deeper into it, I invite you to read his article, coauthored with Lewis and published in *Current Anthropology*.[691]

Without going into detail, let's retain one of the major ideas of Knight and Lewis: the extraordinary flexibility of our vocal apparatus allows us to produce sounds aimed at deceiving the individual receiver (let's note, by the way, that Darwin had already raised this idea: the high priest of evolution was a visionary[692]). The primary receivers are other animals. When we were vulnerable beings with limited weaponry, as we were until very recently, we kept predators at bay by increasing the range and diversity of our vocalizations. And predators were plentiful in the Pleistocene savanna. More than 12 species of saber-toothed tigers, 9 species of hyenas, not to mention all the others—I will spare you the list. According to Knight and Lewis, this type of antipredator strategy would explain the polyphonic songs sung, especially on moonless nights, in certain hunter-gatherer societies (among the Hadza and Ba Yaka of sub-Saharan Africa, for example).[693] We find singing in the dark reassuring, that's for sure. And if each one of us pitches in with some polyphonic variations, the predator will have the impression that these humans are particularly numerous and will not dare approach. On the other hand, when it comes to hunting, we imitate the animals' calls to attract them. This is a strategy practiced daily in hunter-gatherer societies. And even in the present day, don't we do everything we can with our voice or our whistles to imitate the animal whose attention we wish to attract?

Scaring away predators, hunting prey—there is another situation where humans use sounds to communicate with wild animals. This is when they wish to cooperate with them. The case of the greater honeyguide, or indicator bird, is inspiring in this respect. The Latin name of this small African bird is doubly insistent—*Indicator indicator*—as if to make it clear that it will do anything to show us the way. Did you know that, curiously enough, the honeyguide loves to eat beeswax, the substance that bees produce to build the combs of their hives? The problem is that the wax combs are normally inaccessible, hidden at the bottom of some tree trunk, and boldly defended by bees that are as determined as they are armed. The honeyguide has understood, apparently for a very long time, that humans are also looking for hives and that they can be valuable allies provided the reverse is true. The clever bird is thus in the habit of leading hive hunters to the object of their common lust. How does it proceed? By calling and flying from one tree to another, patiently waiting for the humans to catch up. When all have arrived at their destination, the hunters, protected from the bees by clothing and equipped with tools, open the hive to extract the honey from it, leaving the wax combs in plain sight, which their winged guide delights in as they leave. In Mozambique, honey hunters produce a special cry that attracts the indicator bird, something like "brrr-hm!" Claire Spottiswoode of Cambridge University and her colleagues at the FitzPatrick Institute of African Ornithology at the University of Cape Town have shown through playback experiments that this "brrr-hm!" prompts the indicator bird to guide the honey hunters to a hive.[694] The bird does not respond to other sounds. Hunters say that this sound code was taught to them by their fathers. As for the honeyguides, no doubt they learn to recognize the "brrr-hm!" by observing experienced honeyguides feasting on the wax combs provided by the honey hunters. A wild animal, in its natural environment, responding to a call for cooperation from humans: This scene is a rare example of *mutualism* between a wild animal and our human species mediated by acoustics.

Mimicking the cries of animals, finding sound codes to which they are sensitive—our vocal flexibility seems to have been primitively selected to facilitate our interactions with species other than our own. It would then have been put at the service of social interactions within our own species, with some success! From animal languages to human language, the loop is complete.[695]

Dear reader, we are coming to the end of our journey among the voices of nature. I hope I've entertained you sometimes and interested you often. I could have talked to you about many more things, as the world of bioacoustics is so vast. Recently, it's been discovered that it even extends beyond the animal kingdom! Plants might be sensitive to sound waves or even themselves produce acoustic signals carrying information.[696, 697] The flowers of the beach evening primrose *Oenothera drummondii* produce a sweeter nectar within three minutes of "hearing" the sound of bee wings.[698] There are certainly plenty of surprises still in store for the younger generations of bioacousticians, and the next few years should be rich in discoveries.

But . . . let's not forget the essential things! Last but not least, let's evoke a last impression. A last ambience. Our last common bioacoustics moment. . . .

ACKNOWLEDGMENTS

Vallon de la Fauge, Vercors Massif, French Alps. It's five o'clock in the morning and the concert is already starting again. It had stopped late yesterday, well after sunset. But the musicians have decided that the night rest had lasted long enough, and the great symphony resumes. Still lying in my tent, I try to identify who is who. The robin is the first to warm up its voice. Closely followed by a wren. With its explosive singing, this one is a first-rate alarm clock. I imagine it perched on its bush unless it is already snooping here and there while singing, restless as it is. The chaffinches are getting under way—there's more than one of them. Three maybe? And here comes the blackcap. This one brings back memories. I can still see it screeching at the top of my loudspeaker, perched on a branch. In the rising hubbub, I watch for the first part of its sentence, this slightly disordered babbling that makes it possible to distinguish it from its garden cousin. A willow warbler is reeling off its river of notes. And there, a Eurasian blackbird? No, the song is too complex, too full of flourishes—probably a song thrush. Getting a little lost in this festival, I let myself bask in the pleasure of listening to the symphony without trying to isolate the instrumentalists. Last evening, things were simple. Tits and warblers were leading the dance. While it was still light, an owl had begun to hoot before waiting for everyone to shut up. Alone, it had tried to fill the silence of a night still too cold for insects, barely helped by the rare barking of rutting deer. No nightingale, alas, to keep it company. Are we too high for the spring herald? In the afternoon, on the flank of the *Grande Moucherolle* mountain, we came across a herd of 70 Alpine ibex. One of them whistled as we approached, signaling its irritation. We had also heard the whistles of the marmots, frightened by a black kite. I had hoped for wolf howls during the night, but the packs are probably far away, on the uplands over there. Two days later we'll see their droppings, full of fur and bones. I crawl out of my sleeping bag and get out of the tent, still sore from the previous day's long hike. The meadow is soaked with dew. The thrush is there, on the tip of the nearest tree, singing its melodies. Blocked by the peaks, the sun will take a long time to reach the valley. I think I still have two more chapters to write, and I light the little stove to make tea. Tonight we must be at the foot of the *Grand Veymont* mountain . . . the book will wait!

Writing a book is a way to explain things in more depth than giving lectures, where time is always short. In particular, it helps to emphasize the human aspect of research. When writing this book, I have tried to remember all the people who have marked my scientific adventures. Still, an unpredictable memory combined with a well-dimensioned ego can play tricks. Might I have omitted someone who had an idea or initiated a project for which I claim to be the bearer? I would blame myself . . . less for fear of offending someone than for the feeling of having failed in an essential duty: to show that researchers do not forge themselves alone, that no one makes discoveries in a vacuum. In this book, none of the personal stories rested just on my shoulders.

Joëlle Ayats is the second pen of this book. I would like to thank her warmly! I loved discovering her uncompromising corrections, their relevance, their intelligence. They added dynamism and humor. They were also accurate, removing heaviness and errors. "She doesn't miss a thing!" Eliane Viennot had warned me, our mutual friend who had put me at Joëlle's mercy.

The third pen is a pencil. My father, Bernard Mathevon, took up the challenge of illustrating each of the animals I worked with (only the flamingo, the eagle owl, and the Antarctic skua did not find their place in the text; that's why they are here!). I thank him wholeheartedly for his meticulous work. Believe me, the original plates are even more beautiful than the printed ones. His wren is my favorite. Which one was yours?

I was deeply honored when Bernie Krause agreed to write the foreword for the English edition of my book. I warmly thank him for writing a text that is both personal and inspiring. I share with him this vision of

Flamingo

science: the most important thing is the path one takes. Krause is a gentleman of great stature. His expertise on the world's soundscapes is unparalleled. It was a real pleasure to read his kind words about my book: "Like a fine novel, this notable work is impossible to put down once you begin."

Bethan Wakeling did a fabulous job of correcting my English. I am very grateful for her generosity and dedication. The fact that she described her commitment as an "enriching experience" and that she found the book "great" warmed my heart.

I am very grateful to David, Eilean, and Elizabeth Reby for translating the animal sound onomatopoeia into English (except for the Savannah sparrow song, translated by Dan Mennill). I heard it wasn't easy to agree. You be the judge.

Eliane Viennot played an important role in the design of this book, first by opening my eyes a few years ago to the thought biases that mark

Eagle owl

all scientific research, then by helping me write the first chapters, and then by proposing Joëlle as a reviewer.

From Olivia Recasens, my French editor, I retain the passionate interest in science and the soft firmness of her "I can't wait to read you," which punctuated her emails. Writing this book was an exciting adventure, and I am deeply grateful to her for giving me the opportunity.

It was my friend Mark Hauber who encouraged me to send my manuscript to Alison Kalett, acquisitions editor at Princeton University Press. I was thrilled when Alison wrote me that she and her colleagues would be excited to publish *The Voices of Nature*. I sincerely thank her and the PUP team for this opportunity to bring my book to the English-speaking readership. I hope that many vocations will be born from this reading.

I am indebted to all those who reviewed all or part of the manuscript, saving me from serious errors or omissions: Olivier Adam, Thierry Aubin, Renaud Boistel, Elodie Briefer, Isabelle Charrier, Charlotte Curé, Etienne Danchin, Sébastien Derégnaucourt, Tudor Draganoiu, Michael Greenfield, Isabelle Horwath, Florence Levréro, Anne Mathevon, Etienne Parizet, David Reby, Oscar Roman, Dominique Rouger, Fanny

Antarctic
skua

Rybak, Frédéric Sèbe, Marc-André Sélosse, Jérôme Sueur, Frédéric Theunissen, and Laurent Villermet, as well as the two PUP referees, Daniel Blumstein and Tecumseh Fitch, whose careful reading and constructive criticism and advice have greatly improved the book. A huge thank-you for their advice, corrections, and encouragement!

Throughout the book, you have discovered the people with whom I have been fortunate enough to experience these scientific adventures. I continue my journey with some of them. With others, trajectories have diverged. They are all dear to me and I am grateful for the moments I have shared here and there, reflecting, discussing, trudging along, or simply waiting patiently for the right moment to do the experiment.

I would also like to thank my students, my colleagues at the ENES Bioacoustics Research Lab, and my collaborators from France and elsewhere, as well as all those who welcome our research (I have a special sense of gratitude for the people of *Crocoparc* in Agadir, Morocco).

Here I must solemnly thank the University of Saint-Etienne and the people who make it what it is today. By agreeing to host my small team more than 20 years ago, this university has made everything possible.

It took a lot of energy to initiate the ENES Bioacoustics Research Lab (eneslab.com), but through stubbornness and luck, too, things gradually improved. Many people helped me on the way. I am especially indebted to three of them: Professor André Giret, a field geologist, at one time responsible for research in the French Southern and Antarctic Territories; Professor Hervé Barré, one of my former professors, a demanding scientist who headed an important research laboratory at the CNRS; and Professor Jean-Marc Jallon, director of a famous CNRS research lab on memory, learning, and animal communications at the University of Paris–Sud.

A bioacoustics laboratory such as the ENES must actively participate in the training of new generations of bioacousticians. The Bioacoustics Winter School (https://www.eneslab.com/bioacoustic-winter-school) and the International Master of Bioacoustics (https://www.masterofbioacoustics.com/) have been born thanks to the enthusiasm of ENES researchers, in particular David Reby and Frédéric Sèbe, and of many bioacoustician colleagues and friends from France and abroad. I thank all of them warmly here.

I also owe a lot to the *Institut universitaire de France*. The IUF is an organization allowing academics to devote more time to research.[699] Getting in for 15 years as a junior and then as a senior member was (and still is) an incredible opportunity. In the French research landscape where new administrative behemoths are constantly being built, the IUF is a benevolent UFO. May it always escape the mania of forms and reports that paralyzes researchers by depriving them of their time—their most precious commodity.

I would like to thank the funding organizations that support my research projects. The University of Saint-Etienne and the *Institut universitaire de France*, again, but also the *Agence nationale de la recherche* (ANR), the *Centre national de la recherche scientifique* (CNRS), the *Institut national de la santé et de la recherche médicale* (Inserm), the city of Saint-Etienne, the *Labex CeLyA*, and the Miller Institute for Basic Research in Science, to name but a few. The research with crocodiles was made possible by the National Geographic Society, which funded several field expeditions, and by the unfailing support of the *Crocoparc* zoological park in Morocco.

Science allows me to work with amazing, brilliant, singular, enthusiastic people who have shaped both my thinking and my behavior. All of them different. Here I want to pay a special tribute to Thierry Aubin. Over the years, his sharp intelligence, his qualities as a scientist, his talent as an observer of nature, his vision of the world and of life, his humor, and even his culinary tastes have shaped me in many ways. I wouldn't be who I *am* today if our paths hadn't crossed. You have seen in these pages that, without him, nothing would have been possible. I hope that he finds this book worthy.

Last but not least, I end by warmly thanking my wife and children, who sometimes find bioacoustics a bit invasive, as well as my parents who, according to the time-honored-but-true formula, have always supported me in my projects.

And now it's time to start your journey among the voices of nature for real! The white-browed warbler, the jacaré caiman, the spotted hyena, the elephant seal, the Atlantic walrus, and the procession of babbling, singing, barking, jabbering, and talking animals will be with you.

GLOSSARY

Note: The definitions in this glossary are deliberately brief and therefore could appear incomplete. They are written to reflect a bioacoustics perspective.

Acoustic adaptation A process that promotes sound signals that are optimally transmitted in a given environment.

Acoustic avoidance A process that increases the differences between sound signals from animals living at the same time in the same area.

Acoustic biodiversity The totality of animal sound productions characteristic of an environment.

Acoustic biodiversity indices Quantitative measures derived from the analysis of sound recordings made in an environment and indicative of the diversity of living things in that environment.

Acoustic signals Mechanical disturbances (vibration waves) propagating in water or air. Vibration signals propagating through solids (e.g., soil) can also be considered as sounds.

Adaptive (or evolutionary) radiation Diversification of several species from a common ancestor. Each new species occupies, and is adapted to, a new ecological niche of its own.

Amplitude modulation The variation in the intensity (weak/strong) of a sound signal over time.

Amplitude of a sound The intensity of a sound. The amplitude (high, medium, low) of a sound is measured in decibels.

Arms race The evolutionary escalation between a communication signal and the resistance of the individual receiver to respond to this signal. The sender produces a signal with increasingly exaggerated characteristics, while the receiver becomes increasingly reluctant to respond to the signal. A common metaphor for this phenomenon is the Red Queen's instruction to Alice that they must run to stay in the same place.

Audience effect When an animal reacts to a communication signal or changes its signal production based on the individuals around it.

Behavioral ontogeny The development of a behavior over the course of an individual's life.

Bioacoustics The scientific discipline that studies animal and human acoustic communications.

Biological adaptation The process by which a population of individuals becomes better able to survive and reproduce in a particular environment.

Categorization (categorical perception) The mechanism by which an animal responds to the continuous variation of a physical stimulus by forming categories.

Chain of information transmission The sender-signal-receiver unit, as well as the processes of encoding information (by the sender), transmitting information (propagation of the signal in the environment), and decoding information (by the receiver). All elements of the information transmission chain can be affected by noise.

Character shift The increase in differences that allow a distinction, for receivers, between animal populations living in the same location.

Communication The transfer of information between one or several senders and one or several receivers.

Communication network The set of individuals that can transmit and receive sound signals between them.

Cooperative rearing The involvement of individuals other than the parents in the care of the young.

Cost of a signal The processes that cause the emission of a signal to decrease the probability of survival of the sender or its ability to have offspring.

Dear-enemy effect When a territorial animal reacts less violently to the acoustic signals of a familiar neighbor than to those of an unknown individual.

Decibel A unit of measurement (abbreviated dB) for the intensity of a sound.

Dialect The geographic variation in the vocalizations of an animal species.

Doppler effect A phenomenon characterized by a sound being higher-pitched when the sound source is approaching and lower-pitched when it is moving away.

Dynamic information Information about the characteristics of the sender that may vary quickly (e.g., emotions or aggression).

Eavesdropping When an individual listens to a sound signal that was not intended for them and obtains information from it.

Ecoacoustics A discipline derived from bioacoustics that uses natural sounds to address issues of ecology.

Ecological selection (or ecological adaptation) A process that favors sound signals that are best propagated in the environment where they are used.

Evolutionarily stable equilibrium A communication behavior that characterizes the majority of individuals in a population and is never superseded by another: any variation is counter-selected. Also referred to as an evolutionarily stable strategy.

Evolutionary causes of communication Processes that explain the development of communication throughout the history of a species. The evolutionary causes, in the long term, concern the processes of natural selection in particular (including sexual selection), as well as genetic and cultural drifts.

Exaptation A mechanism occurring during evolution where the function of a preexisting trait is redirected to another function (e.g., the drumming behavior of woodpeckers, originally used to find food, has become a means of emitting an acoustic communication signal for recognition within the species).

Formants Reinforced voice frequencies when sound waves travel through the vocal tract.

Frequency modulation The change in pitch (high/low) of a sound signal over time.

Frequency spectrum The set of frequencies present in a sound. A pure sound has a single frequency; a complex sound has several.

Fundamental frequency The lowest frequency of a complex sound (called F0 or "first harmonic"). It determines the pitch of a sound.

Graded vocalizations Vocalizations that vary along an acoustic continuum, with this variation accompanying changes in the information encoded by the sound signal.

Habituation The gradual loss of a receiver's responsiveness to a signal when it is repeatedly perceived.

Handicap theory The theory that certain features that are heavy to bear are evidence of the quality of those who bear them; for example, signals that impose a cost to survival.

Harmonic frequencies Multiples of the fundamental frequency.

Hertz A unit of measurement (abbreviated Hz) for the frequency of a sound. One hertz corresponds to one cycle of vibration per second.

Honest communication When a signal carries reliable information about certain characteristics of the sender (e.g., body size). This signal cannot manipulate the receiver.

Infrasounds Sound waves too low-pitched to be heard by humans (< 20 Hz approximately).

Kin selection Process favoring altruistic (cooperative) behaviors between genetically close individuals, with a reciprocity all the stronger as they are more closely related.

Language Fundamentally, any communication system based on the production of sound signals. However, this term is often reserved for human-articulated language.

Lek Area where males gather to court females.

Mathematical theory of communication The theory that all communication is based on the sender-signal-receiver triad, and is subject to noise. The coding of information by the signal can be optimized to meet the constraint imposed by noise. Also known as information theory.

Morton's principle (or motivation-structural rules) The correspondence between the structure of an acoustic signal and the context of the sender's emission and motivation. For example, and in general, a dull rumble is more indicative of aggressive intentions, while small, high-pitched, low-intensity calls are typical of a submissive attitude.

Natural selection A process that favors the survival and offspring of individuals best adapted to the environment in which they live. In the context of bioacoustics, natural selection favors communication behaviors and sound signals that increase the survival and/or genetic contribution to the next generation (direct and/or indirect progeny) of the sender and/or receiver.

Noise Any process that results in the loss of information as it is transmitted from the sender to the receiver.

Phylogenetic tree A representation of relationships between species (or between populations, individuals, etc.).

Pitch The property of a sound as determined by the fundamental frequency. Ranges from low to high.

Proximal causes of communication Morphological, anatomical, and physiological bases of communication behavior. Proximal causes, in the short term, concern all the mechanisms of production and perception of sound signals.

Quality index (or index signal) A signal that carries reliable information about the sender because its production is subject to an unavoidable physical constraint that reflects its anatomy, morphology, physiological capabilities, or health (e.g., the lowest fundamental frequency that can be emitted by a mammal is constrained by the size of its vocal cords).

Referential communication The use of signals that designate an environmental feature (such as the presence of a particular predator).

Ritualization The modification of a behavior (aggression, search for food, etc.) that becomes a communication signal by stereotypical simplification.

Sensory bias (or sensory exploitation) A situation in which a receiver's preexisting sensitivity to a signal may be exploited by the sender. This sensitivity has emerged through evolution for reasons other than communication.

Sexual dimorphism Differences (e.g., in size or color) between females and males.

Sexual selection Cases of natural selection imposed by individuals of the same sex (intrasexual selection) or of the opposite sex (intersexual selection).

Social grooming Behavior consisting of searching the coat or plumage of a congener for parasites, which plays an important role in maintaining and strengthening social bonds in primates.

Social intelligence The ability to analyze the relationships between members of one's social group and the social status of others.

Sound frequency The number of cycles of vibration per second. A low-frequency sound is low-pitched; a high-frequency sound is high-pitched.

Sound library A collection of sound recordings.

Sound waves Pressure variations that propagate through air or water.

Source-filter theory The theory that the production of sound signals in mammals (and most four-legged vertebrates) is based on an anatomical structure producing a sound wave (the source: larynx in mammals, syrinx in birds), and a system of pipes and cavities modifying this wave (the filter: vocal tract).

Spatial unmasking The ability to locate the origin of a sound signal despite ambient noise when the source of the signal and the source of the noise are not in the same location.

Spectrogram A graphical representation of a sound in which the time scale is given by the horizontal axis, the frequency scale by the vertical axis, and the intensity scale by a palette of colors.

Static information Information relating to the stable characteristics of the sender; the constituents of a sender's vocal signature.

Timbre The sound of the voice as determined by a combination of many characteristics of sound, including the distribution of energy in the frequency spectrum.

Ultrasounds Sound waves that are too high-pitched to be heard by humans (> 20,000 Hz approximately).

Vocal learning The mechanisms by which an animal learns to produce vocalization by imitation.

Vocal plasticity The ability for an individual to rapidly modulate its sound signals.

Vocal repertoire The set of sound signals that an animal can produce.

Vocal signature The information encoded in an audible signal indicating the identity of the sender.

Wavelength The distance between two amplitude maxima of the sound wave. The wavelength is inversely proportional to the frequency of the sound: when it is short, the sound is high-pitched; when it is long, the sound is low-pitched.

NOTES

Chapter 1

1. Bradbury J. W., and Vehrencamp S. L., *Principles of Animal Communication*, Sinauer, 2011.

2. Lorenz K., "Der kumpan in der umwelt des vogels," *J Ornithol*, vol. 83, 1935, 137–289.

3. Von Frisch K., *The Dance Language and Orientation of Bees*, Harvard University Press, 1967.

4. Tinbergen N., "On aims and methods of ethology," *Zeit Tierpsychol*, vol. 20, 1963, 410–433.

5. Bateson P., and Laland K. N., "Tinbergen's four questions: An appreciation and an update," *Trends Ecol Evol*, vol. 28, 2013, 712–718.

6. On dinosaurs' vocalizations, see Weishampel D. B., "Dinosaurian cacophony: Inferring function in extinct organisms," *BioScience*, vol. 47, 1997, 150–159.

7. Hsieh S., and Plotnick R. E., "The representation of animal behaviour in the fossil record," *Anim Behav*, vol. 169, 2020, 65–80; Senter P., "Voices of the past: A review of Paleozoic and Mesozoic animal sounds," *Hist Biol*, vol. 20, 2008, 255–287.

Chapter 2

8. Maurice S., et al., "In situ recording of Mars soundscape," *Nature*, vol. 605, 2022, 653–658.

9. Some animals are able to produce very loud sounds, up to 125 dB! See Jakobsen L., et al., "How loud can you go? Physical and physiological constraints to producing high sound pressures in animal vocalizations," *Front Ecol Evol*, vol. 9, 2021, 657254.

10. To deepen your knowledge of the physics of sound waves, I advise you to read the following chapters, all written for bioacousticians: Larsen O. N., "To shout or to whisper? Strategies for encoding public and private information in sound signals," in Aubin T., and Mathevon N. (eds.), *Coding Strategies in Vertebrate Acoustic Communication*, Springer, 2020, 11–44; Larsen O. N., and Wahlberg M., "Sound and sound sources," in Brown C. and Riede T. (eds.), *Comparative Bioacoustics: An Overview*, Bentham Science, 2017, 3–62; Wahlberg M., and Larsen O. N., "Propagation of sound," in Brown and Riede (eds.), *Comparative Bioacoustics*, 63–121.

Chapter 3

11. Marler P., and Slabbekoorn H. (eds.), *Nature's Music: The Science of Birdsong*, Elsevier, 2004; Catchpole C. K., and Slater P.J.B., *Bird Song: Biological Themes and Variations*, Cambridge University Press, 2008. For a refreshing and interesting perspective on birdsong, see this review: Rose E. M., et al., "The singing question: Re-conceptualizing birdsong," *Biol Rev*, vol. 97, 2022, 326–342.

12. Species recognition is an important process for mating purposes or territorial defense, and many animal signals support information about species identity. For a nonbird example, look at this paper: Fonseca P. J., and Revez M. A., "Song discrimination by male cicadas *Cicada barbara lusitanica* (Homoptera, Cicadidae)," *J Exp Biol*, vol. 205, 2002, 1285–1292.

13. Draganoiu T. I., et al., "Song stability and neighbour recognition in a migratory songbird, the black redstart," *Behaviour*, vol. 151, 2014, 435–453. There are many other articles dealing with the dear-enemy effect; see, for example, Briefer E., Rybak F., and Aubin T., "When to be a dear enemy: Flexible acoustic relationships of neighbouring skylarks, *Alauda arvensis*," *Anim Behav*, vol. 76, 2008, 1319–1325; Briefer E., et al., "How to identify dear enemies: The group signature in the complex song of the skylark *Alauda arvensis*," *J Exp Biol*, vol. 211, 2008, 317–326; Tumulty J. P., et al., "Ecological and social drivers of neighbor recognition and the dear enemy effect in a poison frog," *Behav Ecol*, vol. 32, 2021, 138–150; and Amorim P. S., et al., "Out of sight, out of mind: Dear enemy effect in the rufous hornero, *Furnarius rufus*," *Anim Behav*, vol. 187, 2022, 167–176. See also our recent paper on hippos: Thévenet J., et al., "Voice-mediated interactions in a megaherbivore," *Cur Biol*, vol. 32, 2022, R55–R71.

14. Aubin T., et al., "How a simple and stereotyped acoustic signal transmits individual information: The song of the white-browed warbler *Basileuterus leucoblepharus*," *An Acad Bras Cienc*, vol. 76, 2004, 335–344.

15. Mathevon N., et al., "Singing in the rain forest: How a tropical bird song transfers information," *PLoS ONE*, vol. 3, 2008, e1580.

16. Slabbekoorn H., "Singing in the wild: The ecology of birdsong," in Marler and Slabbekoorn (eds.), *Nature's Music*, 178–205.

17. Their Latin names, respectively, are as follows: *Crypturellus obsoletus, Myrmeciza squamosa, Basileuterus leucoblepharus, Pyriglena leucoptera, Chiroxiphia caudata, Batara cinerea, Trogon surrucura,* and *Carpornis cucullatus*.

18. Brémond J.-C., "Specific value of syntax in the territorial defense signal of the troglodyte (*Troglodytes troglodytes*)," *Behaviour*, vol. 30, 1968, 66–75; Brémond, "Role of the carrier frequency in the territorial songs of oscines," *Ethology*, vol. 73, 1968, 128–135; Brémond, "Acoustic competition between the song of the wren (*Troglodytes troglodytes*) and the songs of other species," *Behaviour*, vol. 65, 1978, 89–98.

19. Kreutzer M., "Stéréotopie et variations dans les chants de proclamation territoriale chez le Troglodyte (*Troglodytes troglodytes*)," *Rev Comport Anim*, vol. 8, no. 2, 1974, 70–286.

20. Mathevon N., and Aubin T., "Reaction to conspecific degraded song by the wren *Troglodytes troglodytes*: Territorial response and choice of song post," *Behav Proc*, vol. 39, 1997, 77–84.

21. Mathevon N., Aubin T., and Dabelsteen T., "Song degradation during propagation: Importance of song post for the wren *Troglodytes troglodytes*," *Ethology*, vol. 102, 1996, 397–412.

22. Mathevon N., et al., "Are high perches in the blackcap *Sylvia atricapilla* song or listening posts? A sound transmission study," *JASA*, vol. 117, 2005, 442–449.

23. Mathevon N., and Aubin T., "Sound-based species-specific recognition in the blackcap *Sylvia atricapilla* shows high tolerance to signal modifications," *Behaviour*, vol. 138, 2001, 511–524.

24. Ey E., and Fischer J., "The 'acoustic adaptation hypothesis'—a review of the evidence from birds, anurans and mammals," *Bioacoustics*, vol. 19, 2009, 21–48.

25. I discuss this in more detail in chapter 17 on ecoacoustics.

26. Balakrishnan R., "Behavioral ecology of insect acoustic communication," in Pollack G. S., Mason A. C., Popper A., and Fay R. R. (eds.), *Insect Hearing*, Springer, 2016, 49–80.

Chapter 4

27. Shannon C. E., and Weaver W., *The Mathematical Theory of Communication*, University of Illinois Press, 1949. Claude Shannon first developed and published this theory in 1948. His paper was republished in book form in 1949, with an introduction by Warren Weaver (which I recommend reading).

28. Mathevon N., and Aubin T., "Acoustic coding strategies through the lens of the mathematical theory of communication," in Aubin and Mathevon (eds.), *Coding Strategies in Vertebrate Acoustic Communication*, 1–10.

29. Charrier I., et al., "Acoustic communication in a black-headed gull colony: How do chicks identify their parents?," *Ethology*, vol. 107, 2001, 961–974.

30. Aubin T., and Jouventin P., "How to vocally identify kin in a crowd: The penguin model," *Adv Stud Behav*, vol. 31, 2002, 243–277.

31. Aubin T., and Jouventin P., "Cocktail-party effect in king penguin colonies," *Proc R Soc B*, vol. 265, 1998, 1665–1673.

32. Lengagne T., et al., "How do king penguins (*Aptenodytes patagonicus*) apply the mathematical theory of information to communicate in windy conditions?," *Proc R Soc B*, vol. 266, 1999, 1623–1628.

33. Jouventin P., Aubin T., and Lengagne T., "Finding a parent in a king penguin colony: The acoustic system of individual recognition," *Anim Behav*, vol. 57, 1999, 1175–1183.

34. The frequencies—the pitch, low or high, of the song—change over time.

35. The intensity—the fact that the song is more or less loud—changes over time.

36. Aubin T., Jouventin P., and Hildebrand C., "Penguins use the two-voice system to recognize each other," *Proc R Soc B*, vol. 267, 2000, 1081–1087.

37. Gomez-Bahamon V., et al., "Sonations in migratory and non-migratory fork-tailed flycatchers (*Tyrannus savana*)," *Integr Comp Biol*, vol. 60, 2020, 1147–1159.

38. Kingsley E. P., et al., "Identity and novelty in the avian syrinx," *PNAS*, vol. 115, 2018, 10209–10217.

39. Riede T., et al., "The evolution of the syrinx," *PLoS Biol*, vol. 17, 2019, e2006507.

40. Robisson P., Aubin T., and Brémond J.-C., "Individuality in the voice of the emperor penguin *Aptenodytes forsteri*: Adaptation to a noisy environment," *Ethology*, vol. 94, 1993, 279–290.

41. Aubin and Jouventin, "Cocktail-party effect in king penguin colonies."

42. Jouventin P., and Aubin T., "Acoustic systems are adapted to breeding ecologies: Individual recognition in nesting penguins," *Anim Behav*, vol. 64, 2002, 747–757.

43. Aubin T., and Jouventin P., "Localisation of an acoustic signal in a noisy environment: The display call of the king penguin *Aptenodytes patagonicus*," *J Exp Biol*, vol. 205, 2002, 3793–3798. See also this paper reporting that king penguins' calls inform receivers about the age of the sender: Kriesell H. J., et al., "How king penguins advertise their sexual maturity," *Anim Behav*, vol. 177, 2021, 253–267.

44. Loesche P., et al., "Signature versus perceptual adaptations for individual vocal recognition in swallows," *Behaviour*, vol. 118, 1991, 15–25; Medvin M. B., Stoddard P. K., and Beecher M. D., "Signals for parent-offspring recognition: Strong sib-sib call similarity in cliff swallows but not barn swallows," *Ethology*, vol. 90, 1992, 17–28; Medvin M. B., Stoddard P. K., and Beecher M. D., "Signals for parent-offspring recognition: A comparative analysis of the begging calls of cliff swallows and barn swallows," *Anim Behav*, vol. 45, 1993, 841–850.

45. Charrier I., "Mother-offspring vocal recognition and social system in pinnipeds," in Aubin and Mathevon (eds.), *Coding Strategies in Vertebrate Acoustic Communication*, 231–246; Charrier I., "Vocal communication in otariids and odobenids," in Campagna C., and Harcourt R. (eds.), *Ethology and Behavioral Ecology of Otariids and the Odobenid*, Springer, 2021, 265–289.

Chapter 5

46. Miller E. H., and Kochnev A. A., "Ethology and behavioral ecology of the walrus (*Odobenus rosmarus*), with emphasis on communication and social behavior," in Campagna and Harcourt (eds.), *Ethology and Behavioral Ecology of Otariids*, 437–488.

47. Charrier I., "Mother-offspring vocal recognition and social system in pinnipeds"; Charrier I., and Casey C., "Social communication in phocids," in Costa D. P., and McHuron E. (eds.), *Ethology and Behavioral Ecology of Phocids*, Springer, 2022, 69–100.

48. Stewart B., "Family Odobenidae," in Wilson D. E., and Mittermeier R. A. (eds.), *Handbook of the Mammals of the World IV, Sea Mammals*, Lynx Edicions, 2014, 102–119.

49. Charrier I., Aubin T., and Mathevon N., "Mother-calf vocal communication in Atlantic walrus: A first field experimental study," *Anim Cogn*, vol. 13, 2010, 471–482.

50. https://videotheque.cnrs.fr/doc=2019.

51. Linossier J., et al., "Maternal responses to pup calls in a high-cost lactation species," *Biol Let*, vol. 17, 2021, 20210469.

52. For a review of recognition mechanisms in the Australian sea lion, see Charrier I., et al., "Mother-pup recognition mechanisms in Australian sea lion (*Neophoca cinerea*) using uni- and multi-modal approaches," *Anim Cogn*, vol. 25, 2022, 1019–1028.

53. Studies show that vocal recognition between mother and pup in mammals can last a long time: Briefer E. F., et al., "Mother goats do not forget their kids' calls," *Proc R Soc B*, vol. 279, 2012, 3749–3755; Charrier I., et al., "Fur seal mothers memorize subsequent versions of developing pups' calls: Adaptation to long-term recognition or evolutionary by-product?," *Biol J Lin Soc*, vol. 80, 2003, 305–312.

54. Charrier I., Mathevon N., and Jouventin P., "Mother's voice recognition by seal pups," *Nature*, vol. 412, 2001, 873.

55. The dynamics of voice recognition between mother and pup has been studied in other mammal species, including humans. See, for instance, Sèbe F., et al., "Establishment of vocal communication and discrimination between ewes and their lamb in the first two days after parturition," *Dev Psychobiol*, vol. 49, 2007, 375–386; and Bouchet H., et al., "Baby cry recognition is independent of motherhood but improved by experience and exposure," *Proc R Soc B*, vol. 287, 2020, 20192499.

56. Charrier I., et al., "The subantarctic fur seal pup switches its begging behaviour during maternal absence," *Can J Zool*, vol. 80, 2002, 1250–1255.

57. Mother-pup vocal recognition in mammals has been studied in other models. See, for instance, Sèbe F., et al., "Early vocal recognition of mother by lambs: Contribution of low- and high-frequency vocalizations," *Anim Behav*, vol. 79, 2010, 1055–1066.

58. Maynard Smith J., and Harper D., *Animal Signals*, Oxford University Press, 2003; Searcy W. A., and Nowicki S., *The Evolution of Animal Communication*, Princeton University Press, 2005.

59. Anderson M. G., Brunton D. H., and Hauber M. E., "Reliable information content and ontogenetic shift in begging calls of grey warbler nestlings," *Ethology*, vol. 116, 2010, 357–365; Caro S. M., West S. A., and Griffin A. S., "Sibling conflict and dishonest signaling in birds," *PNAS*, vol. 113, 2016, 13803–13808.

60. Kedar H., et al., "Experimental evidence for offspring learning in parent-offspring communication," *Proc R Soc B*, vol. 267, 2000, 1723–1727.

61. McCarty J. P., "The energetic cost of begging in nestling passerines," *The Auk*, vol. 113, 1996, 178–188; Leech S. M., and Leonard M. L., "Is there an energetic cost to begging in nestling tree swallows (*Tachycineta bicolor*)?," *Proc R Soc B*, vol. 263, 1996, 983–987.

62. Bachman G. C., and Chappell M. A., "The energetic cost of begging behaviour in nestling house wrens," *Anim Behav*, vol. 55, 1998, 1607–1618.

63. Husby M., "Nestling begging calls increase predation risk by corvids," *Anim Biol*, vol. 69, 2019, 137–155.

64. Magrath R. D., et al., "Calling in the face of danger: Predation risk and acoustic communication by parent birds and their offspring," *Adv Stud Behav*, vol. 41, 2010, 187–253; Husby, "Nestling begging calls increase predation risk."

65. Magrath et al., "Calling in the face of danger."

66. Briskie J. V., Martin P. R., and Martin T. E., "Nest predation and the evolution of nestling begging calls," *Proc R Soc B*, vol. 266, 1999, 2153–2159.

67. Haff T. M., and Magrath R. D., "Calling at a cost: Elevated nestling calling attracts predators to active nests," *Biol Let*, vol. 7, 2011, 493–495.

68. Anderson M. G., Brunton D. H., and Hauber M. E., "Species specificity of grey warbler begging solicitation and alarm calls revealed by nestling responses to playbacks," *Anim Behav*, vol. 79, 2010, 401–409.

69. Kilner R. M., and Hinde C. A., "Information warfare and parent–offspring conflict," *Adv St Behav*, vol. 38, 2008, 283–336.

70. Caro, West, and Griffin, "Sibling conflict and dishonest signaling in birds"; Bowers E. K., et al., "Condition-dependent begging elicits increased parental investment in a wild bird population," *Am Nat*, vol. 193, 2019, 725–737.

71. Price K., "Begging as competition for food in yellow-headed blackbirds," *The Auk*, vol. 4, 1996, 963–967.

72. Mathevon N., and Charrier I., "Parent-offspring conflict and the coordination of siblings in gulls," *Proc R Soc B*, vol. 271, 2004, S145–S147; Blanc A., et al., "Coordination de la quémande entre les jeunes de mouette rieuse," *CR Biol*, vol. 333, 2010, 688–693.

73. Leonard M., and Horn A., "Provisioning rules in tree swallows," *Behav Ecol Sociobiol*, vol. 38, 1996, 341–347; Leonard M. L., Horn A. G., and Mukhida A., "False alarms and begging in nestling birds," *Anim Behav*, vol. 69, 2005, 701–708.

74. Roulin A., "The sibling negotiation hypothesis," in Wright J., and Leonard M. L. (eds.), *The Evolution of Begging*, Kluwer, 2002, 107–126; Ducouret P., et al., "The art of diplomacy in vocally negotiating barn owl siblings," *Front Ecol Evol*, vol. 7, 2019, 351.

75. Dreiss A. N., et al., "Social rules govern vocal competition in the barn owl," *Anim Behav*, vol. 102, 2015, 95–107.

76. Dreiss A. N., et al., "Vocal communication regulates sibling competition over food stock," *Behav Ecol Sociobiol*, vol. 70, 2016, 927–937.

77. Ligout S., et al., "Not for parents only: Begging calls allow nest-mate discrimination in juvenile zebra finches," *Ethology*, vol. 122, 2016, 193–206.

78. Draganoiu T. I., et al., "Parental care and brood division in a songbird, the black redstart," *Behaviour*, vol. 142, 2005, 1495–1514; Draganoiu T. I., et al., "In a songbird, the black redstart, parents use acoustic cues to discriminate between their different fledglings," *Anim Behav*, vol. 71, 2006, 1039–1046.

79. Bugden S. C., and Evans R. M., "Vocal solicitation of heat as an integral component of the developing thermoregulatory system in young domestic chickens," *Can J Zool*, vol. 75, 1997, 1949–1954.

80. Mariette M. M., and Buchanan K. L., "Prenatal acoustic communication programs offspring for high posthatching temperatures in a songbird," *Science*, vol. 353, 2016, 812–814; Mariette M. M., "Acoustic cooperation: Acoustic communication regulates conflict and cooperation within the family," *Front Ecol Evol*, vol. 7, 2019, 445; Pessato A., et al., "A prenatal acoustic signal of heat affects thermoregulation capacities at adulthood in an arid-adapted bird," *Sc Rep*, vol. 12, 2022, 5842.

81. Elie J. E., et al., "Vocal communication at the nest between mates in wild zebra finches: A private vocal duet?," *Anim Behav*, vol. 80, 2010, 597–605; Boucaud I.C.A., et al., "Vocal negotiation over parental care? Acoustic communication at the nest predicts partners' incubation share," *Biol J Lin Soc*, vol. 117, 2016, 322–336.

82. Boucaud I.C.A., et al., "Incubating females signal their needs during intrapair vocal communication at the nest: A feeding experiment in great tits," *Anim Behav*, vol. 122, 2016, 77–86.

83. Dawkins wrote, among other books, *The Selfish Gene*, *The Blind Watchmaker*, and *The Extended Phenotype*.

84. Dawkins R., and Krebs J. R., "Animal signals: Mind-reading and manipulation," in Krebs J. R., and Davies N. B. (eds.), *Behavioural Ecology: An Evolutionary Approach*, Blackwell, 1978, 381–402.

85. This "arms race" can be illustrated by the metaphor of the Red Queen in Lewis Carroll's novel *Through the Looking Glass*, in which Alice and the queen have to run constantly to stay in the same place.

86. Davis N. B., *Cuckoos, Cowbirds and Other Cheats*, T & A Poyser, 2000.

87. Soler M. (ed.), *Avian Brood Parasitism*, Springer, 2017.

88. Anderson M. G., et al., "Begging call matching between a specialist brood parasite and its host: A comparative approach to detect coevolution," *Biol J Lin Soc*, vol. 98, 2009, 208–216; Ursino C. A., et al., "Host provisioning behavior favors mimetic begging calls in a brood-parasitic cowbird," *Behav Ecol*, vol. 29, 2018, 328–332.

89. Kilner R. M., et al., "Signals of need in parent-offspring communication and their exploitation by the common cuckoo," *Nature*, vol. 397, 1999, 667–672.

90. Samas P., et al., "Nestlings of the common cuckoo do not mimic begging calls of two closely related *Acrocephalus* hosts," *Anim Behav*, vol. 161, 2020, 89–94; Jamie G. A., and Kilner R. M., "Begging call mimicry by brood parasite nestlings: Adaptation, manipulation and development," in Soler (ed.), *Avian Brood Parasitism*, 517–538.

91. Langmore N. E., Hunt S., and Kilner R. M., "Escalation of a coevolutionary arms race through host rejection of brood parasitic young," *Nature*, vol. 422, 2003, 157–160; Langmore N. E., and Kilner R. M., "The coevolutionary arms race between Horsfield's bronze-cuckoos and superb fairy-wrens," *Emu*, vol. 110, 2010, 32–38.

92. Wright and Leonard (eds.), *Evolution of Begging*; Royle N., et al., *The Evolution of Parental Care*, Oxford University Press, 2012.

93. Colombelli-Négrel D., et al., "Embryonic learning of vocal passwords in superb fairy-wrens reveals intruder cuckoo nestlings," *Cur Biol*, vol. 22, 2012, 2155–2160.

94. For a review on prenatal acoustic communication and its consequences on developmental processes, see Mariette M. M., et al., "Acoustic developmental programming: A mechanistic and evolutionary framework," *Trends Ecol Evol*, vol. 36, 2021, 722–736.

95. Ames A. E., et al., "Pre- and post-partum whistle production of a bottlenose dolphin (*Tursiops truncatus*) social group," *Int J Comp Psychol*, vol. 32, 2019.

96. King S. L., et al., "Maternal signature whistle use aids mother-calf reunions in a bottlenose dolphin, *Tursiops truncatus*," *Behav Proc*, vol. 126, 2016, 64–70.

Chapter 6

97. Charrier I., Mathevon N., and Aubin T., "Bearded seal males perceive geographic variation in their trills," *Behav Ecol Sociobiol*, vol. 67, 2013, 1679–1689.

98. Montgomery J. C., and Radford C. A., "Marine bioacoustics," *Cur Biol*, vol. 27, 2017, R502–R507.

99. And even birds underwater. A recent paper shows that penguins vocalize underwater: Thiebault A., et al., "First evidence of underwater vocalisations in hunting penguins," *PeerJ*, vol. 7, 2019, e8240.

100. Tolimieri N., et al., "Ambient sound as a navigational cue for larval reef fish," *Bioacoustics*, vol. 12, 2002, 214–217.

101. Lillis A., and Mooney T. A., "Sounds of a changing sea: Temperature drives acoustic output by dominant biological sound-producers in shallow water habitats," *Front Mar Sci*, vol. 9, 2022, 960881.

102. Versluis M., et al., "How snapping shrimp snap: Through cavitating bubbles," *Science*, vol. 289, 2000, 2114–2117; Dinh J. P., and Radford C., "Acoustic particle motion detection in the snapping shrimp (*Alpheus richardsoni*)," *J Comp Physiol A*, vol. 207, 2021, 641–655; Kingston A.C.N., et al., "Snapping shrimp have helmets that protect their brains by dampering shock waves," *Cur Biol*, vol. 32, 2022, 3576–3583.

103. Simpson S. D., et al., "Homeward sound," *Science*, vol. 308, 2005, 221.

104. Gordon T.A.C., et al., "Acoustic enrichment can enhance fish community development on degraded coral reef habitat," *Nat Com*, vol. 10, 2019, 5414.

105. Learn more about whales and dolphins: Whitehead H., and Rendell L., *The Cultural Lives of Whales and Dolphins*, University of Chicago Press, 2015.

106. Wilson and Mittermeier (eds.), *Handbook of the Mammals of the World* IV.

107. Huelsmann M., et al., "Genes lost during the transition from land to water in cetaceans highlight genomic changes associated with aquatic adaptations," *Sc Adv*, vol. 5, 2019, eaaw6671.

108. Adam O., et al., "New acoustic model for humpback whale sound production," *App Acoust*, vol. 74, 2013, 1182–1190; Damien J., et al., "Anatomy and functional morphology of the mysticete rorqual whale larynx: Phonation positions of the U-fold," *Anat Rec*, vol. 302, 2019, 703–717.

109. Things are actually a little more complex. The vocal cords of mysticetes are membranes covering the two cartilages that hold the laryngeal sac to the trachea. When the whale makes a sound, it opens these two cartilages (which are parallel to each other), and the flow of air from the lungs to the laryngeal sacs makes the membranes vibrate. By orienting these cartilages differently, the whale can choose which surface of the membranes starts to vibrate, thus controlling the pitch (low or high frequencies) of the sounds produced. See Damien et al., "Anatomy and functional morphology of the mysticete rorqual whale larynx."

110. Reidenberg J. S., and Laitman J. T., "Anatomy of underwater sound production with a focus on ultrasonic vocalization in toothed whales including dolphins and porpoises," in Brudzynski S. (ed.), *Handbook of Ultrasonic Vocalization* XXV, Elsevier, 2018, 509–519.

111. Ames A. E., Beedholm K., and Madsen P. T., "Lateralized sound production in the beluga whale (*Delphinapterus leucas*)," *J Exp Biol*, vol. 223, 2020, jeb226316.

112. Clark C. W., and Garland E. C. (eds.), *Ethology and Behavioural Ecology of Mysticetes*, Springer, 2022.

113. Stafford K. M., et al., "Extreme diversity in the songs of Spitsbergen's bowhead whales," *Biol Let*, vol. 14, 2018, 20180056; Stafford K. M., "Singing behavior in the bowhead whale," in Clark and Garland (eds.), *Ethology and Behavioural Ecology of Mysticetes*, 277–295.

114. Payne R. S., and McVay S., "Songs of humpback whales," *Science*, vol. 173, 1971, 587–597; Rothenberg D., *Thousand Mile Song: Whale Music in a Sea of Sound*, Basic Books, 2010.

115. Payne K., and Payne R. S., "Large-scale changes over 19 years in songs of humpback whales in Bermuda," *Zeit Psychol*, vol. 68, 1985, 89–114.

116. Noad M. J., et al., "Cultural revolution in whale songs," *Nature*, vol. 408, 2000, 537.

117. Herman L. M., "The multiple functions of male song within the humpback whale (*Megaptera novaeangliae*) mating system: Review, evaluation, and synthesis," *Biol Rev*, vol. 92, 2017, 1795–1818.

118. Tyack P., "Differential response of humpback whales, *Megaptera novaeangliae*, to playback of song or social sounds," *Behav Ecol Sociobiol*, vol. 13, 1983, 49–55.

119. Mercado III E., "The sonar model for humpback whale song revised," *Front Psychol*, vol. 9, 2018, 1156; Mercado III E., "Spectral interleaving by singing humpback whales: Signs of sonar," *J Acoust Soc Am*, vol. 149, 2021, 800–806.

120. Kuna V. M., and Nabelek J. L., "Seismic crustal imaging using fin whale songs," *Science*, vol. 371, 2021, 731–735.

121. See the film by Antonio Fischetti (in French): https://www.youtube.com/watch?v=hbPRRl0Q7hw.

122. King S. L., et al., "Evidence that bottlenose dolphins can communicate with vocal signals to solve a cooperative task," *R Soc Open Sc*, vol. 8, 2021, 202073; King S. L., et al., "Cooperation-based concept formation in male bottlenose dolphins," *Nat Com*, vol. 12, 2021, 2373; Barluet de Beauchesne L., et al., "Friend or foe: Risso's dolphins eavesdrop on conspecific sounds to induce or avoid intra-specific interaction," *Anim Cogn*, vol. 25, 2022, 287–296; Ames A. E., et al., "Evidence of stereotyped contact call use in narwhal (*Monodon monoceros*) mother-calf communication," *PLoS ONE*, vol. 16, 2021, e0254393; Chereskin E., et al., "Allied male dolphins use vocal exchanges to 'bond at a distance,'" *Cur Biol*, vol. 32, 2022, 1657–1663.

123. Janik V. M., Sayigh L. S., and Wells R. S., "Signature whistle shape conveys identity information to bottlenose dolphins," *PNAS*, vol. 103, 2006, 8293–8297.

124. King S. L., and Janik V. M., "Bottlenose dolphins can use learned vocal labels to address each other," *PNAS*, vol. 110, 2013, 13216–13221. The belugas, the famous white whales, also imitate their vocal signatures: Morisaka T., Nishimoto S., et al., "Exchange of 'signature' calls in captive belugas (*Delphinapterus leucas*)," *J Ethol*, vol. 31, 2013, 141–149.

125. Janik V. M., and Sayigh L. S., "Communication in bottlenose dolphins: 50 years of signature whistle research," *J Comp Physiol*, vol. 199, 2013, 479–489.

126. King S. L., et al., "Bottlenose dolphins retain individual vocal labels in multi-level alliances," *Cur Biol*, vol. 28, 2018, 1993–1999.

127. King et al., "Maternal signature whistle use aids mother-calf reunions."

128. Rice A., et al., "Spatial and temporal occurrence of killer whale ecotypes off the outer coast of Washington State, USA," *Mar Ecol Prog Series*, vol. 572, 2017, 255–268.

129. Barrett-Lennard L., "Killer whale evolution: Populations, ecotypes, species, Oh my!," *J Amer Cetac Soc*, vol. 40, 2011, 48–53.

130. Ford J.K.B., "Vocal traditions among resident killer whales (*Orcinus orca*) in coastal waters of British Columbia," *Can J Zool*, vol. 69, 1991, 1454–1483.

131. Deecke V. B., Ford J.K.B., and Spong P., "Dialect change in resident killer whales: Implications for vocal learning and cultural transmission," *Anim Behav*, vol. 40, 2000, 629–638.

132. Poupart M., et al., "Intra-group orca call rate modulation estimation using compact four hydrophones array," *Front Mar Sci*, vol. 8, 2021, 681036.

133. *Centre d'études et d'expertise sur les risques, l'environnement, la mobilité et l'Aménagement* (Center of Studies and Expertise on Risks, Environment, Mobility and Development).

134. Benti B., et al., "Indication that the behavioural responses of humpback whales to killer whale sounds are influenced by trophic relationships," *Mar Ecol Prog Series*, vol. 660, 2021, 217–232.

135. Curé C., et al., "Evidence for discrimination between feeding sounds of familiar fish and unfamiliar mammal-eating killer whale ecotypes by long-finned pilot whales," *Anim Cogn*, vol. 22, 2019, 863–882.

136. Isojunno S., et al., "Sperm whales reduce foraging effort during exposure to 1–2 kHz sonar and killer whale sounds," *Ecol Appl*, vol. 26, 2016, 77–93.

137. Curé C., et al., "Responses of male sperm whales (*Physeter macrocephalus*) to killer whale sounds: Implications for anti-predator strategies," *Sc Rep*, vol. 3, 2013, 1579.

138. Clarke M. R., "Function of the spermaceti organ of the sperm whale," *Nature*, vol. 228, 1970, 873–874.

139. See Joy Reidenberg's animated explanation: https://www.youtube.com/watch?v =sW7o5IC2io0.

140. Huggenberg S., et al., "The nose of the sperm whale: Overviews of functional design, structural homologies and evolution," *J Mar Biol Assoc*, vol. 96, 2016, 783–806.

141. Oliveira C. et al., "The function of male sperm whale slow clicks in a high latitude habitat: Communication, echolocation, or prey debilitation?," *JASA*, vol. 133, 2013, 3135–3144; Tonnesen P. et al., "The long-range echo scene of the sperm whale biosonar," *Biol Let*, vol. 16, 2020, 20200134.

142. For a full explanation of clicks and codas and their roles in the lives of sperm whales, see Whitehead H., and Rendell L., *The Cultural Lives of Whales and Dolphins*, University of Chicago Press, 2015, 146–158.

143. Gordon J.C.D., "Evaluation of a method for determining the length of sperm whales (*Physeter catodon*) from their vocalizations," *J Zool*, vol. 224, 1991, 301–314; Møhl B., et al., "Sperm whale clicks: Directionality and source levels revisited," *JASA*, vol. 107, 2000, 638–648; Møhl B., "Sound transmission in the nose of the sperm whale *Physeter catodon*: A post mortem study," *J Comp Physiol A*, vol. 187, 2001, 335–340.

144. Growcott A., et al., "Measuring sperm whales from their clicks: A new relationship between IPIs and photogrammetrically measured lengths," *JASA*, vol. 130, 2011, 568–573.

145. Rendell L. E., and Whitehead H., "Vocal clans in sperm whales (*Physeter macrocephalus*)," *Proc R Soc B*, vol. 270, 2003, 225–231.

146. Gero S., Whitehead H., and Rendell L., "Individual, unit and vocal clan level identity cues in sperm whale codas," *R Soc Op Sc*, vol. 3, 2016, 150372.

147. Sperm whales are also able to produce nonclick sounds. These include squeals, with a possible communicative social function; pips; short trumpets; and trumpets. For details, see Pace D. S., et al., "Trumpet sounds emitted by male sperm whales in the Mediterranean Sea," *Sc Rep*, vol. 11, 2021, 5867.

Chapter 7

148. Le Bœuf B. J., and Laws R. M., *Elephant Seals, Population Ecology, Behavior and Physiology*, University of California Press, 1994.

149. Dr. Colleen Reichmuth, senior researcher at the Long Marine Lab, University of California, Santa Cruz.

150. Elephant seal males are not the only pinnipeds to emit rhythmic sounds during their reproductive displays. Walrus males also produce some: Larsen O. N., and Reichmuth C., "Walruses produce intense impulse sounds by clap-induced cavitation during breeding displays," *R Soc Open Sc*, vol. 8, 2021, 210197.

151. Casey C., et al., "Rival assessment among northern elephant seals: Evidence of associative learning during male-male contests," *R Soc Open Sc*, vol. 2, 2015, 150228.

152. Mathevon N., et al., "Northern elephant seals memorize the rhythm and timbre of their rivals' voices," *Cur Biol*, vol. 27, 2017, 2352–2356.

153. Casey C., et al., "The genesis of giants: Behavioural ontogeny of male northern elephant seals," *Anim Behav*, vol. 166, 2020, 247–259.

Chapter 8

154. To know everything about crocodilians: Grigg G., and Kirshner D., *Biology and Evolution of Crocodylians*, CSIRO, 2015.

155. Vergne A. L., Pritz M. B., and Mathevon N., "Acoustic communication in crocodilians: From behaviour to brain," *Biol Rev*, vol. 84, 2009, 391–411.

156. Reber S. A., "Crocodilia communication," in Vonk J., and Shackelford T. K. (eds.), *Encyclopedia of Animal Cognition and Behavior*, Springer, 2018, 1–10.

157. Marquis O., et al., "Observations on breeding site, bioacoustics and biometry of hatchlings of *Paleosuchus trigonatus* (Schneider, 1801) from French Guiana (Crocodylia: Alligatoridae)," *Herp Notes*, vol. 13, 2020, 513–516.

158. Vergne A. L., and Mathevon N., "Crocodile egg sounds signal hatching time," *Cur Biol*, vol. 18, 2008, R513–R514.

159. Hatching synchronization seems to occur in other animals, even insects: Tanaka S., "Embryo-to-embryo communication facilitates synchronous hatching in grasshoppers," *J Orthop Res*, vol. 30, 2021, 107–115.

160. Vergne A. L., et al., "Acoustic signals of baby black caimans," *Zoology*, vol. 114, 2011, 313–320.

161. Mathevon N., et al., "The code size: Behavioural response of crocodile mothers to offspring calls depends on the emitter's size, not on its species identity," in *Crocodiles, Proceedings of the 24th Working Meeting of the Crocodile Specialist Group IUCN*, 2016, 79–85.

162. Vergne A. L., et al., "Acoustic communication in crocodilians: Information encoding and species specificity of juvenile calls," *Anim Cogn*, vol. 15, 2012, 1095–1109.

163. Vergne A. L., et al., "Parent-offspring communication in the Nile crocodile *Crocodylus niloticus*: Do newborns' calls show an individual signature?," *Naturwissen*, vol. 94, 2007, 49–54.

164. Papet L., et al., "Influence of head morphology and natural postures on sound localization cues in crocodilians," *R Soc Open Sc*, vol. 6, 2019, 190423; Papet L., et al., "Crocodiles use both interaural level differences and interaural time differences to locate a sound source," *JASA*, vol. 148, 2020, EL307.

165. For a comparative perspective on sound localizations, see this paper: Carr C. E., and Christensen-Dalsgaard J., "Sound localization strategies in three predators," *Brain Behav Evol*, vol. 86, 2015, 17–27. I also talk about sound localization in chapter 9, "Hear, at all costs."

166. Thévenet J., et al., "Spatial release from masking in crocodilians," *Com Biol*, vol. 5, 2022, 869.

Chapter 9

167. Some snakes do produce sounds, however (scraping scales, rattling the end of their tails, whistling sounds). The royal cobra *Ophiophagus hannah* produces a whistling sound at a frequency of about 600 Hz thanks to diverticula in its trachea, which act as resonators. See Young B. A., "The comparative morphology of the larynx in snakes," *Acta Zool*, vol. 81, 2000, 177–193; Young B. A., et al., "The morphology of sound production in *Pituophis melanoleucus* (Serpentes, Colubridae) with the first description of a vocal cord in snakes," *J Exp Zool*, vol. 273, 1995, 472–481; Young B. A., "Snake bioacoustics: Toward a richer understanding of the behavioral

ecology of snakes,"*Q Rev Biol*, vol. 78, 2003, 303–325; and Gans C., "Sound producing mechanisms in recent reptiles: Review and comment," *Amer Zool*, vol. 13, 1973, 1195–1203.

168. Greenfield M. D., "Evolution of acoustic communication in insects," in Pollack G. S., Mason A. C., Popper A., and Fay R. R. (eds.), *Insect Hearing*, Springer, 2016, 17–47.

169. Vermeij G. J., "Sound reasons for silence: Why do molluscs not communicate acoustically?," *Biol J Lin Soc*, vol. 100, 2010, 485–493. The snail emits sounds, but we do not know if they are used to communicate: Breure A.S.H., "The sound of a snail: Two cases of acoustic defence in gastropods," *J Mollusc Stud*, vol. 81, 2015, 290–293. The mussel perceives sound vibrations: Charifi M., et al., "The sense of hearing in the Pacific oyster, *Magallana gigas*," *PLoS ONE*, vol. 12, 2017, e0185353.

170. Bradbury J. W., and Vehrencamp S. L., *Principles of Animal Communication*, Sinauer, 2011.

171. Some internal mechanoreceptors serve a completely different purpose, like measuring blood pressure or the amount of food currently distending our stomachs.

172. These "short-distance" ears are sensitive to variations in the speed of air molecules.

173. "Long-distance" ears are sensitive to variations in air pressure.

174. Larsen and Wahlberg, "Sound and sound sources," in Brown and Riede (eds.), *Comparative Bioacoustics*, 3–61. This chapter provides very detailed explanations about these notions, which, as you might have guessed, are far more complex than what I've told you here. . . . Allergic to physics? Best sit this one out.

175. Budelmann B. U., "Hearing in crustacea," in Webster D. B., and Fay R. R. (eds.), *Evolutionary Biology of Hearing*, Springer, 1992, 131–139; Jezequel Y., et al., "Sound detection by the American lobster (*Homarus americanus*)," *J Exp Biol*, vol. 224, 2021, jeb240747.

176. Kaifu K., et al., "Underwater sound detection by cephalopod statocyst," *Fish Sc*, vol. 74, 2008, 781–786.

177. Ladich F., and Winkler H., "Acoustic communication in terrestrial and aquatic vertebrates," *J Exp Biol*, vol. 220, 2017, 2306–2317.

178. Clack J. A., Fay R. R., and Popper A. N., *Evolution of the Vertebrate Ear*, Springer, 2016; Montealegre-Z F., Robert D., et al., "Convergent evolution between insect and mammalian audition," *Science*, vol. 338, 2012, 968–971.

179. Römer H., "*Acoustic communication*," in Córdoba-Aguilar A., González-Tokman D., and González-Santoyo I. (eds.), *Insect Behavior: From Mechanisms to Ecological and Evolutionary Consequences*, Oxford University Press, 2018, 174–188; Balakrishnan R., "Behavioral ecology of insect acoustic communication," in Pollack, Mason, Popper, and Fay (eds.), *Insect Hearing*, 49–80; Göpfert M. C., and Hennig R. M., "Hearing in insects," *Ann Rev Entomol*, vol. 61, 2016, 257–276.

180. Matthews R. W., and Matthews J. R., *Insect Behavior*, Springer, 2010.

181. Since ears are derived from preexisting mechanoreceptors, and since these receptors are present throughout the bodies of insects, finding ears in different positions is not surprising. This is evidence that the ear has been "invented" several times, independently, throughout the history of insects.

182. The bodies of insects have three successive parts: head, thorax, and abdomen. Each part is a succession of rings, the metameres. The thorax has three metameres, each with a pair of legs. The abdomen has a greater number of metameres.

183. Göpfert M. C., et al., "Tympanal and atympanal 'mouth-ears' in hawkmoths (Sphingidae)," *Proc R Soc B*, vol. 269, 2002, 89–95.

184. The respiratory system of insects is a genuine piping system (made up of the trachea), bringing air everywhere in the animal's body.

185. A scolopidium is a fairly complex structure consisting of one or more sensory neurons, each having an extension (a dendrite) ending in a cilium (connected to the eardrum in the case of the tympanic ear), and two protective cells. Scolopidia are grouped into chordotonal organs.

186. Brown E. E., et al., "Quantifying the completeness of the bat fossil record," *Palaeontology*, vol. 62, 2019, 757–776.

187. Greenfield M. D., "Evolution of acoustic communication in insects," in Pollack, Mason, Popper, and Fay (eds.), *Insect Hearing*, 17–47.

188. Gerhardt H. C., and Huber F., *Acoustic Communication in Insects and Anurans: Common Problems and Diverse Solutions*, University of Chicago Press, 2002.

189. Yager D. D., and Svenson G. J., "Patterns of praying mantis auditory system evolution based on morphological, molecular, neurophysiological, and behavioural data," *Biol J Lin Soc*, vol. 94, 2008, 541–568.

190. The ears of bush crickets (katydids) are also amazing, with specific anatomical features that enhance the ability of these animals to pinpoint a sound source. Have a look at this recent study: Veitch D., et al., "A narrow ear canal reduces sound velocity to create additional acoustic inputs in a microscale insect ear," *PNAS*, vol. 118, 2021, e2017281118.

191. Albert J. T., and Kozlov A. S., "Comparative aspects of hearing in vertebrates and insects with antennal ears," *Cur Biol*, vol. 26, 2016, R1050–R1061.

192. On hearing by mosquitoes, see Feugere L., et al., "Behavioural analysis of swarming mosquitoes reveals high hearing sensitivity in *Anopheles coluzzii*," *J Exp Biol*, vol. 225, 2022, jeb243535; and Feugere L., et al., "The role of hearing in mosquito behaviour," in Hill S., Ignell R., Lazzari C., and Lorenzo M. (eds.), *Sensory Ecology of Disease Vectors*, Wageningen Academic Publishers, forthcoming.

193. Baker C. A., et al., "Neural network organization for courtship-song feature detection in *Drosophila*," *Cur Biol*, vol. 32, 2022, 3317–3333.

194. Kerwin P., et al., "Female copulation song is modulated by seminal fluid," *Nat Com*, vol. 11, 2020, 1430.

195. Rybak F., Sureau G., and Aubin T., "Functional coupling of acoustic and chemical signals in the courtship behaviour of the male *Drosophila melanogaster*," *Proc R Soc B*, vol. 269, 2002, 695–701.

196. Jackson J. C., and Robert D., "Nonlinear auditory mechanism enhances female sounds for male mosquitoes," *PNAS*, vol. 103, 2006, 16734–16739.

197. Gibson G., and Russell I., "Flying in tune: Sexual recognition in mosquitoes," *Cur Biol*, vol. 16, 2006, 1311–1316.

198. Manley G. A., "Cochlear mechanisms from a phylogenetic viewpoint," *PNAS*, vol. 97, 2000, 11736–11743.

199. Moffat A.J.M., and Capranica R. R., "Auditory sensitivity of the saccule in the American toad (*Bufo americanus*)," *J Comp Physiol*, vol. 105, 1976, 1–8.

200. Although the near sound field of some elephant vocalizations can extend to 20 meters. We humans can't hear them with our ears, but we do feel the particle oscillations they produce.

201. Sisneros J. A., *Fish Hearing and Bioacoustics*, Springer, 2016.

202. Parmentier E., and Diogo R., "Evolutionary trends of swimbladder sound mechanisms in some teleost fishes," in Ladich F., Collin P. S., Moller P., and Kapoor B. G. (eds.), *Communication in Fishes*, Science Publishers, 2006, 43–68.

203. Ladich F., and Schulz-Mirbach T., "Diversity in fish auditory systems: One of the riddles of sensory biology," *Front Ecol Evol*, vol. 4, 2016, 28; Chapuis L., and Collin S. P., "The auditory system of cartilaginous fishes," *Rev Fish Biol Fisheries*, vol. 32, 2022, 521–554.

204. van Hemmen J. L., et al., "Animals and ICE: Meaning, origin, and diversity," *Biol Cybern*, vol. 110, 2016, 237–246.

205. Vedurmudi A. P., et al., "How internally coupled ears generate temporal and amplitude cues for sound localization," *Phys Rev Let*, vol. 116, 2016, 028101; Klump G. M., "Sound localization in birds," in Dooling R. J., Fay R. R., and Popper A. N. (eds.), *Comparative Hearing: Birds and Reptiles*, Springer, 2000, 249–307.

206. Schnupp J.W.H., and Carr C. E., "On hearing with more than one ear: Lessons from evolution," *Nat Neurosc*, vol. 12, 2009, 692–697.

207. Robert D., "Directional hearing in insects," in Fay R. R. (ed.), *Sound Source Localization*, Springer, 2005, 6–35; Montealegre-Z F., et al., "Convergent evolution between insect and mammalian audition," *Science*, vol. 338, 2012, 968–971.

208. Lee N., et al., "Lung mediated auditory contrast enhancement improves the signal-to-noise ratio for communication in frogs," *Cur Biol*, vol. 31, 2021, 1488–1498.

209. Boistel R., Aubry J.-F., et al., "How minute sooglossid frogs hear without a middle ear," *PNAS*, vol. 17, 2013, 15360–15364.

210. For a very interesting, lively, and still modern introduction to frog acoustic communication, see Narins P., "Frog communication," *Sc Amer*, vol. 273, 1995, 78–83.

211. On the origin of the avian ear, see Hanson M., et al., "The early origin of a birdlike inner ear and the evolution of dinosaurian movement and vocalization," *Science*, vol. 372, 2021, 601–609.

212. Zeyl J. N., et al., "Aquatic birds have middle ears adapted to amphibious lifestyles," *Sc Rep*, vol. 12, 2022, 5251.

213. Supin A. Y., et al., *The Sensory Physiology of Aquatic Mammals*, Springer, 2001.

214. Hemilä S., Nummela S., and Reuter T., "Anatomy and physics of the exceptional sensitivity of dolphin hearing (Odontoceti: Cetacea)," *J Comp Physiol A*, vol. 196, 2010, 165–179.

215. Churchill M., et al., "The origin of high-frequency hearing in whales," *Cur Biol*, vol. 26, 2016, 2144–2149.

216. Au W.W.L., and Fay R. R., *Hearing by Whales and Dolphins*, Springer, 2000; Cranfor T., and Krysl P., "Fin whale sound reception mechanisms: Skull vibration enables low-frequency hearing," *PLoS ONE*, vol. 10, 2015, e0116222; Park T., et al., "Low-frequency hearing preceded the evolution of giant body size and filter feeding in baleen whales," *Proc R Soc B*, vol. 284, 2017, 20162528.

217. Nelson D. A., and Marler P., "Categorical perception of a natural stimulus continuum: Birdsong," *Science*, vol. 244, 1989, 976–978.

218. May B., et al., "Categorical perception of conspecific communication sounds by Japanese macaques, *Macaca fuscata*," *J Acoust Soc Am*, vol. 85, 1989, 837–847.

219. Baugh A. T., Akre K. L., and Ryan M. J., "Categorical perception of a natural, multivariate signal: Mating call recognition in túngara frogs," *PNAS*, vol. 105, 2008, 8985–8988.

220. Greenfield M. D., "Mechanisms and evolution of communal sexual displays in arthropods and anurans," *Adv Stud Behav*, vol. 35, 2005, 1–62.

221. Paul A., et al., "Behavioral discrimination and time-series phenotyping of birdsong performance," *PLoS Comput Biol*, vol. 17, 2021, e1008820.

Chapter 10

222. Hippos have a vocal repertoire, with several calls. During a recent field trip in the Maputo Special Reserve in Mozambique, we started to study hippo acoustic communication: Thévenet et al., "Voice-mediated interactions in a megaherbivore," *Cur Biol*, vol. 32, 2022, R55–R71.

223. Vergne A. L., et al., "Parent-offspring communication in the Nile crocodile *Crocodylus niloticus*: Do newborns' calls show an individual signature?," *Naturwissen*, vol. 94, 2007, 49–54.

224. Chabert T., et al., "Size does matter: Crocodile mothers react more to the voice of smaller offspring," *Sc Rep*, vol. 5, 2015, 15547.

225. For an extensive review on sound production in nonavian reptiles, see Russell A. P., and Bauer A. M., "Vocalization by extant nonavian reptiles: A synthetic overview of phonation and the vocal apparatus," *Anat Rec*, vol. 304, 2021, 1478–1528.

226. Taylor A. M., et al., "Vocal production by terrestrial mammals: Source, filter, and function," in Suthers R. A., Fitch W. T., Fay R. R., and Popper A. N. (eds.), *Vertebrate Sound Production and Acoustic Communication*, Springer, 2016, 229–259.

227. The Adam's apple is larger, and therefore more visible, in most men. But everyone has one. Put your finger to your throat and swallow; you'll feel your Adam's apple move.

228. Or when the air comes in. But it's harder. Try to talk while you breathe, just to see.

229. Bowling D. L., et al., "Body size and vocalization in primates and carnivores," *Sc Rep*, vol. 7, 2017, 41070.

230. Garcia M., et al., "Acoustic allometry revisited: Morphological determinants of fundamental frequency in primate vocal production," *Sc Rep*, vol. 7, 2017, 10450. On the evolution of the larynx, see also Bowling D. L., et al., "Rapid evolution of the primate larynx?," *PLoS Biol*, vol. 18, 2020, e3000764.

231. For an in-depth discussion of the concept of timbre, see Piazza E. A., et al., "Rapid adaptation to the timbre of natural sounds," *Sc Rep*, vol. 8, 2018, 13826; and Elliott T. M., et al., "Acoustic structure of the five perceptual dimensions of timbre in orchestral instrument tones," *JASA*, vol. 133, 2013, 389–404.

232. Garcia M., et al., "Honest signaling in domestic piglets (*Sus scrofa domesticus*): Vocal allometry and the information content of grunt calls," *J Exp Biol*, vol. 219, 2016, 1913–1921; Garcia-Navas V., and Blumstein D. T., "The effect of body size and habitat on the evolution of alarm vocalizations in rodents," *Biol J Lin Soc*, vol. 118, 2016, 745–751.

233. Reber S. A., et al., "A Chinese alligator in heliox: Formant frequencies in a crocodilian," *J Exp Biol*, vol. 218, 2015, 2442–2447; Reber S. A., et al., "Formants provide honest acoustic cues to body size in American alligators," *Sc Rep*, vol. 7, 2017, 1816.

234. Charlton B. D., and Reby D., "The evolution of acoustic size exaggeration in terrestrial mammals," *Nat Com*, vol. 7, 2016, 12739.

235. Reby D., and McComb K., "Vocal communication and reproduction in deer," *Adv St Behav*, vol. 33, 2003, 231–264; Frey R., et al., "Roars, groans and moans: Anatomical correlates of vocal diversity in polygenous deer," *J Anat*, vol. 239, 2021, 1336–1369.

236. See this paper on birds: Francis C. D., and Wilkins M. R., "Testing the strength and direction of selection on vocal frequency using metabolic scaling theory," *Ecosphere*, vol. 12, 2021, e03733.

237. Charlton B. D., Reby D., and McComb K., "Female red deer prefer the roars of larger males," *Biol Let*, vol. 3, 2007, 382–385.

238. McComb K., "Female choice for high roaring rates in red deer, *Cervus elaphus*," *Anim Behav*, vol. 41, 1991, 79–88.

239. Reby D., et al., "Red deer (*Cervus elaphus*) hinds discriminate between the roars of their current harem-holder stag and those of neighbouring stags," *Ethology*, vol. 107, 2001, 951–959.

240. In reality, the koala is a marsupial mammal, closer to the kangaroo than to the polar bear.

241. Charlton B. D., et al., "Koalas use a novel vocal organ to produce unusually low-pitched mating calls," *Cur Biol*, vol. 23, 2013, R1035–R1036.

242. Fitch W. T., "Vertebrate vocal production: An introductory overview," in Suthers, Fitch, Fay, and Popper (eds.), *Vertebrate Sound Production and Acoustic Communication*, 1–18.

243. Bradbury J. W., "Lek mating behavior in the hammer-headed bat," *Ethology*, vol. 45, 1977, 225–255.

244. Frey R., and Gebler A., "Mechanisms and evolution of roaring-like vocalization in mammals," in Brudzynski S. (ed.), *Handbook of Mammalian Vocalization*, Elsevier, 2010, 439–450.

245. Mathevon N., and Viennot É., "Avant-propos," in Mathevon N., and Viennot É. (eds.), *La différence des sexes*, Belin, 2017, 7–28.

246. Titze I. R., *Fascinations with the Human Voice*, National Center for Voice and Speech, 2010; Dunn J. C., et al., "Evolutionary trade-off between vocal tract and testes dimensions in howler monkeys," *Cur Biol*, vol. 25, 2015, 2839–2844.

247. Grawunder S., et al., "Higher fundamental frequency in bonobos is explained by larynx morphology," *Cur Biol*, vol. 28, 2018, R1171–R1189.

248. Greenfield M., "Honesty and deception in animal signals," in Lucas J., and Simmons L. (eds.), *Essays in Animal Behaviour*, Elsevier, 2006, 279–300.

249. Interspecific signals also follow the *honesty principle*. Look at this amazing paper: Forsthofer M., et al., "Frequency modulation of rattlesnake acoustic display affects acoustic distance perception in humans," *Cur Biol*, vol. 31, 2021, 4367–4372.

250. Bee M. A., et al., "Male green frogs lower the pitch of acoustic signals in defense of territories: A possible dishonest signal of size?," *Behav Ecol*, vol. 11, 2000, 169–177.

251. Zahavi A., "Mate selection—a selection for a handicap," *J Theor Biol*, vol. 53, 1975, 205–214; Bradbury J. W., and Vehrencamp S. L., *Principles of Animal Communication*, Sinauer, 2011.

252. Vocalizations are not the only sound signals that can inform about the size of the sender. In mountain gorillas, for example, a recent study shows that the acoustic characteristics of the chest beat are correlated with body size: larger males have significantly lower peak frequencies than smaller ones: Wright E., et al., "Chest beats as an honest signal of body size in male mountain gorillas (*Gorilla beringei beringei*)," *Sc Rep*, vol. 11, 2021, 6879.

253. Pisanski K., and Reby D., "Efficacy in deceptive vocal exaggeration of human body size," *Nat Com*, vol. 12, 2021, 968. See also Raine J., et al., "Human listeners can accurately judge strength and height relative to self from aggressive roars and speech," *iScience*, vol. 4, 2018, 273–280.

254. Vallet E., and Kreutzer M., "Female canaries are sexually responsive to special song phrases," *Anim Behav*, vol. 49, 1995, 1603–1610.

255. Vallet E., et al., "Two-note syllables in canary songs elicit high levels of sexual display," *Anim Behav*, vol. 55, 1998, 291–297.

256. Oberweger K., and Goller F., "The metabolic cost of birdsong production," *J Exp Biol*, vol. 204, 2001, 3379–3388.

257. Casagrande S., Pinxten R., and Eens M., "Honest signaling and oxidative stress: The special case of avian acoustic communication," *Front Ecol Evol*, vol. 4, 2016, 52; Spencer K. A., et al., "Parasites affect song complexity and neural development in a songbird," *Proc R Soc B*, vol. 272, 2005, 2037–2043.

258. Gil D., and Gahr M., "The honesty of bird song: Multiple constraints for multiple traits," *Trends Ecol Evol*, vol. 17, 2002, 133–141.

259. Buchanan K. L., "Stress and the evolution of condition-dependent signals," *Trends Ecol Evol*, vol. 15, 2000, 156–160.

260. Buchanan K. L., et al., "Song as an indicator of parasitism in the sedge warbler," *Anim Behav*, vol. 57, 1999, 307–314.

261. Whether and how acoustic features drive mate choice is still an active field of research. For instance, see Wang D., et al., "Is female mate choice repeatable across males with nearly identical songs?," *Anim Behav*, vol. 181, 2021, 137–149.

262. Dalziall A. H., et al., "Male lyrebirds create a complex acoustic illusion of a mobbing flock during courtship and copulation," *Cur Biol*, vol. 31, 2021, 1970–1976.

263. Maynard Smith J., and Harper D., *Animal Signals*, Oxford University Press, 2003.

264. See this paper dealing with sound production in fossil insects: Schubnel T., et al., "Sound vs. light: Wing-based communication in Carboniferous insects," *Com Biol*, vol. 4, 2021, 794.

265. Tinbergen N., "'Derived' activities: Their causation, biological significance, origin, and emancipation during evolution," *Q Rev Biol*, vol. 27, 1952, 1–32.

266. Clark C. J., "Locomotion-induced sounds and sonations: Mechanisms, communication function, and relationship with behavior," in Suthers, Fitch, Fay, and Popper (eds.), *Vertebrate Sound Production and Acoustic Communication*, 83–117; Clark C. J., "Ways that animal wings produce sound," *Integr Comp Biol*, vol. 61, 2021, 696–709.

267. Murray T. G., Zeil J., and Magrath R. D., "Sounds of modified flight feathers reliably signal danger in a pigeon," *Cur Biol*, vol. 27, 2017, 3520–3525.

268. Clark C. J., "Signal or cue? Locomotion-induced sounds and the evolution of communication," *Anim Behav*, vol. 143, 2018, 83–91.

269. Gould S. J., *The Structure of Evolutionary Theory*, Harvard University Press, 2002.

270. Montealegre-Z F., et al., "Sound radiation and wing mechanics in stridulating field crickets (Orthoptera: Gryllidae)," *J Exp Biol*, vol. 214, 2011, 2105–2117.

271. Fonseca P. J., "Cicada acoustic communication," in Hedwig B. (ed.), *Insect Hearing and Acoustic Communication*, Springer, 2014, 101–121.

272. Godthi V., et al., "The mechanics of acoustic signal evolution in field crickets," *J Exp Biol*, vol. 225, 2022, jeb243374.

273. Parmentier E., Diogo R., and Fine M. L., "Multiple exaptations leading to fish sound production," *Fish & Fisheries*, vol. 18, 2017, 958–966.

274. Looby A., et al., "A quantitative inventory of global soniferous fish diversity," *Rev Fish Biol Fisheries*, vol. 32, 2022, 581–595; Rice A. N., et al., "Evolutionary patterns in sound production across fishes," *Ichthyology & Herpetology*, vol. 110, 2022, 1–12.

275. Parmentier E., Herrel A., et al., "Sound production in the clownfish *Amphiprion clarkii*," *Science*, vol. 316, 2007, 1006.

276. Millot S., Vandewalle P., and Parmentier E., "Sound production in red-bellied piranhas (*Pygocentrus nattereri*, Kner): An acoustical, behavioural and morphofunctional study," *J Exp Biol*, vol. 214, 2011, 3613–3618.

277. Amorim P., "Diversity of sound production in fish," in Ladich, Collin, Moller, and Kapoor (eds.), *Communication in Fishes*, 71–105.

278. Amorim P., "Fish sounds and mate choice," in Ladich F. (ed.), *Sound Communication in Fishes*, Springer, 2015, 1–33.

279. Bertucci F., et al., "Sounds produced by the cichlid fish *Metriaclima zebra* allow reliable estimation of size and provide information on individual identity," *J Fish Biol*, vol. 80, 2012, 752–766.

280. Bertucci F., et al., "Sounds modulate males' aggressiveness in a cichlid fish," *Ethology*, vol. 116, 2010, 1179–1188; Bertucci F., et al., "The relevance of temporal cues in a fish sound: A first experimental investigation using modified signals in cichlids," *Anim Cogn*, vol. 16, 2013, 45–54.

281. Amorim M.C.P., et al., "Mate preference in the painted goby: The influence of visual and acoustic courtship signals," *J Exp Biol*, vol. 216, 2013, 3996–4004.

282. Wyttenbach R. A., May M. L., and Hoy R. R., "Categorical perception of sound frequency by crickets," *Science*, vol. 273, 1996, 1542–1544.

283. Robillard T., Grandcolas P., and Desutter-Grandcolas L., "A shift toward harmonics for high-frequency calling shown with phylogenetic study of frequency spectra in Eneopterinae crickets (Orthoptera, Grylloidea, Eneopteridae)," *Can J Zool*, vol. 85, 2007, 1264–1274.

284. Hofstede H. M., et al., "Evolution of a communication system by sensory exploitation of startle behavior," *Cur Biol*, vol. 25, 2015, 3245–3252.

285. Benavides-Lopez J. L., Ter Hofstede H., and Robillard T., "Novel system of communication in crickets originated at the same time as bat echolocation and includes male-male multimodal communication," *Sci Nat*, vol. 107, 2020, 9.

286. Römer H., "Insect acoustic communication: The role of transmission channel and the sensory system and brain of receivers," *Func Ecol*, vol. 34, 2020, 310–321.

287. Wilkins M. R., Seddon N., and Safran R. J., "Evolutionary divergence in acoustic signals: Causes and consequences," *Trends Ecol Evol*, vol. 28, 2013, 156–166.

288. To give you an idea, the common ancestor of the human species and the great apes of today goes back 6 or 7 million years. Our species *Homo sapiens* was distinguished from other human species only 300,000 years ago. See Hublin J.-J., et al., "New fossils from Jebel Irhoud, Morocco and the pan-African origin of *Homo sapiens*," *Nature*, vol. 546, 2017, 289–292.

289. Prum R. O., et al., "A comprehensive phylogeny of birds (Aves) using targeted next-generation DNA sequencing," *Nature*, vol. 526, 2015, 569–573.

290. Mason N. A., et al., "Song evolution, speciation, and vocal learning in passerine birds," *Evolution*, vol. 71, 2016, 786–796.

291. Amezquita A., et al., "Acoustic interference and recognition space within a complex assemblage of dendrobatid frogs," *PNAS*, vol. 108, 2011, 17058–17063; Tobias J. A., et al., "Species interactions and the structure of complex communication networks," *PNAS*, vol. 111, 2014, 1020–1025.

292. See, for example, regarding the evolution of mammalian sound signals, Charlton B. D., et al., "Coevolution of vocal signal characteristics and hearing sensitivity in forest mammals," *Nat Com*, vol. 10, 2019, 2778. See also Leighton G. M., and Birmingham T., "Multiple factors affect the evolution of repertoire size across birds," *Behav Ecol*, vol. 32, 2021, 380–385.

293. Hyacinthe C., Attia J., and Rétaux S., "Evolution of acoustic communication in blind cavefish," *Nat Com*, vol. 10, 2019, 4231.

294. My colleagues included Frédéric Theunissen and Frédéric Sèbe, whom I will tell you more about, as well as Maxime Garcia, who was a postdoctoral researcher on my team at the time, and a few other scientists.

295. Garcia M., et al., "Evolution of communication signals and information during species radiation," *Nat Com*, vol. 11, 2020, 4970. See also the post on the *Nature Ecology & Evolution* blog: https://natureecoevocommunity.nature.com/posts/information-tinkering-in-an-animal-communication-system.

296. Freeman B. G., et al., "Faster evolution of a premating reproductive barrier is not associated with faster speciation rates in New World passerine birds," *Proc R Soc B*, vol. 289, 2022, 20211514; Arato J., and Fitch W. T., "Phylogenetic signal in the vocalizations of vocal learning and vocal non-learning birds," *Phil Trans R Soc B*, vol. 376, 2021, 20200241.

297. Podos J., "Correlated evolution of morphology and vocal signal structure in Darwin's finches," *Nature*, vol. 409, 2001, 185–188; Podos J., and Nowicki S., "Beaks, adaptation, and vocal evolution in Darwin's finches," *Bioscience*, vol. 54, 2004, 501–510; Servedio M. R., et al., "Magic traits in speciation: 'Magic' but not rare?," *Trends Ecol Evol*, vol. 26, 2011, 389–397.

298. An interesting paper on the evolution of signals when evolutionary constraints are relaxed is Rayner J. G., et al., "The persistence and evolutionary consequences of vestigial behaviours," *Biol Rev*, vol. 97, 2022, 1389–1407.

Chapter 11

299. McGregor P. K. (ed.), *Animal Communication Networks*, Cambridge University Press, 2005; Reichert M. S., Enriquez M. S., and Carlson N. V., "New dimensions for animal communication networks: Space and time," *Integ Comp Biol*, vol. 61, 2021, 814–824.

300. Zann R. A., *The Zebra Finch*, Oxford University Press, 1996.

301. Elie J. E., and Theunissen F. E., "The vocal repertoire of the domesticated zebra finch: A data-driven approach to decipher the information-bearing acoustic features of communication signals," *Anim Cogn*, vol. 19, 2015, 285–315.

302. Vignal C., et al., "Mate recognition by female zebra finch: Analysis of individuality in male call and first investigations on female decoding process," *Behav Process*, vol. 77, 2008, 191–198.

303. Vignal C., et al., "Audience drives male songbird response to partner's voice," *Nature*, vol. 430, 2004, 448–451.

304. Perry S., et al., "White-faced capuchin monkeys show triadic awareness in their choice of allies," *Anim Behav*, vol. 67, 2004, 165–170; Jolly A., "Lemur social behavior and primate intelligence," *Science*, vol. 153, 1966, 501–506; Tomasello M., and Call J., *Primate Cognition*, Oxford University Press, 1997.

305. Elie J. E., et al., "Dynamics of communal vocalizations in a social songbird, the zebra finch (*Taeniopygia guttata*)," *JASA*, vol. 129, 2011, 4037–4046.

306. Elie J. E., et al., "Same-sex pair-bonds are equivalent to male-female bonds in a life-long socially monogamous songbird," *Behav Ecol Sociobiol*, vol. 65, 2011, 2197–2208.

307. Geberzhan N., and Gahr M., "Undirected (solitary) birdsong in female and male blue-capped cordon-bleus (*Uraeginthus cyanocephalus*) and its endocrine correlates," *PLoS ONE*, vol. 6, 2011, e26485; Ota N., et al., "Tap dancing birds: The multimodal mutual courtship display of males and females in a socially monogamous songbird," *Sc Rep*, vol. 5, 2015, 16614; Ota N., et al., "Songbird tap dancing produces non-vocal sounds," *Bioacoustics*, vol. 26, 2017, 161–168.

308. Ota N., et al., "Couples showing off: Audience promotes both male and female multimodal courtship display in a songbird," *Sc Adv*, vol. 4, 2018, eaat4779.

309. Zuberbühler K., "Audience effects," *Cur Biol*, vol. 18, 2008, R189–R190.

310. Marler P., et al., "Vocal communication in the domestic chicken: II. Is a sender sensitive to the presence and nature of a receiver?," *Anim Behav*, vol. 34, 1986, 194–198.

311. Slocombe K. E., and Zuberbühler K., "Chimpanzees modify recruitment screams as a function of audience composition," *PNAS*, vol. 104, 2007, 17228–17233.

312. Semple S., et al., "Bystanders affect the outcome of mother-infant interactions in rhesus macaques," *Proc R Soc B*, vol. 276, 2009, 2257–2262.

313. Wich S. A., and de Vries H., "Male monkeys remember which group members have given alarm calls," *Proc R Soc B*, vol. 273, 2006, 735–740.

314. Cheney D. L., and Seyfarth R. M., *How Monkeys See the World: Inside the Mind of Another Species*, Chicago University Press, 1990. Look also at this recent paper, which shows that marmosets socially evaluate vocal exchanges between congeners and are able to distinguish between cooperative and non-cooperative conspecifics: Brugger R. K., et al., "Do marmosets understand others' conversations? A thermography approach," *Sc Adv*, vol. 7, 2021, eabc8790.

315. Grosenick L., et al., "Fish can infer social rank by observation alone," *Nature*, vol. 445, 2007, 429–432.

316. McGregor P., and Doutrelant C., "Eavesdropping and mate choice in female fighting fish," *Behaviour*, vol. 137, 2000, 1655–1668. See also Doutrelant C., et al., "The effect of an audience on intrasexual communication in male siamese fighting fish, *Betta splendens*," *Behav Ecol*, vol. 12, 2001, 283–286.

317. Clotfelter E. D., and Paolino A. D., "Bystanders to contests between conspecifics are primed for increased aggression in male fighting fish," *Anim Behav*, vol. 66, 2003, 343–347.

318. Mennill D. J., et al., "Female eavesdropping on male song contests in songbirds," *Science*, vol. 296, 2002, 873; Otter K., et al., "Do female great tits (*Parus major*) assess males by eavesdropping? A field study using interactive song playback," *Proc R Soc B*, vol. 266, 1999, 1305–1309.

319. Templeton C. N., et al., "Allometry of alarm calls: Black-capped chickadees encode information about predator size," *Science*, vol. 308, 2005, 1934–1937.

320. Templeton C. N., and Greene E., "Nuthatches eavesdrop on variations in heterospecific chickadee mobbing alarm calls," *PNAS*, vol. 104, 2007, 5479–5482; Carlson N. V., et al., "Nuthatches vary their alarm calls based upon the source of the eavesdropped signals," *Nat Com*, vol. 11, 2020, 526.

321. Another example of eavesdropping is when animals listen to the vocalizations of their predators to assess the risk they represent: Hettena A. M., et al., "Prey responses to predator's sounds: A review and empirical study," *Ethology*, vol. 120, 2014, 427–452.

322. Demartsev V., et al., "Signalling in groups: New tools for the integration of animal communication and collective movement," *Met Ecol Evol*, forthcoming.

323. An interesting research question is the brain basis of communication networks. An excellent example of research in this field is given in this paper, coauthored by Julie Elie: Rose M. C., et al., "Cortical representation of group social communication in bats," *Science*, vol. 374, 2021, eaba9584.

Chapter 12

324. Marler P., and Tamura M., "Culturally transmitted patterns of vocal behavior in sparrows," *Science*, vol. 146, 1964, 1483–1486.

325. Baker M. C., "Bird song research: The past 100 years," *Bird Behav*, vol. 14, 2001, 3–50; Barrington D., "Experiments and observations on the singing of birds," *Phil Trans R Soc*, vol. 63, 1773, 249–291.

326. Thorpe W. H., "The learning of song patterns by birds, with especial reference to the song of the chaffinch *Fringilla coelebs*," *Ibis*, vol. 100, 1958, 535–570; Poulsen H., "Inheritance and learning in the song of the chaffinch (*Fringilla coelebs* L.)," *Behaviour*, vol. 3, 1951, 216–228.

327. Actually, it was Darwin who first announced it. In *The Expression of Emotions in Man and Animals*, he wrote that, among birds, it is the father who teaches the young to sing.

328. Searcy W. A., et al., "Variation in vocal production learning across songbirds," *Phil Trans R Soc B*, vol. 376, 2021, 20200257.

329. Marler P., and Slabbekoorn H. (eds.), *Nature's Music: The Science of Birdsong*, Elsevier, 2004; Araya-Sala M., and Wright T., "Open-ended song learning in a hummingbird," *Biol Let*,

vol. 9, 2013, 2013062; Johnson K. E., and Clark C. J., "Ontogeny of vocal learning in a hummingbird," *Anim Behav*, vol. 167, 2020, 139–150.

330. For a recent review on vocal learning in nonoscine birds, see Ten Cate C., "Re-evaluating vocal prodution learning in non-oscine birds," *Phil Trans R Soc B*, vol. 376, 2021, 20200249.

331. On hummingbirds, see this review: Duque F. G., and Carruth L. L., "Vocal communication in hummingbirds," *Brain Behav Evol*, vol. 97, 2022, 241–252.

332. It should be noted that the songs of songbirds, although culturally transmitted, still have a strong genetic component: Arato J., and Fitch W. T., "Phylogenetic signal in the vocalizations of vocal learning and vocal non-learning birds," *Phil Trans R Soc B*, vol. 376, 2021, 20200241.

333. This statement needs to be tempered a bit since ducks show some ability to imitate sounds: Ten Cate C., and Fullagar P. J., "Vocal imitations and production learning by Australian musk ducks (*Biziura lobata*)," *Phil Trans R Soc B*, vol. 376, 2021, 20200243.

334. Kroodsma D. E., and Konishi M., "A suboscine bird (eastern phoebe, *Sayornis phoebe*) develops normal song without auditory feedback," *Anim Behav*, vol. 42, 1991, 477–487.

335. Liu W., et al., "Rudimentary substrates for vocal learning in a suboscine," *Nat Com*, vol. 4, 2013, 2082.

336. Kroodsma D., et al., "Behavioral evidence for song learning in the suboscine bellbirds (*Procnias* spp.; Cotingidae)," *Wilson J Orn*, vol. 125, 2013, 1–14.

337. Marler P., "Three models of song learning: Evidence from behavior," *J Neurobiol*, vol. 33, 1997, 501–516; Tchernichovski O., et al., "Dynamics of the vocal imitation process: How a zebra finch learns its song," *Science*, vol. 291, 2001, 2564–2569; Carousio-Peck S., and Goldstein M. H., "Female social feedback reveals non-imitative mechanisms of vocal learning in zebra finches," *Cur Biol*, vol. 29, 2019, 631–636.

338. A recent study suggests that vocal learning may begin earlier, when the embryo is still in its egg, at least in certain bird species: Colombelli-Negrel D., et al., "Prenatal auditory learning in avian vocal learners and non-learners," *Phil Trans R Soc*, vol. 376, 2021, 20200247.

339. Bloomfield T. C., et al., "What birds have to say about language," *Nat Neurosci*, vol. 14, 2011, 947–948; Lipkind D., et al., "Stepwise acquisition of vocal combinatorial capacity in songbirds and human infants," *Nature*, vol. 498, 2013, 104–108; Brainard M. S., and Doupe A. J., "What songbirds teach us about learning," *Nature*, vol. 417, 2002, 351–358; Doupe A. J., and Kuhl P. K., "Birdsong and human speech: Common themes and mechanisms," *Ann Rev Neurosc*, vol. 22, 1999, 567–631.

340. Mennill D. J., et al., "Wild birds learn songs from experimental vocal tutors," *Cur Biol*, vol. 28, 2018, 3273–3278.

341. If you want to know more about the Savannah sparrow's story, check out this paper: Williams H., et al., "Cumulative cultural evolution and mechanisms for cultural selection in wild bird songs," *Nat Com*, vol. 13, 2022, 4001.

342. Mets D. G., and Brainard M. S., "Genetic variation interacts with experience to determine interindividual differences in learned song," *PNAS*, vol. 115, 2018, 421–426.

343. Carouso-Peck S., and Goldstein M. H., "Female social feedback reveals non-imitative mechanisms of vocal learning in zebra finches," *Cur Biol*, vol. 29, 2019, 631–636.

344. For a recent study on the effect of visual cues on song learning, see Varkevisser J. M., et al., "Multimodality during live tutoring is relevant for vocal learning in zebra finches," *Anim Behav*, vol. 187, 2022, 263–280.

345. Note that there is still a critical period during the first year.

346. Kroodsma D., "The diversity and plasticity of birdsong," in Marler and Slabbekoorn (eds.), *Nature's Music*, 108–131.

347. Klatt D. H., and Stefanski R. A., "How does a mynah bird imitate human speech?," *JASA*, vol. 55, 1974, 822–832.

348. On the flexibility of vocal learning, see this paper, which demonstrates that Bengalese finches can learn to rapidly modify the order of syllables in their song: Veit L., et al., "Songbirds can learn flexible contextual control over syllable sequencing," *eLife*, vol. 10, 2021, e61610.

349. Nowicki S., and Searcy W. A., "The evolution of vocal learning," *Cur Opin Neurobiol*, vol. 28, 2014, 48–53; Beecher M. D., and Brenowitz E. A., "Functional aspects of song learning in songbirds," *Trends Ecol Evol*, vol. 20, 2005, 143–149.

350. See also Osiejuk T. S., et al., "Songbird presumed to be age-limited learner may change repertoire size and composition throughout their life," *J Zool*, vol. 309, 2019, 231–240.

351. Kroodsma, "Diversity and plasticity of birdsong."

352. Odom K. J., et al., "Female song is widespread and ancestral in songbirds," *Nat Com*, vol. 5, 2014, 3379; Riebel K., et al., "New insights from female bird song: Towards an integrated approach to studying male and female communication roles," *Biol Let*, vol. 15, 2019, 20190059.

353. Bolhuis J. J., and Gahr M., "Neural mechanisms of birdsong memory," *Nat Rev Neurosci*, vol. 7, 2006, 347–357.

354. Sakata J. T., et al., *The Neuroethology of Birdsong*, Springer Handbook of Auditory Research, vol. 71, Springer, 2020.

355. Mooney R., "The neurobiology of innate and learned vocalizations in rodents and songbirds," *Cur Opin Neurobiol*, vol. 64, 2020, 24–31; Theunissen F. E., and Shaevitz S. S., "Auditory processing of vocal sounds in birds," *Cur Opin Neurobiol*, vol. 16, 2006, 400–407; Van Ruijssevelt L., et al., "fMRI reveals a novel region for evaluating acoustic information for mate choice in a female songbird," *Cur Biol*, vol. 28, 2018, 711–721.

356. Nottebohm F., and Arnold A. P., "Sexual dimorphism in vocal control areas of the songbird brain," *Science*, vol. 194, 1976, 211–213.

357. Jarvis E. D., et al., "Behaviourally driven gene expression reveals song nuclei in hummingbird brain," *Nature*, vol. 406, 2000, 628–632.

358. Theunissen F. E., and Elie J. E., "Neural processing of natural sounds," *Nature Rev Neurosc*, vol. 15, 2014, 355–366.

359. For an interesting perspective on song production in birds, see During D. N., and Elemans C.P.H., "Embodied motor control of avian vocal production," in Suthers, Fitch, Fay, and Popper (eds.), *Vertebrate Sound Production and Acoustic Communication*, 119–157.

360. Jarvis E. D., "Neural systems for vocal learning in birds and humans: A synopsis," *J Ornithol*, 2007, 35–44; Prather J. F., et al., "Brains for birds and babies: Neural parallels between birdsong and speech acquisition," *Neurosc Biobehav Rev*, vol. 81, 2017, 225–237.

361. Suthers R. A., "How birds sing and why it matters," in Marler and Slabbekoorn (eds.), *Nature's Music*, 272–295; Goller F., and Larsen O. N., "A new mechanism of sound generation

in songbirds," *PNAS*, vol. 94, 1997, 14787–14791; Riede T., et al., "The evolution of the syrinx: An acoustic theory," *PLoS Biol*, vol. 17, 2019, 1–22.

362. Moorman S., et al., "Human-like brain hemispheric dominance in birdsong learning," *PNAS*, vol. 109, 2012, 12782–12787.

363. Amin N., et al., "Development of selectivity for natural sounds in the songbird auditory forebrain," *J Neurophysiol*, vol. 97, 2007, 3517–3531.

364. Derégnaucourt S., et al., "How sleep affects the developmental learning of bird song," *Nature*, vol. 433, 2005, 710–716.

365. Stickgold R., "Sleep-dependent memory consolidation," *Nature*, vol. 437, 2005, 1272–1278; Nusbaum H. C., et al., "Consolidating skill learning through sleep," *Cur Opin Behav Sc*, vol. 20, 2018, 174–182; Saletin J. M., "Memory: Necessary for deep sleep?," *Cur Biol*, vol. 30, 2020, R234–R236.

366. Lachlan R. F., et al., "Cultural conformity generates extremely stable traditions in bird song," *Nat Com*, vol. 9, 2018, 2417.

367. Kroodsma D., *The Singing Life of Birds*, Houghton Mifflin, 2005; Price T., *Speciation in Birds*, Roberts & Co., 2008. See also this recent paper, which investigates the dynamics of vocal cultural transmission in a songbird: Tchernichovski O., Eisenberg-Edidin S., and Jarvis E. D., "Balanced imitation sustains song culture in zebra finches," *Nat Com*, vol. 12, 2021, 2562.

368. Wang D., et al., "Machine learning reveals cryptic dialects that explain mate choice in a songbird," *Nat Com*, vol. 13, 2022, 1630.

369. Irwin D. E., et al., "Speciation in a ring," *Nature*, vol. 409, 2001, 333–337.

370. Crates R., et al., "Loss of vocal culture and fitness costs in a critically endangered songbird," *Proc R Soc B*, vol. 288, 2021, 20210225.

371. Burroughs E. R., *Tarzan of the Apes*, A. C. McClurg & Co., 1914.

372. Hauber M. E., et al., "A password for species recognition in a brood-parasitic bird," *Proc R Soc B*, vol. 268, 2001, 1041–1048.

373. Louder M.I.M., et al., "An acoustic password enhances auditory learning in juvenile brood parasitic cowbirds," *Cur Biol*, vol. 29, 2019, 4045–4051.

374. Tyack P. L., "A taxonomy for vocal learning," *Phil Trans R Soc B*, vol. 375, 2019, 20180406; Nieder A., and Mooney R., "The neurobiology of innate, volitional and learned vocalizations in mammals and birds," *Phil Trans R Soc B*, vol. 375, 2019, 20190054.

375. For a recent perspective on vocal learning in mammals, see Janik V. M., and Knörnschild M., "Vocal production learning in mammals," *Phil Trans R Soc B*, vol. 376, 2021, 20200244.

376. This paper presents an interesting perspective on the different facets of vocal learning: Vernes S. C., et al., "The multi-dimensional nature of vocal learning," *Phil Trans R Soc B*, vol. 376, 2021, 20200236.

377. See this special issue on vocal learning: Vernes S. C., et al., "Vocal learning in animals and humans," *Phil Trans R Soc B*, vol. 376, 2021, 20200234.

378. Payne R., *Among Whales*, Simon & Schuster, 1995.

379. Whitehead H., and Rendell L., *The Cultural Lives of Whales and Dolphins*, University of Chicago Press, 2015; Payne K., and Payne R. S., "Large-scale changes over 19 years in songs of humpback whales in Bermuda," *Zeit Psychol*, vol. 68, 1985, 89–114; Noad M. J., et al., "Cultural revolution in whale songs," *Nature*, vol. 408, 2000, 537; Garland E. C., et al., "Dynamic horizontal cultural transmission of humpback whale song at the ocean basin scale," *Cur Biol*, vol. 21, 2011, 687–691.

380. This view has been recently challenged. See the following papers: Mercado III E., and Perazio C. E., "Similarities in composition and transformations of songs by humpback whales (*Megaptera novaeangliae*) over time and space," *J Comp Psychol*, vol. 135, 2021, 28–50; Mercado III E., "Song morphing by humpback whales: Cultural or epiphenomenal?," *Front Psychol*, vol. 11, 2021, 574403; and Mercado III E., "The humpback's new songs: Diverse and convergent evidence against vocal culture via copying in humpback whales," *Anim Behav Cogn*, vol. 9, 2022, 196–206.

381. Zandberg L., et al., "Global cultural evolutionary model of humpback whale song," *Phil Trans R Soc B*, vol. 376, 2021, 20200242.

382. Richards D. G., et al., "Vocal mimicry of computer-generated sounds and vocal labeling of objects by a bottlenosed dolphin, *Tursiops truncatus*," *J Comp Psychol*, vol. 98, 1984, 10–28; Janik V. M., and Sayigh L. S., "Communication in bottlenose dolphins: 50 years of signature whistle research," *J Comp Physiol*, vol. 199, 2013, 479–489; Luis A. R., et al., "Vocal universals and geographic variations in the acoustic repertoire of the common bottlenose dolphin," *Sc Rep*, vol. 11, 2021, 11847.

383. Eaton R. L., "A beluga whale imitates human speech," *Carnivore*, vol. 2, 1979, 22–23; Murayama T., et al., "Vocal imitation of human speech, synthetic sounds and beluga sounds, by a beluga (*Delphinapterus leucas*)," *Int J Comp Psychol*, vol. 27, 2014, 369–384.

384. Janik V. M., "Cetacean vocal learning and communication," *Cur Opin Neurobiol*, vol. 28, 2014, 60–65; Ridgway S., et al., "Spontaneous human speech mimicry by a cetacean," *Cur Biol*, vol. 22, 2012, R860–R861; Abramson J. Z., et al., "Imitation of novel conspecific and human speech sounds in the killer whale (*Orcinus orca*)," *Proc R Soc B*, vol. 285, 2018, 20172171.

385. Bernie Krause made a recording of an orca, *Orcinus orca*, imitating pinniped vocalizations, likely to attract the prey or at least get close enough to attack. Bernie sent me this recording, and the imitation is indeed truly amazing.

386. Reichmuth C., and Casey C., "Vocal learning in seals, sea lions, and walruses," *Cur Opin Neurobiol*, vol. 28, 2014, 66–71.

387. Rawls K., Fiorelli P., and Gish S., "Vocalizations and vocal mimicry in captive harbor seals, *Phoca vitulina*," *Can J Zool*, vol. 63, 1985, 1050–1056.

388. Stansbury A. L., and Janik V. M., "Formant modification through vocal production learning in gray seals," *Cur Biol*, vol. 29, 2019, 2244–2249.

389. Not everyone agrees. See, for example, Schusterman R., "Vocal learning in mammals with special emphasis on pinnipeds," in Oller D. K., and Griebel U. (eds.), *Evolution of Communicative Flexibility: Complexity, Creativity, and Adaptability in Human and Animal Communication*, MIT Press, 2008, 41–70.

390. Stoeger A. S., and Manger P., "Vocal learning in elephants: Neural bases and adaptive context," *Cur Opin Neurobiol*, vol. 28, 2014, 101–107.

391. Stoeger A. S., and Baotic A., "Operant control and call usage learning in African elephants," *Phil Trans R Soc B*, vol. 376, 2021, 20200254.

392. Poole J. H., et al., "Elephants are capable of vocal learning," *Nature*, vol. 434, 2005, 455–456.

393. Stoeger A. S., et al., "An Asian elephant imitates human speech," *Cur Biol*, vol. 22, 2012, 2144–2148. See also this recent paper reporting that elephants can use lip buzzing (like human

brass players) to produce high-pitched sounds: Beeck V. C., et al., "A novel theory of Asian elephant high-frequency squeak production," *BMC Biol*, vol. 19, 2021, 121.

394. Vernes S. C., and Wilkinson G. S., "Behaviour, biology and evolution of vocal learning in bats," *Phil Trans R Soc B*, vol. 375, 2019, 20190061; Knörnschild M., "Vocal production learning in bats," *Cur Opin Neurobiol*, vol. 28, 2014, 80–85.

395. Jones G., and Ransome R. D., "Echolocation calls of bats are influenced by maternal effects and change over a lifetime," *Proc R Soc B*, vol. 252, 1993, 125–128.

396. Prat Y., et al., "Vocal learning in a social mammal: Demonstrated by isolation and playback experiments in bats," *Sc Adv*, vol. 1, 2015, e1500019. See also Prat Y., et al., "Crowd vocal learning induces vocal dialects in bats: Playback of conspecifics shapes fundamental frequency usage by pups," *PLoS Biol*, vol. 15, 2017, e2002556.

397. Lattenkamp E. Z., et al., "The vocal development of the pale spear-nosed bat is dependent on auditory feedback," *Phil Trans R Soc*, vol. 376, 2021, 20200253.

398. Fernandez A. A., et al., "Babbling in a vocal learning bat resembles human infant babbling," *Science*, vol. 373, 2021, 923–926.

399. Knörnschild M., et al., "Complex vocal imitation during ontogeny in a bat," *Biol Let*, vol. 6, 2010, 156–159.

400. For instance, look at the works of Mirjam Knörnschild at the Museum of Natural History in Berlin, and Sonja Catherine Vernes at the University of St. Andrews, Scotland.

401. Barker A. J., et al., "Cultural transmission of vocal dialect in the naked mole-rat," *Science*, vol. 371, 2021, 503–507.

402. Taylor D., et al., "Vocal functional flexibility: What it is and why it matters." *Anim Behav*, vol. 186, 2022, 93–100.

403. Brockelman W. Y., and Schilling D., "Inheritance of stereotyped gibbon calls," *Nature*, vol. 312, 1984, 634–636.

404. Hayes K. J., and Hayes C., "Imitation in a home-raised chimpanzee," *J Comp Physiol Psychol*, vol. 45, 1952, 450–459.

405. Takahashi D. Y., et al., "The developmental dynamics of marmoset monkey vocal production," *Science*, vol. 349, 2015, 734–738; Gultekin Y. B., and Hage S. R., "Limiting parental feedback disrupts vocal development in marmoset monkeys," *Nat Com*, vol. 8, 2017, 14046.

406. Levrero F., et al., "Social shaping of voices does not impair phenotype matching of kinship in mandrills," *Nat Com*, vol. 6, 2015, 7609.

407. Fischer J., et al., "Vocal convergence in a multi-level primate society: Insights into the evolution of vocal learning," *Proc R Soc B*, vol. 287, 2020, 20202531.

408. Lameira A. R., et al., "Sociality predicts orangutan vocal phenotype," *Nat Ecol Evol*, vol. 6, 2022, 644–652.

409. Ruch H., et al., "The function and mechanism of vocal accommodation in humans and other primates," *Biol Rev*, vol. 93, 2018, 996–1013; Watson S. K., et al., "Vocal learning in the functionally referential food grunts of chimpanzees," *Cur Biol*, vol. 25, 2015, 495–499; Fischer J., et al., "Vocal convergence in a multi-level primate society."

410. On the interactions between genetics and learning, see Mets and Brainard, "Genetic variation interacts with experience."

Chapter 13

411. Céline Rochais is conducting studies on the cognitive performance of striped mice in the wild. See, for example, Rochais C., et al., "How does cognitive performance change in relation to seasonal and experimental changes in blood glucose levels?," *Anim Behav*, vol. 158, 2019, 149–159.

412. Schradin C., "Seasonal changes in testosterone and corticosterone levels in four social classes of a desert dwelling sociable rodent," *Horm Behav*, vol. 53, 2008, 573–579.

413. Schradin C., and Pillay N., "Intraspecific variation in the spatial and social organization of the African striped mouse," *J Mammal*, vol. 86, 2005, 99–107.

414. Holy T. E., and Guo Z., "Ultrasonic songs of male mice," *PLoS Biol*, vol. 3, 2005, e386.

415. Griffin D. R., and Galambos R., "The sensory basis of obstacle avoidance by flying bats," *J Exp Zool*, vol. 86, 1941, 481–506; Galambos R., and Griffin D. R., "Obstacle avoidance by flying bats: The cries of bats," *J Exp Zool*, vol. 89, 1942, 475–490; Griffin D. R., "Supersonic cries of bats," *Nature*, vol. 158, 1946, 46–48.

416. For a review of our knowledge of rodents' vocalizations (including production mechanisms), look at Fernandez-Vargas M., et al., "Mechanisms and constraints underlying acoustic variation in rodents," *Animal Behav*, vol. 184, 2021, 135–147.

417. Simola N., and Brudzynski S. M., "Repertoire and biological function of ultrasonic vocalizations in adolescent and adult rats," in Brudzynski S. (ed.), *Handbook of Ultrasonic Vocalization XXV*, Elsevier, 2018, 177–186.

418. Ehret G., "Characteristics of vocalization in adult mice," in Brudzynski (ed.), *Handbook of Ultrasonic Vocalization*, 187–196.

419. Fischer J., and Hammerschmidt K., "Ultrasonic vocalizations in mouse models for speech and socio-cognitive disorders: Insights into the evolution of vocal communication," *Genes Brain Behav*, vol. 10, 2010, 17–27.

420. Portfors C. V., and Perkel D. J., "The role of ultrasonic vocalizations in mouse communication," *Cur Opin Neurobiol*, vol. 28, 2014, 115–120; Miranda R., et al., "Altered social behavior and ultrasonic communication in the dystrophin-deficient mdx mouse model of Duchenne muscular dystrophy," *Mol Autism*, vol. 6, 2015, 60; Faure A., et al., "Dissociated features of social cognition altered in mouse models of schizophrenia: Focus on social dominance and acoustic communication," *Neuropharmacol*, vol. 159, 2019, 107334.

421. Pierce G. W., *The Songs of Insects*, Harvard University Press, 1948.

422. Pierce G. W., and Griffin D. R., "Experimental determination of supersonic notes emitted by bats," *J Mammal*, vol. 19, 1938, 454–455.

423. Fenton M. B., Grinnell A. D., Popper A. N., and Fay R. R. (eds.), *Bat Bioacoustics*, Springer, 2016.

424. Ulanovsky N., and Moss C. F., "What the bat's voice tells the bat's brain," *PNAS*, vol. 105, 2008, 8491–8498.

425. Leiser-Miller L. B., and Santana S. E., "Functional differences in echolocation call design in an adaptive radiation of bats," *Ecol Evol*, vol. 11, 2021, 16153–16164.

426. Bats are not the only animals to echolocate. In the book, we have already seen that dolphins and other sperm whales are capable of it. It is the same for some birds, and even . . . humans! Indeed, it has been shown that trained blind people can locate objects in their environment

by producing "clicks" and listening to their echoes. See Brinklov S., "Echolocation in oilbirds and swiftlets," *Front Physiol*, vol. 4, 2013, 123; and Thaler L., et al., "Human echolocators adjust loudness and number of clicks for detection of reflectors at various azimuth angles," *Proc R Soc B*, vol. 285, 2018, 20172735. Recently, researchers have discovered another lineage of echolocating mammals: He K., et al., "Echolocation in soft-furred tree mice," *Science*, vol. 372, 2021, eaay1513.

427. Carew T. J., *Behavioral Neurobiology*, Sinauer, 2000.

428. Simmons J. A., et al., "Echo-delay resolution in sonar images of the big brown bat, *Eptesicus fuscus*," *PNAS*, vol. 95, 1998, 12647–12652.

429. Schnitzler H. U., "Control of Doppler shift compensation in the greater horseshoe bat, *Rhinolophus ferrumequinum*," *J Comp Physiol*, vol. 82, 1973, 79–92; Metzner W., "A possible neuronal basis for Doppler-shift compensation in echo-locating horseshoe bats," *Nature*, vol. 341, 1989, 529–532; Metzner W., et al., "Doppler-shift compensation behavior in horseshoe bats revisited: Auditory feedback controls both a decrease and an increase in call frequency," *J Exp Biol*, vol. 205, 2002, 1607–1616.

430. Simmons J. A., "Response of the Doppler echolocation system in the bat, *Rhinolophus ferrumequinum*," *JASA*, vol. 56, 1974, 672–682; Zhang Y., et al., "Performance of Doppler shift compensation in bats varies with species rather than with environmental clutter," *Anim Behav*, vol. 158, 2019, 109–120.

431. Schnitzler H.-U., and Denzinger A., "Auditory fovea and Doppler shift compensation: Adaptations for flutter detection in echolocating bats using CF-FM signals," *J Comp Physiol A*, vol. 197, 2011, 541–559.

432. Yack J. E., and Fullard J. H., "Ultrasonic hearing in nocturnal butterflies," *Nature*, vol. 403, 2000, 265–266.

433. Greenfield M. D., "Evolution of acoustic communication in insects," in Pollack G. S., Mason A. C., Popper A., and Fay R. R. (eds.), *Insect Hearing*, Springer, 2016, 17–47.

434. Yack J. E., et al., "Neuroethology of ultrasonic hearing in nocturnal butterflies (*Hedyloidea*)," *J Comp Physiol A*, vol. 193, 2007, 577–590.

435. Kawahara A. Y., and Barber J. R., "Tempo and mode of antibat ultrasound production and sonar jamming in the diverse hawkmoth radiation," *PNAS*, vol. 112, 2015, 6407–6412; Barber J. R., et al., "Anti-bat ultrasound production in moths is globally and phylogenetically widespread," *PNAS*, vol. 119, 2022, e2117485119.

436. Ancillotto L., et al., "Bats mimic hymenopteran insect sounds to deter predators," *Cur Biol*, vol. 32, 2022, R399–R413.

437. Conner W. E., and Corcoran A. J., "Sound strategies: The 65-million-year-old battle between bats and insects," *Ann Rev Entomol*, vol. 57, 2012, 21–39; Ter Hofstede H. M., and Ratcliffe J. M., "Evolutionary escalation: The bat-moth arms race," *J Exp Biol*, vol. 219, 2016, 1589–1602.

438. von Helversen D., and von Helversen O., "Acoustic guide in bat-pollinated flower," *Nature*, vol. 398, 1999, 759–760.

439. The bat ultrasonic world is an active field of research. Look at this paper, reporting that differences among bats in hearing abilities contribute to the diversity in foraging strategies: Geipel I., et al., "Hearing sensitivity: An underlying mechanism for niche differentiation in gleaning bats," *PNAS*, vol. 118, 2021, e2024943118.

440. Feng A. S., and Narins P. M., "Ultrasonic communication in concave-eared torrent frogs (*Amolops tormotus*)," *J Comp Physiol*, vol. 194, 2008, 159–167.

441. Feng A. S., et al., "Ultrasonic communication in frogs," *Nature*, vol. 440, 2006, 333–336; Shen J.-X., et al., "Ultrasonic frogs show hyperacute phonotaxis to female courtship calls," *Nature*, vol. 453, 2008, 914–916; Gridi-Papp M., et al., "Active control of ultrasonic hearing in frogs," *PNAS*, vol. 105, 2008, 11014–11019.

442. Narins P. M., et al., "Plant-borne vibrations modulate calling behaviour in a tropical amphibian," *Cur Biol*, vol. 23, 2018, R1333–R1334; Lewis E. R., and Narins P. M., "Do frogs communicate with seismic signals?," *Science*, vol. 227, 1985, 187–189; Narins P. M., "Seismic communication in anuran amphibians," *Bioscience*, vol. 40, 1990, 268–274.

443. A recent paper explores the richness of the vibrational environment that can be perceived and exploited by insects: Sturm R., et al., "Hay meadow vibroscape and interactions within insect vibrational community," *iScience*, vol. 24, 2021, 103070. Another paper deals with crocodiles: Grap N. J., et al., "Stimulus discrimination and surface wave source localization in crocodilians," *Zoology*, vol. 139, 2020, 125743.

444. Hager F. A., et al., "Vibrational behavior in termites (Isoptera)," in Hill P., Lakes-Harlan R., Mazzoni V., Narins P., Virant-Doberlet M., and Wessel A. (eds.), *Biotremology: Studying Vibrational Behavior*, Springer, 2019, 309–327.

445. Hager F. A., and Kirchner W. H., "Vibrational long-distance communication in the termites *Macrotermes natalensis* and *Odontotermes* sp.," *J Exp Biol*, vol. 216, 2013, 3249–3256.

446. Similar patterns can be observed in bees; see, for example, Grüter C., et al., "Propagation of olfactory information within the honeybee hive," *Behav Ecol Sociobiol*, vol. 60, 2006, 707–715.

447. Hager F. A., and Kirchner W. H., "Directional vibration sensing in the termite *Macrotermes natalensis*," *J Exp Biol*, vol. 217, 2014, 2526–2530.

448. Spiders can develop amazing abilities to detect vibrations and sounds. In the orb-weaving spider *Larinioides sclopetarius*, the web acts as an acoustic antenna that captures the sound-induced air particle movements: Zhou J., et al., "Outsourced hearing in an orb-weaving spider that uses its web as an auditory sensor," *PNAS*, vol. 119, 2022, e2122789119.

449. Narins P. M., et al., "Infrasonic and seismic communication in the vertebrates with special emphasis on the Afrotheria: An update and future directions," in Suthers, Fitch, Fay, and Popper (eds.), *Vertebrate Sound Production and Acoustic Communication*, 191–227.

450. Schleich C., and Francescoli G., "Three decades of subterranean acoustic communication studies," in Dent M. L., Fay R. R., and Popper A. N. (eds.), *Rodent Bioacoustics*, Springer, 2018, 43–69.

451. Elephants also emit high-pitched sounds: Beeck et al., "Novel theory of Asian elephant high-frequency squeak production."

452. O'Connell-Rodwell C. E., "Keeping an 'ear' to the ground: Seismic communication in elephants," *Physiology*, vol. 22, 2007, 287–294; O'Connell-Rodwell C. E., et al., "Wild African elephants (*Loxodonta africana*) discriminate between familiar and unfamiliar conspecific seismic alarm calls," *JASA*, vol. 122, 2007, 823–830.

453. Parihar D. S., et al., "Seismic signal analysis for the characterization of elephant movements in a forest environment," *Ecol Infor*, vol. 64, 2021, 101329.

454. O'Connell C. E., et al., "Vibrational communication in elephants: A case for bone conduction," in Hill, Lakes-Harlan, Mazzoni, Narins, Virant-Doberlet, and Wessel (eds.), *Biotremology: Studying Vibrational Behavior*, 259–276.

455. McComb K., et al., "Unusually extensive networks of vocal recognition in African elephants," *Anim Behav*, vol. 59, 2000, 1103–1109.

456. The importance of infrasound in the animal world may have been underestimated. For instance, although they do not communicate in the infrasound range, some birds are able to hear it: Zeyl J. N., et al., "Infrasonic hearing in birds: A review of audiometry and hypothesized structure-function relationships," *Biol Rev*, vol. 95, 2020, 1036–1054.

Chapter 14

457. Holekamp K. E., and Kolowski J. M., "Family Hyaenidae (hyenas)," in Wilson D. E., and Mittermeier R. A. (eds.), *Handbook of the Mammals of the World* I, Carnivores, Lynx Edicions, 2009, 234–251.

458. Kruuk H., *The Spotted Hyena: A Study of Predation and Social Behaviour*, Chicago University Press, 1972.

459. Watts H. E., and Holekamp K. E., "Hyena societies," *Cur Biol*, vol. 17, 2007, R657–R660.

460. Yalcinkaya T. M., et al., "A mechanism for virilization of female spotted hyenas in utero," *Science*, vol. 260, 1993, 1929–1931; Glickman S. E., et al., "Androstenedione may organize or activate sex-reversed traits in female spotted hyenas," *PNAS*, vol. 84, 1987, 3444–3447.

461. Holekamp K. E., and Smale L., "Dominance acquisition during mammalian social development: The 'inheritance' of maternal rank," *Amer Zool*, vol. 31, 1991, 306–317; Engh A. L., et al., "Mechanisms of maternal rank 'inheritance' in the spotted hyaena, *Crocuta crocuta*," *Anim Behav*, vol. 60, 2000, 323–332.

462. Smith J. E., and Holekamp K. E., "Spotted hyenas," in Breed M. D., and Moore J. (eds.), *Encyclopedia of Animal Behavior*, 2nd ed., Academic Press, 2019, 190–208.

463. On the importance of individual recognition in animal social networks, see Gokcekus S., et al., "Recognizing the key role of individual recognition in social networks," *Trends Ecol Evol*, vol. 36, 2021, 1024–1035.

464. Elie J. E., and Theunissen F. E., "Zebra finches identify individuals using vocal signatures unique to each call type," *Nat Com*, vol. 9, 2018, 4026.

465. Frédéric Theunissen's team recently showed that the zebra finches have fast and high-capacity auditory memory for vocalizer identity. They can remember a mean number of 42 different vocalizers based solely on the individual signatures found in their songs and distance calls. The learning is very efficient, taking only a few trials, and is maintained for up to a month. See Yu K., et al., "High-capacity auditory memory for vocal communication in a social songbird," *Sc Adv*, vol. 6, 2020, eabe0440.

466. Christiansen P., and Adolfseen J. S., "Bite forces, canine strength and skull allometry in carnivores (Mammalia, Carnivora)," *J Zool*, vol. 266, 2005, 133–151.

467. Mathevon N., et al., "What the hyena's laugh tells: Sex, age, dominance and individual signature in the giggling call of *Crocuta crocuta*," *BMC Ecol*, vol. 10, 2010, 9.

468. Ross-Gillespie A., and Griffin A. S., "Meerkats," *Cur Biol*, vol. 17, 2007, R442–R443; Gilchrist J. S., et al., "Family Herpestidae (mongooses)," in Wilson and Mittermeier (eds.), *Handbook of the Mammals of the World* I, 262–328.

469. Manser M. B., "The acoustic structure of suricates' alarm calls varies with predator type and the level of response urgency," *Proc R Soc B*, vol. 268, 2001, 2315–2324; Manser M. B., et al., "The information that receivers extract from alarm calls in suricates," *Proc R Soc B*, vol. 268, 2001, 2485–2491.

470. Townsend S. W., et al., "Acoustic cues to identity and predator context in meerkat barks," *Anim Behav*, vol. 94, 2014, 143–149.

471. Manser M. B., "Semantic communication in vervet monkeys and other animals," *Anim Behav*, vol. 86, 2013, 491–496; Rauber R., Kranstauber B., and Manser M. B., "Call order within vocal sequences of meerkats contains temporary contextual and individual information," *BMC Biol*, vol. 18, 2020, 119.

472. Hollén L. I., and Manser M. B., "Ontogeny of alarm call responses in meerkats, *Suricata suricatta*: The roles of age, sex and nearby conspecifics," *Anim Behav*, vol. 72, 2006, 1345–1353.

473. Darwin C., *The Descent of Man and Selection in Relation to Sex*, John Murray, 1871.

474. Struhsaker T. T., "Auditory communication among vervet monkeys (*Cercopithecus aetthiops*)," in Altmann S. A. (ed.), *Social Communication among Primates*, University of Chicago Press, 1967, 281–324.

475. Seyfarth R. M., et al., "Monkey responses to three different alarm calls: Evidence of predator classification and semantic communication," *Science*, vol. 210, 1980, 801–803; Price T., et al., "Vervets revisited: A quantitative analysis of alarm call structure and context specificity," *Sc Rep*, vol. 5, 2015, 13220.

476. Cheney D. L., and Seyfarth R. M., *How Monkeys See the World*, University of Chicago Press, 1990.

477. Premack D., "Concordant preferences as a precondition for affective but not for symbolic communication (or how to do experimental anthropology)," *Cognition*, vol. 1, 1972, 251–264.

478. Zuberbühler K., "Referential labelling in Diana monkeys," *Anim Behav*, vol. 59, 2000, 917–927.

479. Zuberbühler K., "A syntactic rule in forest monkey communication," *Anim Behav*, vol. 63, 2002, 293–299.

480. Rendall D., et al., "The meaning and function of grunt variants in baboons," *Anim Behav*, vol. 57, 1999, 583–592.

481. Caesar C., et al., "The alarm call system of wild black-fronted titi monkeys, *Callicebus nigrifrons*," *Behav Ecol Sociobiol*, vol. 66, 2012, 653–667.

482. Slocombe K. E., and Zuberbühler K., "Functionally referential communication in a chimpanzee," *Cur Biol*, vol. 15, 2005, 1779–1784.

483. Pereira M. E., and Macedonia J. M., "Ringtailed lemur anti-predator calls denote predator class, not response urgency," *Anim Behav*, vol. 26, 1991, 760–777.

484. Schel A. M., and Zuberbühler K., "Predator and non-predator long-distance calls in Guereza colobus monkeys," *Behav Process*, vol. 91, 2012, 41–49.

485. Ouattara K., et al., "Campbell's monkeys concatenate vocalizations into context-specific call sequences," *PNAS*, vol. 106, 2009, 22026–22031.

486. Zuberbühler K., et al., "Diana monkey long-distance calls: Messages for conspecifics and predators," *Anim Behav*, vol. 53, 1997, 589–604.

487. Slocombe K. E., and Zuberbühler K., "Functionally referential communication in a chimpanzee," *Cur Biol*, vol. 15, 2005, 1779–1784; Slocombe K. E., et al., "Chimpanzee vocal communication: What we know from the wild," *Cur Opin Behav Sc*, vol. 46, 2022, 101171.

488. Clay Z., and Zuberbühler K., "Food-associated calling sequences in bonobos," *Anim Behav*, vol. 77, 2009, 1387–1396; Shorland G., et al., "Bonobos assign meaning to food calls based on caller food preferences," *PLoS ONE*, vol. 17, 2022, e0267574.

489. Understanding how the sequences of calls produced by primates are organized and what they mean is an active field of research. See, for example, Girard-Buttoz C., et al., "Chimpanzees produce diverse vocal sequences with ordered and recombinatorial properties," *Com Biol*, vol. 5, 2022, 410.

490. Townsend S. W., and Manser M. B., "Functionally referential communication in mammals: The past, present and the future," *Ethology*, vol. 119, 2013, 1–11.

491. Kiriazis J., and Slobodchikoff C. N., "Perceptual specificity in the alarm calls of Gunnison's prairie dogs," *Behav Process*, vol. 73, 2006, 29–35.

492. Smith C. L., "Referential signalling in birds: The past, present and future," *Anim Behav*, vol. 124, 2017, 315–323.

493. Marler P., et al., "Semantics of an avian alarm call system: The male domestic fowl, *Gallus domesticus*," *Behaviour*, vol. 102, 1987, 15–40; Evans C. S., and Evans L., "Chicken food calls are functionally referential," *Anim Behav*, vol. 58, 1999, 307–319; Evans C. S., and Evans L., "Representational signalling in birds," *Biol Let*, vol. 3, 2007, 8–11.

494. Evans C. S., et al., "Effects of apparent size and speed on the response of chickens, *Gallus gallus*, to computer-generated simulations of aerial predators," *Anim Behav*, vol. 46, 1993, 1–11.

495. Gill S. A., and Bierema A.M.K., "On the meaning of alarm calls: A review of functional reference in avian alarm calling," *Ethology*, vol. 119, 2012, 449–461.

496. Magrath R., et al., "Interspecific communication: Gaining information from heterospecific alarm calls," in Aubin T., and Mathevon N. (eds.), *Coding Strategies in Vertebrate Acoustic Communication*, Springer, 2020, 287–314.

497. Templeton C. N., et al., "Allometry of alarm calls: Black-capped chickadees encode information about predator size," *Science*, vol. 308, 2005, 1934–1937.

498. Suzuki T. N., "Communication about predator type by a bird using discrete, graded and combinatorial variation in alarm calls," *Anim Behav*, vol. 87, 2014, 59–65; Suzuki T. N., "Referential mobbing calls elicit different predator-searching behaviours in Japanese great tits," *Anim Behav*, vol. 84, 2012, 53–57; Suzuki T. N., et al., "Experimental evidence for compositional syntax in bird calls," *Nat Com*, vol. 7, 2016, 10986.

499. Suzuki T. N., "Alarm calls evoke a visual search image of a predator in birds," *PNAS*, vol. 115, 2018, 1541–1545.

500. For a review on bird calls and their interest for the understanding of language evolution, see Suzuki T. N., "Animal linguistics: Exploring referentiality and compositionality in bird calls," *Ecol Res*, vol. 36, 2021, 221–231.

501. Engesser S., et al., "Meaningful call combinations and compositional processing in the southern pied babbler," *PNAS*, vol. 113, 2016, 5976–5981.

502. Engesser S., et al., "Chestnut-crowned babbler calls are composed of meaningless shared building blocks," *PNAS*, vol. 116, 2019, 19579–19584.

503. On songbird syntax, see Searcy W. A., et al., "Long-distance dependencies in birdsong syntax," vol. 289, 2022, 20212473; and Backhouse F., et al., "Higher-order sequences of vocal mimicry performed by male Albert's lyrebirds are socially transmitted and enhance acoustic contrast," *Proc R Soc B*, vol. 289, 2022, 20212498.

504. In Molière's famous comedy *Le bourgeois gentilhomme*, Mr. Jourdain learns from his philosophy teacher that all language is classified according to the way it is spoken, as poetry or prose (the ordinary form of language). Mr. Jourdain is delighted to realize that he has been speaking in prose for a very long time without even knowing it.

505. Seyfarth R. M., and Cheney D. L., "Precursors to language: Social cognition and pragmatic inference in primates," *Psychon Bull Rev*, vol. 24, 2017, 79–84.

506. Cheney D. L., and Seyfarth R. M., *Baboon Metaphysics*, University of Chicago Press, 2007.

Chapter 15

507. Mathevon N., et al., "The code size: Behavioural response of crocodile mothers to offspring calls depends on the emitter's size, not on its species identity," in *Crocodiles, Proceedings of the 24th Working Meeting of the Crocodile Specialist Group IUCN*, 2016, 79–85.

508. Darwin C., *The Expression of Emotions in Man and Animals*, Penguin Classics, 2009.

509. Briefer E. F., "Coding for 'dynamic' information: Vocal expression of emotional arousal and valence in non-human animals," in Aubin and Mathevon (eds.), *Coding Strategies in Vertebrate Acoustic Communication*, 137–162.

510. Briefer E. F., et al., "Segregation of information about emotional arousal and valence in horse whinnies," *Sc Rep*, vol. 5, 2015, 9989.

511. Things are in fact even more complex, because each of these frequencies, F0 and G0, is accompanied by its harmonic series. For details, see Briefer et al., "Segregation of information about emotional arousal."

512. Briefer E. F., "Vocal expression of emotions in mammals: Mechanisms of production and evidence," *J Zool*, vol. 288, 2012, 1–20.

513. Briefer E. F., "Classification of pig calls produced from birth to slaughter according to their emotional valence and context of production," *Sc Rep*, vol. 12, 2022, 3409.

514. Darwin C., *Descent of Man*.

515. Morton E. S., "On the occurrence and significance of motivation-structural rules in some bird and mammal sounds," *Am Nat*, vol. 111, 1977, 855–869.

516. Koutseff A., et al., "The acoustic space of pain: Cries as indicators of distress recovering dynamics in pre-verbal infants," *Bioacoustics*, vol. 27, 2018, 313–325.

517. Filippi P., et al., "Humans recognize emotional arousal in vocalizations across all classes of terrestrial vertebrates: Evidence for acoustic universals," *Proc R Soc B*, vol. 284, 2017, 20170990.

518. Lingle S., and Riede T., "Deer mothers are sensitive to infant distress vocalizations of diverse mammalian species," *Am Nat*, vol. 184, 2014, 510–522.

519. Kelly T., et al., "Adult human perception of distress in the cries of bonobo, chimpanzee, and human infants," *Biol J Lin Soc*, vol. 120, 2017, 919–930.

520. Briefer E. F., "Vocal contagion of emotions in non-human animals," *Proc R Soc B*, vol. 285, 2018, 20172783.

521. If you want to learn a lot about primates' acoustic communication, I advise you to read Fischer J., *Monkeytalk: Inside the Worlds and Minds of Primates*, University of Chicago Press, 2017.

522. The response of chacma baboons to barks of conspecifics is not easy to understand. See these two papers: Fischer J., et al., "Baboon responses to graded bark variants," *Anim Behav*, vol. 61, 2001, 925–931; and Cheney D., et al., "The function and mechanisms underlying baboon 'contact' barks," *Anim Behav*, vol. 52, 1996, 507–518.

523. Vignal C., et al., "Audience drives male songbird response to partner's voice," *Nature*, vol. 430, 2004, 448–451.

524. Perez E. C., et al., "The acoustic expression of stress in a songbird: Does corticosterone drive isolation-induced modifications of zebra finch calls?," *Horm Behav*, vol. 61, 2012, 573–581.

525. Perez E. C., et al., "Physiological resonance between mates through calls as possible evidence of empathic processes in songbirds," *Horm Behav*, vol. 75, 2015, 130–141.

526. Ben-Aderet T., et al., "Dog-directed speech: Why do we use it and do dogs pay attention to it?," *Proc R Soc B*, vol. 284, 2017, 20162429. See also Massenet M., et al., "Nonlinear vocal phenomena affect human perceptions of distress, size and dominance in puppy whines," *Proc R Soc B*, vol. 289, 2022, 20220429.

527. On human-dog vocal communication, see, for instance, Gabor A., et al., "The acoustic basis of human voice identity processing in dogs," *Anim Cogn*, vol. 25, 2022, 905–916; and Massenet et al., "Nonlinear vocal phenomena affect human perceptions of distress."

528. Azhari F., "Bird strike case study at airport level to include take off, landing and taxiways," *Adv J Tech Voc Educ*, vol. 1, 2017, 364–374.

529. Aubin T., and Brémond J.-C., "Parameters used for recognition of distress calls in two species: *Larus argentatus* and *Sturnus vulgaris*," *Bioacoustics*, vol. 2, 1999, 22–33; Aubin T., and Brémond J.-C., "Perception of distress call harmonic structure by the starling (*Sturnus vulgaris*)," *Behaviour*, vol. 120, 1992, 3–4; Brémond J.-C., and Aubin T., "Responses to distress calls by black-headed gulls, *Larus ridibundus*: The role of non-degraded features," *Anim Behav*, vol. 39, 1990, 503–511; Maigrot A. L., et al., "Cross-species discrimination of vocal expression of emotional valence by Equidae and Suidae," *BMC Biol*, vol. 20, 2022, 106.

530. Aubin T., "Why do distress calls evoke interspecific responses? An experimental study applied to some species of birds," *Behav Proc*, vol. 23, 1991, 103–111.

Chapter 16

531. Dentressangle F., et al., "Males use time whereas females prefer harmony: Individual call recognition in the dimorphic blue-footed booby," *Anim Behav*, vol. 84, 2012, 413–420.

532. Janicke T., et al., "Darwinian sex roles confirmed across the animal kingdom," *Sc Adv*, vol. 2, 2016, e1500983.

533. Vignal C., "Biologie du comportement animal," in Mathevon N., and Viennot É. (eds.), *La différence des sexes*, Belin, 2017, 53–80.

534. Green K. K., and Madjidian J. A., "Active males, reactive females: Stereotypic sex roles in sexual conflict research?," *Anim Behav*, vol. 81, 2011, 901–907.

535. Bourgeois K., et al., "Morphological versus acoustic analysis: What is the most efficient method for sexing yelkouan shearwaters *Puffinus yelkouan*?," *J Ornithol*, vol. 148, 2007, 261–269.

536. Curé C., et al., "Sex discrimination and mate recognition by voice in the yelkouan shearwater *Puffinus yelkouan*," *Bioacoustics*, vol. 20, 2011, 235–250; Curé C., et al., "Mate vocal recognition in the Scopoli's shearwater *Calonectris diomedea*: Do females and males share the same acoustic code?," *Behav Process*, vol. 128, 2016, 96–102.

537. Curé C., et al., "Acoustic cues used for species recognition can differ between sexes and sibling species: Evidence in shearwaters," *Anim Behav*, vol. 84, 2012, 239–250; Curé C., et al., "Acoustic convergence and divergence in two sympatric burrowing nocturnal seabirds," *Biol J Lin Soc*, vol. 96, 2009, 115–134; Curé C., et al., "Intra-sex vocal interactions in two hybridizing seabird species (*Puffinus* sp.)," *Behav Ecol Sociobiol*, vol. 64, 2010, 1823–1837.

538. Mouterde S. C., "From vocal to neural encoding: A transversal investigation of information transmission at long distance in birds," in Aubin and Mathevon (eds.), *Coding Strategies in Vertebrate Acoustic Communication*, 203–229.

539. Mouterde S. C., et al., "Acoustic communication and sound degradation: How do the individual signatures of male and female zebra finch calls transmit over distance?," *PLoS ONE*, vol. 9, 2014, e102842.

540. Mouterde S. C., et al., "Learning to cope with degraded sounds: Female zebra finches can improve their expertise in discriminating between male voices at long distances," *J Exp Biol*, vol. 217, 2014, 3169–3177.

541. Interestingly, unlike the call, the zebra finch's song seems to be more of a "short-distance" signal: Loning H., et al., "Zebra finch song is a very short-range signal in the wild: Evidence from an integrated approach," *Behav Ecol*, vol. 33, 2022, 37–46.

542. Mouterde S. C., et al., "Single neurons in the avian auditory cortex encode individual identity and propagation distance in naturally degraded communication calls," *J Neuro*, vol. 37, 2017, 3491–3510.

543. Darwin C., *On the Origin of Species by Means of Natural Selection, or the Preservation of Favoured Races in the Struggle for Life*, John Murray, 1859.

544. Odom K. J., et al., "Female song is widespread and ancestral in songbirds," *Nat Com*, vol. 5, 2014, 3379.

545. Oliveros C. H., et al., "Earth history and the passerine superradiation," *PNAS*, vol. 116, 2019, 7916–7925.

546. Gahr M., "Seasonal hormone fluctuations and song structure of birds," in Aubin and Mathevon (eds.), *Coding Strategies in Vertebrate Acoustic Communication*, 163–201.

547. On this topic, see this recent paper reporting a 17-year study on the vocal behavior of female and male rufous-and-white wrens, *Thryophilus rufalbus*, where both the female and the male sing. For both sexes, the authors found variation in vocal behaviors with time of day and time of year. However, some behavioral differences were noted between sexes, e.g., female wrens change song types more often in areas with more neighbors, whereas the male vocal behavior did not change with the number of neighbors. This result suggests that female wrens may be

using song-type switching in territorial defense against conspecifics more than males: Owen K. C., and Mennill D. J., "Singing in a fragmented landscape: Wrens in a tropical dry forest show sex differences in the effects of neighbours, time of day, and time of year," *J Ornithol*, vol. 162, 2021, 881–893.

548. Odom J. K., et al., "Differentiating the evolution of female song and male-female duets in the New World blackbirds: Can tropical natural history traits explain duet evolution?," *Evolution*, vol. 69, 2015, 839–847; Elie J. E., et al., "Vocal communication at the nest between mates in wild zebra finches: A private vocal duet?," *Anim Behav*, vol. 80, 2010, 597–605; Lemazina A., et al., "The multifaceted vocal duets of white-browed sparrow weavers are based on complex duetting rules," *J Avian Biol*, vol. 52, 2021, e02703.

549. Fortune E. S., et al., "Neural mechanisms for the coordination of duet singing in wrens," *Science*, vol. 334, 2011, 666–670; Coleman M., and Fortune E., "Duet singing in plain-tailed wrens," *Cur Biol*, vol. 28, 2018, R1–R3; Coleman M., et al., "Neurophysiological coordination of duet singing," *PNAS*, vol. 118, 2021, e2018188118.

550. Another study, conducted in the wild, looked at the vocal coordination mechanisms in the white-browed sparrow-weaver, *Plocepasser mahali*: Hoffmann S., et al., "Duets recorded in the wild reveal that interindividually coordinated motor control enables cooperative behavior," *Nat Com*, vol. 10, 2019, 2577.

551. Levréro F., "Éthologie des primates non humains," in Mathevon and Viennot (eds.), *La différence des sexes*, 277–304.

552. Brunton P. J., and Russel J. A., "The expectant brain: Adapting for motherhood," *Nat Rev Neurosci*, vol. 9, 2008, 11–25; Hoekzema E., et al., "Pregnancy leads to long-lasting changes in human brain structure," *Nat Neurosci*, vol. 20, 2017, 287–296; Kim P., et al., "The maternal brain and its plasticity in humans," *Horm Behav*, vol. 77, 2016, 113–123.

553. Hrdy S. B., "The neurobiology of paternal care: Cooperative breeding and the paradox of facultative fathering," in Bridges R. S. (ed.), *Neurobiology of the Parental Brain*, Elsevier, 2008, 407–416.

554. Wiesenfeld A. R., et al., "Differential parental response to familiar and unfamiliar infant distress signals," *Infant Behav Dev*, vol. 4, 1981, 281–295; Green J. A., and Gustafson G. E., "Individual recognition of human infants on the basis of cries alone," *Dev Psychobiol*, vol. 16, 1983, 485–493.

555. Gustafsson E., et al., "Fathers are just as good as mothers at recognizing the cries of their baby," *Nat Com*, vol. 4, 2013, 1698.

556. Bouchet H., et al., "Baby cry recognition is independent of motherhood but improved by experience and exposure," *Proc R Soc B*, vol. 287, 2020, 20192499.

557. Corvin S., et al., "Adults learn to identify pain in babies' cries," *Cur Biol*, vol. 32, 2022, R807–R827.

558. Reby D., et al., "Sex stereotypes influence adults' perception of babies' cries," *BMC Psychol*, vol. 4, 2016, 19.

Chapter 17

559. Marin-Cudraz T., et al., "Acoustic monitoring of rock ptarmigan: A multi-year comparison with point-count protocol," *Ecol Indic*, vol. 101, 2019, 710–719. See also Guibard A., et al., "Influence of meteorological conditions and topography on the active space of mountain birds assessed by a wave-based sound propagation model," *JASA*, vol. 151, 2022, 3703.

560. Ulloa J. S., et al., "Screening large audio datasets to determine the time and space distribution of screaming piha birds in a tropical forest," *Ecol Infor*, vol. 31, 2016, 91–99; Ulloa J. S., et al., "Explosive breeding in tropical anurans: Environmental triggers, community composition and acoustic structure," *BMC Ecol*, vol. 19, 2019, 28; Ducrettet M., et al., "Acoustic monitoring of the white-throated toucan (*Ramphastos tucanus*) in disturbed tropical landscapes," *Biol Conserv*, vol. 245, 2020, 108574; Perez-Granados C., and Schuchmann K. L., "Passive acoustic monitoring of the diel and annual vocal behavior of the black and gold howler monkey," *Am J Primatol*, vol. 83, 2021, e23241; Stowell D., et al., "Automatic acoustic detection of birds through deep learning: The first bird audio detection challenge," *Meth Ecol Evol*, vol. 10, 2018, 368–380; Stowell D., et al., "Automatic acoustic identification of individuals in multiple species: Improving identification across recording conditions," *J R Soc Interface*, vol. 16, 2019, 20180940; Desjonqueres C., "Passive acoustic monitoring as a potential tool to survey animal and ecosystem processes in freshwater environments," *Fresh Biol*, vol. 65, 2020, 7–19; Poupard M., et al., "Passive acoustic monitoring of sperm whales and anthropogenic noise using stereophonic recordings in the Mediterranean Sea, North West Pelagos Sanctuary," *Sc Rep*, vol. 12, 2022, 2007.

561. Parmentier E., et al., "How many fish could be vocal? An estimation from a coral reef (Moorea Island)," *Belg J Zool*, vol. 151, 2021, 1–29.

562. For a discussion on the definition of *soundscape*, see Grinfeder E., et al., "What do we mean by 'soundscape'? A functional description," *Front Ecol Evol*, vol. 10, 2022, 894232.

563. Sethi S. S., et al., "Characterizing soundscapes across diverse ecosystems using a universal acoustic feature set," *PNAS*, vol. 117, 2020, 17049–17055.

564. Lin T. H., et al., "Exploring coral reef biodiversity via underwater soundscapes," *Biol Conserv*, vol. 253, 2021, 108901; Raick X., et al., "From the reef to the ocean: Revealing the acoustic range of the biophony of a coral reef (Moorea Island, French Polynesia)," *J Mar Sc Eng*, vol. 9, 2021, 420.

565. Morrison C. A., et al., "Bird population declines and species turnover are changing the acoustic properties of spring soundscapes," *Nat Com*, vol. 12, 2021, 6217.

566. The CRIOBE is a laboratory bringing together researchers from the CNRS, the University of Perpignan, and the École pratique des hautes études (EPHE). Founded in 1868, the EPHE has counted among its professors the famous Claude Bernard and Claude Lévi-Strauss.

567. Bertucci F., et al., "Local sonic activity reveals potential partitioning in a coral reef fish community," *Oecologia*, vol. 193, 2020, 125–134.

568. Gross M., "Can science rescue coral reefs?," *Cur Biol*, vol. 26, 2016, R481–R492.

569. Lamont T. A., et al., "The sound of recovery: Coral reef restoration success is detectable in the soundscape," *J App Ecol*, vol. 59, 2022, 742–756.

570. Aran Mooney T., et al., "Listening forward: Approaching marine biodiversity assessments using acoustic methods," *R Soc Open Sc*, vol. 7, 2020, 201287.

571. Sueur J., and Farina A., "Ecoacoustics: The ecological investigation and interpretation of environmental sound," *Biosemiotics*, vol. 8, 2015, 493–502.

572. I invite you to watch these few interviews and lectures by Bernie Krause: https://www
.ted.com/talks/bernie_krause_the_voice_of_the_natural_world/transcript?language=fr;
https://www.youtube.com/watch?v=osgERQKVrhA; https://www.youtube.com/watch?v
=RRM51DPgQXg.

573. Krause B., *The Great Animal Orchestra: Finding the Origins of Music in the World's Wild Places*, Profile Books, 2012.

574. *Anthropos*: Greek word for "human."

575. Duarte C. M., et al., "The soundscape of the Anthropocene ocean," *Science*, vol. 371, 2021, eaba4658; Grinfeder E., et al., "Soundscape dynamics of a cold protected forest: Dominance of aircraft noise," *Lands Ecol*, vol. 37, 2022, 567–582.

576. Sueur J., Krause B., and Farina A., "Climate change is breaking Earth's beat," *Trends Ecol Evol*, vol. 34, 2019, 971–973. In addition, natural soundscapes have been shown to have beneficial effects on human health; on this topic, see Buxton R. T., et al., "A synthesis of health benefits of natural sounds and their distribution in national parks," *PNAS*, vol. 118, 2021, e2013097118. In addition, anthropogenic noise can act in synergy with artificial light: Wilson A. A., et al., "Artificial night light and anthropogenic noise interact to influence bird abundance over a continental scale," *Glob Change Biol*, vol. 27, 2021, 3987–4004.

577. Brumm H., *Animal Communication and Noise*, Springer, 2013.

578. Kunc H. P., and Schmidt R., "The effects of anthropogenic noise on animals: A meta-analysis," *Biol Let*, vol. 15, 2019, 20190649; Raboin M., and Elias D. O., "Anthropogenic noise and the bioacoustics of terrestrial invertebrates," *J Exp Biol*, vol. 222, 2019, jeb178749.

579. The rise of temperature due to climate change may also impact acoustic communication in animals. See Coomes, C. M., and Derryberry, E. P., "High temperatures reduce song production and alter signal salience in songbirds," *Anim Behav*, vol. 180, 2021, 13–22.

580. Faria A., et al., "Boat noise impacts early life stages in the Lusitanian toadfish: A field experiment," *Sc Total Environ*, vol. 811, 2022, 151367.

581. Recent research shows that noise pollution interferes with cognitive functions in birds: Osbrink A., et al., "Traffic noise inhibits cognitive performance in a songbird," *Proc R Soc B*, vol. 288, 2021, 20202851. Another study demonstrates that bats use more sonar pulses when hunting in a noisy environment: Allen L. C., et al., "Noise distracts foraging bats," *Proc R Soc B*, vol. 288, 2021, 20202689. See this study, which demonstrates that birds and bats avoid areas with high sound levels: Gomes D.G.E., et al., "Phantom rivers filter birds and bats by acoustic niche," *Nat Com*, vol. 12, 2021, 3029. Also have a look at this study, which shows that while urbanization affects territorial and vocal behaviors in southern house wrens, *Troglodytes aedon musculus*, noise does not seem to alter birds' vocal behavior: Diniz P., and Duca C., "Anthropogenic noise, song, and territorial aggression in southern house wrens," *J Avian Biol*, 2021, e02846. Finally, this review examines the effect of anthropogenic noise on a range of vertebrates: Gomes L., et al., "Influence of anthropogenic sounds on insect, anuran and bird acoustic signals: A meta-analysis," *Front Ecol Evol*, vol. 10, 2022, 827440.

582. Wale M. A., et al., "From DNA to ecological performance: Effects of anthropogenic noise on a reef-building mussel," *Sc Total Environ*, vol. 689, 2019, 126–132.

583. Buxton et al., "A synthesis of health benefits of natural sounds."

584. Popper A. N., et al., "Taking the animals' perspective regarding anthropogenic underwater sound," *Trends Ecol Evol*, vol. 35, 2020, 787–794; Tougaard, J., "Thresholds for noise induced hearing loss in marine mammals: Background note to revision of guidelines from the Danish Energy Agency," Technical report no. 28, Aarhus University, DCE—Danish Centre for Environment and Energy, 2021.

585. Wensveen P. J., et al., "Northern bottlenose whales in a pristine environment respond strongly to close and distant navy sonar signals," *Proc R Soc B*, vol. 286, 2019, 20182592.

586. For a review on the effect of marine noise pollution, see Di Franco E., et al., "Effects of marine noise pollution on Mediterranean fishes and invertebrates: A review," *Mar Pollut Bull*, vol. 159, 2020, 111450.

587. A recent study shows by using playback experiments that cetacean species that are more responsive to predator presence (orcas) are also those that react the most to anthropogenic noise: Miller P.J.O., et al., "Behavioral responses to predatory sounds predict sensitivity of cetaceans to anthropogenic noise within a soundscape of fear," *PNAS*, vol. 119, 2022, e2114932119.

588. Ferrier-Pages C., et al., "Noise pollution on coral reefs? A yet underestimated threat to coral reef communities," *Mar Pollut Bull*, vol. 165, 2021, 112129; Van der Knaap I., et al., "Effects of a seismic survey on movement of free-ranging Atlantic cod," *Cur Biol*, vol. 31, 2021, 1555–1562; Leduc A.O.H., et al., "Land-based noise pollution impairs reef fish behavior: A case study with a Brazilian carnival," *Biol Conserv*, vol. 253, 2021, 108910; Vieira M., et al., "Boat noise affects meagre (*Argyrosomus regius*) hearing and vocal behaviour," *Mar Pollut Bulletin*, vol. 172, 2021, 112824.

589. Hatch L. T., et al., "Quantifying loss of acoustic communication space for right whales in and around a US national marine sanctuary," *Conserv Biol*, vol. 26, 2012, 983–994. For an example in fish, see Alves D., et al., "Boat noise interferes with Lusitanian toadfish acoustic communication," *J Exp Biol*, vol. 224, 2021, jeb234849.

590. McCauley R. D., et al., "Widely used marine seismic survey air gun operations negatively impact zooplankton," *Nat Ecol Evol*, vol. 1, 2017, 0195.

591. A study suggests that anthropogenic noise may negatively impact the seagrass *Posidonia*: Sole M., et al., "Seagrass *Posidonia* is impaired by human-generated noise," *Com Biol*, vol. 4, 2021, 743.

592. Klingbeil B. T., et al., "Geographical associations with anthropogenic noise pollution for North American breeding birds," *Glob Ecol Biogeog*, vol. 29, 2020, 148–158.

593. Brumm H., "The impact of environmental noise on song amplitude in a territorial bird," *J Anim Ecol*, vol. 73, 2004, 434–440.

594. Guazzo R. A., et al., "The Lombard effect in singing humpback whales: Source levels increase as ambient ocean noise levels increase," *JASA*, vol. 148, 2020, 542.

595. Fournet M.E.H., et al., "Limited vocal compensation for elevated ambient noise in bearded seals: Implications for an industrializing Arctic Ocean," *Proc R Soc B*, vol. 288, 2021, 20202712.

596. Zhao L., et al., "Differential effect of aircraft noise on the spectral-temporal acoustic characteristics of frog species," *Anim Behav*, vol. 182, 2021, 9–18.

597. Slabbekoorn H., and Peet M., "Birds sing at a higher pitch in urban noise," *Nature*, vol. 424, 2003, 267.

598. Nemeth E., et al., "Bird song and anthropogenic noise: Vocal constraints may explain why birds sing higher-frequency songs in cities," *Proc R Soc B*, vol. 280, 2013, 20122798.

599. Zollinger S. A., et al., "Higher songs of city birds may not be an individual response to noise," *Proc R Soc B*, vol. 284, 2017, 20170602.

600. Derryberry E. P., et al., "Singing in a silent spring: Birds respond to a half-century soundscape reversion during the COVID-19 shutdown," *Science*, vol. 370, 2020, 575–579.

601. Gallego-Abenza M., et al., "Experience modulates an insect's response to anthropogenic noise," *Behav Ecol*, vol. 31, 2020, 90–96.

602. Bent A. M., et al., "Anthropogenic noise disrupts mate choice behaviors in female *Gryllus bimaculatus*," *Behav Ecol*, vol. 32, 2021, 201–210.

603. Hanache P., et al., "Noise-induced reduction in the attack rate of a planktivorous freshwater fish revealed by functional response analysis," *Fresh Biol*, vol. 65, 2020, 75–85; Rojas E., et al., "From distraction to habituation: Ecological and behavioural responses of invasive fish to anthropogenic noise," *Fresh Biol*, vol. 66, 2021, 1606–1618.

604. See, however, Francis C. D., et al., "Noise pollution alters ecological services: Enhanced pollination and disrupted seed dispersal," *Proc R Soc B*, vol. 279, 2012, 2727–2735; Francis C. D., et al., "Noise pollution changes avian communities and species interactions," *Cur Biol*, vol. 19, 2009, 1415–1419; and Francis C. D., et al., "Noise pollution filters bird communities based on vocal frequency," *PLoS ONE*, vol. 6, 2011, e27052.

605. Gordon T.A.C., et al., "Habitat degradation negatively affects auditory settlement behavior of coral reef fishes," *PNAS*, vol. 115, 2018, 5193–5198.

606. Lecchini D., et al., "Boat noise prevents soundscape-based habitat selection by coral planulae," *Sc Rep*, vol. 8, 2018, 9283.

607. See, for instance, Morrison C. A., et al., "Bird population declines and species turnover are changing the acoustic properties of spring soundscapes," *Nat Com*, vol. 12, 2022, 6217; and Rappaport D. I., et al., "Animal soundscapes reveal key markers of Amazon forest degradation from fire and logging," *PNAS*, vol. 119, 2022, e2102878119.

608. Farina A., *Soundscape Ecology: Principles, Patterns, Methods and Applications*, Springer, 2014; Mullet T. C., et al., "The acoustic habitat hypothesis: An ecoacoustics perspective on species habitat selection," *Biosemiotics*, vol. 10, 2017, 319–336.

609. Begon M., Townsend C. R., and Harper J. L., *Ecology, from Individuals to Ecosystems*, 4th ed., Blackwell, 2006.

610. Krause B. L., "Niche hypothesis: A virtual symphony of animal sounds, the origins of musical expression and the health of habitats," *Sound Newslet*, vol. 6, 1993, 6–10.

611. Sueur J., et al., "Acoustic indices for biodiversity assessment and landscape investigation," *Acta Acust Acust*, vol. 100, 2014, 772–781.

612. Brémond J.-C., "Acoustic competition between the song of the wren (*Troglodytes troglodytes*) and the songs of other species," *Behaviour*, vol. 65, 1978, 89–98.

613. Amezquita A., et al., "Acoustic interference and recognition space within a complex assemblage of dendrobatid frogs," *PNAS*, vol. 108, 2011, 17058–17063; Tobias J.A., et al., "Species interactions and the structure of complex communication networks," *PNAS*, vol. 111, 2014, 1020–1025; Sueur J., "Cicada acoustic communication: Potential sound partitioning in a multispecies community from Mexico (Hemiptera: Cicadomorpha: Cicadidae)," *Biol J Lin Soc*, vol. 75, 2002, 379–394; Luther D., "The influence of the acoustic community on songs of birds in a neotropical rain forest," *Behav Ecol*, vol. 20, 2009, 864–871; Schmidt A.K.D., and Balakrishnan R., "Ecology of acoustic signalling and the problem of masking interference in insects," *J Comp Physiol A*, vol. 201, 2014, 133–142; Ruppé L., et al., "Environmental constraints drive

the partitioning of the soundscape in fishes," *PNAS*, vol. 12, 2015, 6092–6097; Schmidt A.K.D., et al., "Spectral niche segregation and community organization in a tropical cricket assemblage," *Behav Ecol*, vol. 24, 2012, 470–480; Balakrishnan R., "Behavioral ecology of insect acoustic communication," in Pollack G. S., Mason A. C., Popper A., and Fay R. R. (eds.), *Insect Hearing*, Springer, 2016, 49–80; Chitnis S. S., et al., "Sympatric wren-warblers partition acoustic signal space and song perch height," *Behav Ecol*, vol. 31, 2020, 559–567; Bertucci F., et al., "Local sonic activity reveals potential partitioning in a coral reef fish community," *Oecologia*, vol. 193, 2020, 125–134; Allen-Ankins S., and Schwarzkopf L., "Spectral overlap and temporal avoidance in a tropical savannah frog community," *Anim Behav*, vol. 180, 2021, 1–11; Allen-Ankins S., and Schwarzkopf L., "Using citizen science to test for acoustic niche partitioning in frogs," *Sc Rep*, vol. 12, 2022, 2447.

614. Boncoraglio G., and Saino N., "Habitat structure and the evolution of bird song: A meta-analysis of the evidence for the acoustic adaptation hypothesis," *Funct Ecol*, vol. 21, 2007, 134–142; Ey E., and Fischer J., "The 'acoustic adaptation hypothesis'—a review of the evidence from birds, anurans and mammals," *Bioacoustics*, vol. 19, 2009, 21–48.

615. Or the sender uses a strategy that reinforces the range of its signals, such as certain insects that position themselves judiciously in the environment. See Montealegre-Z F., et al., "Generation of extreme ultrasonics in rainforest katydids," *J Exp Biol*, vol. 209, 2006, 4923–4937.

616. Morton E. S., "Ecological sources of selection on avian sounds," *Am Nat*, vol. 109, 1975, 17–34; Morton E. S., "On the occurrence and significance of motivation-structural rules in some bird and mammal sounds," *Am Nat*, vol. 111, 1977, 855–869; Marten K., and Marler P., "Sound transmission and its significance for animal vocalization," *Behav Ecol Sociobiol*, vol. 2, 1977, 271–290; Wiley R. H., and Richards D. G., "Physical constraints on acoustic communication in the atmosphere: Implications for the evolution of animal vocalizations," *Behav Ecol Sociobiol*, vol. 3, 1978, 69–94; Richards D. G., and Wiley R. H., "Reverberations and amplitude fluctuations in the propagation of sound in a forest: Implications for animal communication," *Am Nat*, vol. 115, 1980, 381–399.

617. Klump G. M., "Bird communication in the noisy world," in Barth F. G., and Schmid A., *Ecology of Sensing—Ecology and Evolution of Acoustic Communication in Birds*, Cornell University Press, 1996, 321–338; Zhao L., et al., "Noise constrains the evolution of call frequency contours in flowing water frogs: A comparative analysis in two clades," *Front Zool*, vol. 18, 2021, 37.

618. Goutte S., et al., "How the environment shapes animal signals: A test of the acoustic adaptation hypothesis in frogs," *J Evol Biol*, vol. 31, 2017, 148–158.

619. Mikula P., et al., "A global analysis of song frequency in passerines provides no support for the acoustic adaptation hypothesis but suggests a role for sexual selection," *Ecol Let*, vol. 24, 2020, 477–486.

620. Riondato I., et al., "Allometric escape and acoustic signal features facilitate high-frequency communication in an endemic Chinese primate," *J Comp Physiol A*, vol. 207, 2021, 327–336.

621. This integrative side of ecoacoustics is reminiscent of ecological monitoring methods, such as biotic indices, where insect larvae present in a watercourse are counted. The quality of the water is deduced from the species present and their density. The advantage of ecoacoustics is that no living organisms are taken from the water.

622. The International Bioacoustics Society (IBAC) organizes an international scientific conference every two years. If you are interested, you can have a look on the society website: https://www.ibac.info/.

623. For a recent review on acoustic indices, see Alcocer I., et al., "Acoustic indices as proxies for biodiversity: A meta-analysis," *Biol Rev*, vol. 97, 2022, 2209–2236.

624. Sueur J., et al., "Acoustic indices for biodiversity assessment and landscape investigation," *Acta Acust Acust*, vol. 100, 2014, 772–781; Sueur J., et al., "Acoustic biodiversity," *Cur Biol*, vol. 31, 2021, R1141–R1224.

625. See, for example, these studies: Depraetere M., et al., "Monitoring animal diversity using acoustic indices: Implementation in a temperate woodland," *Ecol Indic*, vol. 13, 2012, 46–54; Gasc A., et al., "Assessing biodiversity with sound: Do acoustic diversity indices reflect phylogenetic and functional diversities of bird communities?," *Ecol Indic*, vol. 25, 2013, 279–287; Linke S., et al., "Freshwater ecoacoustics as a tool for continuous ecosystem monitoring," *Front Ecol Environ*, vol. 16, 2018, 231–238; Ulloa J. S., et al., "Estimating animal acoustic diversity in tropical environments using unsupervised multiresolution analysis," *Ecol Indic*, vol. 90, 2018, 346–355; Van der Lee G. H., et al., "Freshwater ecoacoustics: Listening to the ecological status of multi-stressed lowland waters," *Ecol Indic*, vol. 113, 2020, 106252; and Flowers C., et al., "Looking for the -scape in the sound: Discriminating soundscapes categories in the Sonoran Desert using indices and clustering," *Ecol Indic*, vol. 127, 2021, 107805.

626. Efforts are currently being made to develop automated monitoring of soundscapes using global ecosystem monitoring. For instance, see Sethi S., et al., "Characterizing soundscapes across diverse ecosystems using a universal acoustic feature set," *PNAS*, vol. 117, 2020, 17049–17055.

627. Ferrier-Pages C., et al., "Noise pollution on coral reefs? A yet underestimated threat to coral reef communities," *Mar Pollut Bull*, vol. 165, 2021, 112129.

628. The acoustic complexity index (ACI) was developed based on the fact that most sounds produced by animals have varying intensities over time, while noises—and in particular anthropogenic noises—are more constant. When biodiversity is poor, the ACI is low. It rises with increasing numbers of sound-producing animal species: Pieretti, N., "A new methodology to infer the singing activity of an avian community: The acoustic complexity index (ACI)," *Ecol Indic*, vol. 11, 2011, 868–873; Bolgan M., et al., "Acoustic complexity of vocal fish communities: A field and controlled validation," *Sc Rep*, vol. 8, 2018, 10559.

629. Bertucci F., et al., "A preliminary acoustic evaluation of three sites in the lagoon of Bora Bora, French Polynesia," *Environ Biol Fish*, vol. 103, 2020, 891–902.

630. Elise S., et al., "An optimised passive acoustic sampling scheme to discriminate among coral reefs' ecological state," *Ecol Indic*, vol. 107, 2019, 105627.

631. https://youtube/WK0sR7e4F0Y.

632. Each index was calculated on five different frequency bands: 0.1–0.5 kHz; 0.5–1 kHz; 1–2 kHz; 2–7 kHz; and the full bandwidth, 0–50 kHz. For details, see Elise S., et al., "Assessing key ecosystem functions through soundscapes: A new perspective from coral reefs," *Ecol Indic*, vol. 107, 2019, 105623.

633. Dimoff S. A., et al., "The utility of different acoustic indicators to describe biological sounds of a coral reef soundscape," *Ecol Indic*, vol. 124, 2021, 107435.

634. The assessment of stress due to anthropogenic noise can lead to noise mitigation measures: Nedelec S. L., et al., "Limiting motorboat noise on coral reefs boosts fish reproductive success," *Nat Com*, vol. 13, 2022, 2822.

635. Gross M., "Listening to the sounds of the biosphere," *Cur Biol*, vol. 28, 2018, R847–R870.

636. Lin T. H., et al., "Exploring coral reef biodiversity via underwater soundscapes," *Biol Conserv*, vol. 253, 2018, 108901.

637. Stowell D., "Computational bioacoustics with deep learning: A review and roadmap," *PeerJ*, vol. 10, 2022, e13152.

638. Such ecoacoustic programs are already in place. See the site http://ear.cnrs.fr/ for examples.

639. The ecoacoustic approach allows us to evaluate functional aspects of ecosystems. See, for example, Folliot A., et al., "Using acoustics and artificial intelligence to monitor pollination by insects and tree use by woodpeckers," *S Tot Envir*, vol. 838, 2022, 155883. Soundscapes may even be useful for assessing soil biodiversity: Maeder M., et al., "Temporal and spatial dynamics in soil acoustics and their relation to soil animal diversity," *PLoS ONE*, vol. 17, 2022, e0263618.

640. Sueur, Krause, and Farina, "Climate change is breaking Earth's beat."

641. Mbu Nyamsi R. G., et al., "On the extraction of some time dependent parameters of an acoustic signal by means of the analytic signal concept: Its application to animal sound study," *Bioacoustics*, vol. 5, 1994, 187–203.

642. True. See De Novion C., et al., "L'impact des concepts de Pierre-Gilles de Gennes sur l'innovation en France dans le domaine des matériaux," *Reflets Phys*, vol. 56, 2018, 10–19.

Chapter 18

643. Hauser M. D., et al., "The faculty of language: What is it, who has it, and how did it evolve?," *Science*, vol. 298, 2002, 1569–1579.

644. Prat Y., "Animals have no language, and humans are animals too," *Pers Psychol Sc*, vol. 14, 2019, 885–893.

645. Recent research is attempting to find large-scale organizational rules in animal vocalizations. See, for example, Markowitz J. E., et al., "Long-range order in canary song," *PLoS Comput Biol*, vol. 9, 2013, e1003052; and Sainburg T., et al., "Parallels in the sequential organization of birdsong and human speech," *Nat Com*, vol. 10, 2019, 3636.

646. I have translated into modern-day English what Montaigne wrote. The original quote is as follows: "*Ce defaut qui empesche la communication d'entre elles et nous, pourquoy n'est il aussi bien à nous qu'à elles? C'est à deviner à qui est la faute de ne nous entendre point: Car nous ne les entendons non plus qu'elles nous. Par ceste mesme raison elles nous peuvent estimer bestes, comme nous les estimons*" (Montaigne 1595, Essais Livre II).

647. Pisanski K., et al., "Voice modulation: A window into the origins of human vocal control?," *Trends Cogn Sci*, vol. 20, 2016, 304–318.

648. Favaro L., et al., "Do penguins' vocal sequences conform to linguistic laws?," *Biol Let*, vol. 16, 2020, 20190589.

649. Huang M., et al., "Male gibbon loud morning calls conform to Zipf's law of brevity and Menzerath's law: Insights into the origin of human language," *Anim Behav*, vol. 160, 2020, 145–155; Valente D., et al., "Linguistic laws of brevity: Conformity in *Indri indri*," *Anim Cogn*, vol. 24, 2021, 897–906.

650. Pougnault L., Levréro F., and Lemasson A., "Conversation among primate species," in Masataka N. (ed.), *The Origins of Language Revisited*, vol. II, Springer, 2020, 73–96.

651. Ravignani A., et al., "Interactive rhythms across species: The evolutionary biology of animal chorusing and turn-taking," *Ann NY Acad Sc*, vol. 1453, 2019, 12–21; Banerjee A., and Vallentin D., "Convergent behavioral strategies and neural computations during vocal turn-taking across diverse species," *Cur Opin Neurobiol*, vol. 73, 2022, 102529.

652. Cornec C., et al., "A pilot study of calling patterns and vocal turn-taking in wild bonobos *Pan paniscus*," *Ethol Ecol Evol*, vol. 34, 2022, 360–377.

653. Pougnault L., et al., "Temporal calling patterns of a captive group of chimpanzees (*Pan troglodytes*)," *Int J Primatol*, vol. 42, 2021, 809–832.

654. Demartsev V., et al., "Vocal turn-taking in meerkat group calling sessions," *Cur Biol*, vol. 28, 2018, 3661–3666.

655. Pougnault L., et al., "Breaking conversational rules matters to captive gorillas: A playback experiment," *Sc Rep*, vol. 10, 2020, 6947.

656. An interesting perspective that reviews linguistic laws and explores the potential relevance of these laws across all biological levels is Semple S., et al., "Linguistic laws in biology," *Trends Ecol Evol*, vol. 37, 2022, 53–66.

657. Schlenker P., et al., "What do monkey calls mean?," *Trends Cogn Sc*, vol. 20, 2016, 894–904; Fischer J., "Primate vocal communication and the evolution of speech," *Cur Dir Psychol Sc*, vol. 30, 2021, 55–60; Pougnault L., et al., "Social pressure drives "conversational rule" in great apes," *Biol Rev*, vol. 97, 2022, 749–765.

658. Fitch W. T., "Animal cognition and the evolution of human language: Why we cannot focus solely on communication," *Phil Trans R Soc B*, vol. 375, 2019, 20190046.

659. Lyn H., et al., "Apes and the evolution of language: Taking stock of 40 years of research," in Vonk J., and Shackelford T. K. (eds.), *Oxford Handbook of Comparative Evolutionary Psychology*, Oxford University Press, 2012, 356–378; Krause M. A., and Beran M. J., "Words matter: Reflections on language projects with chimpanzees and their implications," *Am J Primat*, vol. 82, 2020, e23187.

660. Hayes K. J., and Hayes C., "The intellectual development of a home-raised chimpanzee," *Proc Amer Phil Soc*, vol. 95, 1951, 105–109.

661. Hayes C., *The Ape in Our House*, Harper, 1951.

662. Gardner R. A., and Gardner B. T., "Teaching sign language to a chimpanzee," *Science*, vol. 165, 1969, 664–672.

663. Terrace H. S., *Why Chimpanzees Can't Learn Language and Only Humans Can*, Columbia University Press, 2019.

664. Terrace H. S., "Can an ape create a sentence?," *Science*, vol. 206, 1979, 891–902.

665. There were also experiments with an orangutan and a gorilla. See Shettleworth S. J., *Cognition, Evolution and Behavior*, Oxford University Press, 2010.

666. Matsuzawa T., "Use of numbers by a chimpanzee," *Nature*, vol. 315, 1985, 57–59.

667. On the different projects to teach apes how to talk, see this review: Ristau C. A., and Robbins D., "Language in the great apes: A critical review," *Adv St Behav*, vol. 12, 1982, 141–255. For a recent point of view, see Bergman T. J., et al., "The speech-like properties of nonhuman primate vocalizations," *Anim Behav*, vol. 151, 2019, 229–237.

668. Pepperberg I. M., *The Alex Studies*, Harvard University Press, 2000.

669. Bodin C., et al., "Functionally homologous representation of vocalizations in the auditory cortex of humans and macaques," *Cur Biol*, vol. 31, 2021, 4839–4844.

670. Lieberman P. H., et al., "Vocal tract limitations on the vowel repertoires of rhesus monkey and other nonhuman primates," *Science*, vol. 164, 1969, 1185–1187.

671. Fitch W. T., et al., "Monkey vocal tracts are speech-ready," *Sc Adv*, vol. 2, 2016, e1600723; Grawunder S., et al., "Chimpanzee vowel-like sounds and voice quality suggest formant space expansion through the hominoid lineage," *Phil Trans R Soc*, vol. 377, 2021, 20200455.

672. Meyer J., "Coding human languages for long-range communication in natural ecological environments: Shouting, whistling, and drumming," in Aubin T., and Mathevon N. (eds.), *Coding Strategies in Vertebrate Acoustic Communication*, Springer, 2020, 91–113.

673. Vargha-Khadem F., et al., "FOXP2 and the neuroanatomy of speech and language," *Nat Rev Neurosc*, vol. 6, 2005, 131–138.

674. Lai C. S., et al., "A forkhead-domain gene is mutated in a severe speech and language disorder," *Nature*, vol. 413, 2002, 519–523; Fisher S. E., et al., "Localisation of a gene implicated in a severe speech and language disorder," *Nat Genet*, vol. 18, 1998, 168–170; Liégeois F., et al., "Language fMRI abnormalities associated with FOXP2 gene mutation," *Nat Neurosc*, vol. 6, 2003, 1230–1237.

675. Enard W., et al., "Molecular evolution of FOXP2, a gene involved in speech and language," *Nature*, vol. 418, 2002, 869–872.

676. Fisher S. E., and Scharff C., "FOXP2 as a molecular window into speech and language," *Trends Gen*, vol. 25, 2009, 166–177.

677. Haesler S., et al., "FOXP2 expression in avian vocal learners and non-learners," *J Neurosc*, vol. 24, 2004, 3164–3175.

678. For instance, a study has just shown that FOXP1, the little brother of FOXP2, plays a role in the cultural transmission of vocalizations in the zebra finch: Garcia-Oscos F., et al., "Autism-linked gene FoxP1 selectively regulates the cultural transmission of learned vocalizations," *Sc Adv*, vol. 7, 2021, eabd2827.

679. Pfenning A. R., et al., "Convergent transcriptional specializations in the brains of humans and song-learning birds," *Science*, vol. 346, 2014, 1256846.

680. Fisher S. E., "Human genetics: The evolving story of FOXP2," *Cur Biol*, vol. 29, 2019, R50–R70.

681. If you wish to delve deeper into the evolutionary history of human spoken language, I recommend Fitch's clear, well-documented, and very accessible article, "The biology and evolution of speech: A comparative analysis," *Annu Rev Linguist*, vol. 4, 2018, 255–279; and the book by the same author, *The Evolution of Language*, Cambridge University Press, 2012.

682. Darwin C., *The Descent of Man and Selection in Relation to Sex*, John Murray, 1871.

683. For a discussion about our understanding of human origins and evolution, see Richerson P. J., et al., "Modern theories of human evolution foreshadowed by Darwin's *Descent of Man*," *Science*, vol. 372, 2021, eaba3776.

684. Nishimura T., et al., "Evolutionary loss of complexity in human vocal anatomy as an adaptation for speech," *Science*, vol. 377, 2022, 760–763; Gouzoules H., "When less is more in the evolution of language," *Science*, vol. 377, 2022, 706–707.

685. A recent study suggests that auditory abilities of Neanderthals were comparable to those of modern humans, allowing a similar vocal communication system: Conde-Valverde M., et al., "Neanderthals and *Homo sapiens* had similar auditory and speech capacities," *Nat Ecol Evol*, vol. 5, 2021, 609–615.

686. Dunbar R.I.M., "Group size, vocal grooming and the origins of language," *Psychon Bull Rev*, vol. 24, 2017, 209–212.

687. Mathevon N., and Aubin T., "Acoustic coding strategies through the lens of the mathematical theory of communication," in Aubin and Mathevon (eds.), *Coding Strategies in Vertebrate Acoustic Communication*, 1–10.

688. Lehmann J., et al., "Group size, grooming and social cohesion in primates," *Anim Behav*, vol. 74, 2007, 1617–1629.

689. Dunbar R., *Grooming, Gossip, and Evolution of Language*, Harvard University Press, 1996.

690. Knight C., and Lewis J., "Towards a theory of everything," in Power C., Finnegan M., and Callan H. (eds.), *Human Origins: Contributions from Social Anthropology*, Berghahn, 2017, 84–102.

691. Knight C., and Lewis J. D., "Wild voices: Mimicry, reversal, metaphor, and the emergence of language," *Cur Anthropol*, vol. 58, 2017, 435–453.

692. Darwin, *Descent of Man*.

693. I won't go into the complex topic of the origin of music. For an interesting perspective, see Leongomez J. D., et al., "Musicality in human vocal communication: An evolutionary perspective," *Phil Trans R Soc B*, vol. 377, 2021, 20200391. Also see this paper: De Gregorio C., et al., "Categorical rhythms in a singing primate," *Cur Biol*, vol. 31, 2021, R1379–R1380.

694. Spottiswoode C. N., et al., "Reciprocal signaling in honeyguide-human mutualism," *Science*, vol. 353, 2016, 387–389; van der Wal J. E., et al., "Awer honey-hunting culture with greater honeyguides in coastal Kenya," *Front Conserv Sc*, vol. 2, 2022, 727479. See also, from the BBC: https://www.bbc.co.uk/sounds/play/b07z43f8.

695. The study of the origins of human language is a very active field because many questions remain unanswered. See, for example, Taylor D., et al., "Vocal functional flexibility: What it is and why it matters," *Anim Behav*, vol. 186, 2022, 93–100.

696. Cardinal S., et al., "The evolution of floral sonification, a pollen foraging behavior used by bees (*Anthophila*)," *Evolution*, vol. 72, 2018, 590–600.

697. Khait I., et al., "Sound perception in plants," *Sem Cell Dev Biol*, vol. 92, 2019, 134–138.

698. Veits M., et al., "Flowers respond to pollinator sound within minutes by increasing nectar sugar concentration," *Ecol Let*, vol. 22, 2019, 1483–1492. But see Pyke G. H., et al., "Changes in floral nectar are unlikely adaptive responses to pollinator flight sound," *Ecol Let*, vol. 23, 2020, 1421–1422.

Acknowledgments

699. French academics devote many hours to teaching (more than twice as many as their American, British, or Swiss colleagues) and are in great demand for administrative tasks.

INDEX